Radon Series on Computational and Applied Mathematics 10

Radon Series on Computational and Applied Mathematics

1 *Lectures on Advanced Computational Methods in Mechanics*
Johannes Kraus and Ulrich Langer (eds.), 2007

2 *Gröbner Bases in Symbolic Analysis*
Markus Rosenkranz and Dongming Wang (eds.), 2007

3 *Gröbner Bases in Control Theory and Signal Processing*
Hyungju Park and Georg Regensburger (eds.), 2007

4 *A Posteriori Estimates for Partial Differential Equations*
Sergey Repin, 2008

5 *Robust Algebraic Multilevel Methods and Algorithms*
Johannes Kraus and Svetozar Margenov, 2009

6 *Iterative Regularization Methods for Nonlinear Ill-Posed Problems*
Barbara Kaltenbacher, Andreas Neubauer and Otmar Scherzer, 2008

7 *Robust Static Super-Replication of Barrier Options*
Jan H. Maruhn, 2009

8 *Advanced Financial Modelling*
Hansjörg Albrecher, Wolfgang J. Runggaldier and Walter Schachermayer (eds.), 2009

9 *Theoretical Foundations and Numerical Methods for Sparse Recovery*
Massimo Fornasier (ed.), 2010

10 *Regularization Methods in Banach Spaces*
Thomas Schuster, Barbara Kaltenbacher, Bernd Hofmann and Kamil S. Kazimierski, 2012

Thomas Schuster
Barbara Kaltenbacher
Bernd Hofmann
Kamil S. Kazimierski

Regularization Methods in Banach Spaces

De Gruyter

Mathematics Subject Classification 2010: Primary: 65J20, 47A52, 47J06, 65J22; Secondary: 65F22, 47J20, 49J40.

ISBN 978-3-11-025524-9
e-ISBN 978-3-11-025572-0
ISSN 1865-3707

Library of Congress Cataloging-in-Publication Data

A CIP catalog record for this book has been applied for at the Library of Congress.

Bibliographic information published by the Deutsche Nationalbibliothek

The Deutsche Nationalbibliothek lists this publication in the Deutsche Nationalbibliografie; detailed bibliographic data are available in the internet at http://dnb.dnb.de.

Typesetting: Da-TeX Gerd Blumenstein, Leipzig, www.da-tex.de
Printing and binding: Hubert & Co. GmbH & Co. KG, Göttingen
⊗Printed on acid-free paper

Printed in Germany

www.degruyter.com

Dedications

Thomas Schuster dedicates his efforts spent on writing this book to
his beloved son ADRIAN (*2010)
and his patient and equally beloved wife PETRA.

The parts by **Barbara Kaltenbacher** are dedicated to
her dear and patient family.

Bernd Hofmann dedicates his chapters to
RENATE and UWE.

Kamil S. Kazimierski dedicates his chapters to
his ever so patient, understanding and loving family.

Preface

Regularization methods aimed at finding stable approximate solutions are a necessary tool to tackle inverse and ill-posed problems. Inverse problems arise in a large variety of applications ranging from medical imaging and non-destructive testing, via finance to systems biology. Many of these problems belong to the class of parameter identification problems in partial differential equations (PDEs) and thus are computationally demanding and mathematically challenging. Hence there is a substantial need for stable and efficient solvers for this class of problems, as well as for a rigorous convergence analysis of these methods. Usually the mathematical model of an inverse problem is expressed by a linear or nonlinear operator equation of the first kind and often the associated forward operator acts between Hilbert spaces. As compared to general Banach spaces, Hilbert spaces have a nicer structure, thus enabling convergence analysis, a fact which has led to a rigorous and comprehensive study of regularization methods in Hilbert spaces during the last three decades.

However, for numerous problems such as X-ray diffractometry, certain inverse scattering problems, and a number of parameter identification problems in PDEs, the reasons for using a Hilbert space setting seem to be based rather on conventions than on an appropriate and realistic model choice, and often a Banach space setting would be closer to reality. Furthermore, sparsity constraints, using general L^p-norms or the BV-norm, have recently become very popular. It is therefore a consequent and quite natural step to extend well-established regularization techniques for linear and nonlinear operator equations from a Hilbert to a Banach space setting and to analyze their convergence behavior. Meanwhile the most well-known methods such as Tikhonov-type methods, requiring the solution of extremal problems, iterative regularization methods like the Landweber and the Gauß - Newton method, as well as the method of approximate inverse, have been investigated for linear and nonlinear operator equations in Banach spaces. Convergence has been proved, the rates of convergence determined, and conditions on the solution smoothness on the structure of nonlinearity have been formulated.

Over the last decade research on regularization methods in Banach spaces has become a very lively and fast growing field. The expansion of standard methods like Landweber's method or the Gauß - Newton method to Banach spaces on the one hand and the establishment of Tikhonov methods with sparsity constraints on the other hand have been the starting point of a fruitful development which could have hardly been foreseen, which provided the motivation for the authors to write a monograph which comprehensively discusses the main progress in the field of Banach space driven regularization theory.

The book consists of five parts. Part I motivates the importance of developing and analyzing regularization methods in Banach spaces by presenting four applications which intrinsically require a Banach space setting and giving a brief glimpse of sparsity constraints (Chapter 1). Part II summarizes all mathematical tools necessary to carry out an analysis in Banach spaces, such as some facts on convex analysis, duality mappings and Bregman distances (Chapter 2). Part II furthermore includes a chapter on ill-posed operator equations and regularization theory in Banach spaces (Chapter 3), which also introduces the reader to modern ingredients of smoothness analysis for ill-posed problems like approximate source conditions and variational inequalities. In view of solution methods for inverse problems we distinguish between Tikhonov-type or variational regularization methods (Part III), iterative techniques (Part IV) and the method of approximate inverse (Part V). Part III represents the current state-of-the-art concerning Tikhonov regularization in Banach spaces. After stating the theory, including error estimates and convergence rates for general convex penalties and nonlinear problems (Chapter 4), we specifically address linear problems and power-type penalty terms, propose parameter choice rules, and present methods for solving the resulting minimization problems in Chapter 5. Part IV, dealing with iterative regularization methods, is divided into two chapters: the first one is concerned with linear operator equations and contains the Landweber method, as well as the numerically accelerated sequential subspace methods and the general framework of split feasibility problems (Chapter 6), and the second one deals with the iterative solution of nonlinear operator equations by gradient type methods and the iteratively regularized Gauß–Newton method (Chapter 7). Finally, part V outlines the method of approximate inverse, which is based on the efficient evaluation of the measured data with reconstruction kernels. After a brief introduction to the method (Chapter 8), we investigate its regularization property and convergence rates in L^p-spaces, as well as in the space of continuous functions on a compact set (Chapter 9). The application of these results to the problem of X-ray diffractometry concludes the work (Chapter 10).

The authors wish to thank several people, whose research has strongly contributed to this book. Thomas Schuster and Barbara Kaltenbacher are deeply indebted to Dr. Frank Schöpfer for the many fruitful discussions. His scientific work was essential to the content of Chapters 6 and parts of Chapter 7. Thomas Schuster furthermore thanks Prof. Dr. Alfred K. Louis for introducing him to the method of approximate inverse a long time ago; he owes him so much. Barbara Kaltenbacher and Bernd Hofmann wish to thank Prof. Dr. Otmar Scherzer and Dr. Christiane Pöschl for initiating their involvement in the exciting field of regularization in Banach spaces, as well as Dr. Elena Resmerita for instructive communication. Bernd Hofmann expresses his gratitude to Dr. Jens Flemming and PD Dr. Radu I. Boţ for many helpful suggestions and discussions.

February 2012

Thomas Schuster, Oldenburg
Barbara Kaltenbacher, Klagenfurt
Bernd Hofmann, Chemnitz
Kamil S. Kazimierski, Bremen

Contents

Preface vii

I Why to use Banach spaces in regularization theory?

1 Applications with a Banach space setting 4

1.1 X-ray diffractometry 4

1.2 Two phase retrieval problems 6

1.3 A parameter identification problem for an elliptic partial differential equation 9

1.4 An inverse problem from finance 13

1.5 Sparsity constraints 18

II Geometry and mathematical tools of Banach spaces

2 Preliminaries and basic definitions 28

2.1 Basic mathematical tools 28

2.2 Convex analysis 31

2.2.1 The subgradient of convex functionals 31

2.2.2 Duality mappings 34

2.3 Geometry of Banach space norms 36

2.3.1 Convexity and smoothness 37

2.3.2 Bregman distance 44

3 Ill-posed operator equations and regularization 49

3.1 Operator equations and the ill-posedness phenomenon 49

3.1.1 Linear problems 50

3.1.2 Nonlinear problems 52

3.1.3 Conditional well-posedness 55

3.2 Mathematical tools in regularization theory 56

3.2.1 Regularization approaches 57

3.2.2 Source conditions and distance functions 63

3.2.3 Variational inequalities 67

3.2.4 Differences between the linear and the nonlinear case 69

III Tikhonov-type regularization

4 Tikhonov regularization in Banach spaces with general convex penalties 81
4.1 Basic properties of regularized solutions 81
4.1.1 Existence and stability of regularized solutions 81
4.1.2 Convergence of regularized solutions 84
4.2 Error estimates and convergence rates 89
4.2.1 Error estimates under variational inequalities 90
4.2.2 Convergence rates for the Bregman distance 95
4.2.3 Tikhonov regularization under convex constraints 99
4.2.4 Higher rates briefly visited 101
4.2.5 Rate results under conditional stability estimates 103
4.2.6 A glimpse of rate results under sparsity constraints 105

5 Tikhonov regularization of linear operators with power-type penalties 108
5.1 Source conditions 108
5.2 Choice of the regularization parameter 113
5.2.1 A priori parameter choice 113
5.2.2 Morozov's discrepancy principle 115
5.2.3 Modified discrepancy principle 116
5.3 Minimization of the Tikhonov functionals 122
5.3.1 Primal method 123
5.3.2 Dual method 135

IV Iterative regularization

6 Linear operator equations 144
6.1 The Landweber iteration 146
6.1.1 Noise-free case 146
6.1.2 Regularization properties 152
6.2 Sequential subspace optimization methods 157
6.2.1 Bregman projections 158
6.2.2 The method for exact data (SESOP) 163
6.2.3 The regularization method for noisy data (RESESOP) 165
6.3 Iterative solution of split feasibility problems (SFP) 177
6.3.1 Continuity of Bregman and metric projections 179
6.3.2 A regularization method for the solution of SFPs 183

7 Nonlinear operator equations 193

7.1 Preliminaries 193
7.1.1 Conditions on the spaces 193
7.1.2 Variational inequalities 194
7.1.3 Conditions on the forward operator 195

7.2 Gradient type methods 199
7.2.1 Convergence of the Landweber iteration with the discrepancy principle 199
7.2.2 Convergence rates for the iteratively regularized Landweber iteration with a priori stopping rule 203

7.3 The iteratively regularized Gauss–Newton method 212
7.3.1 Convergence with a priori parameter choice 215
7.3.2 Convergence with a posteriori parameter choice 225
7.3.3 Numerical illustration 230

V The method of approximate inverse

8 Setting of the method 236

9 Convergence analysis in $L^p(\Omega)$ and $\mathcal{C}(K)$ 239

9.1 The case $X = L^p(\Omega)$ 239

9.2 The case $X = \mathcal{C}(K)$ 244

9.3 An application to X-ray diffractometry 248

10 A glimpse of semi-discrete operator equations 253

Bibliography 265

Index 280

Part I

Why to use Banach spaces in regularization theory?

The purpose of this first part of the book is to give reasons why Banach space settings play such an important role in the past decade of research in the area of regularization theory for inverse and ill-posed problems, and serve as an appropriate framework for such applied problems. The research on regularization methods in Banach spaces was driven by different mathematical viewpoints: on the one hand, there are indeed numerous practical applications where models that use Hilbert spaces, for example by formulating the problem as an operator equation in L^2-spaces, are not realistic or appropriate. The nature of such applications requires Banach space models working in L^p-spaces, non-Hilbertian–Sobolev spaces, or spaces of continuous functions. In this context, sparse solutions of linear and nonlinear ill-posed operator equations are often to be determined. On the other hand, mathematical tools and techniques typical of Banach spaces can help to overcome the limitations of Hilbert space models.

One reason why Banach spaces became popular in regularization theory, is the fact that they represent a framework that allows us to formulate a model for a specific application in a more general setting than can be done based on Hilbert spaces alone. To demonstrate this, we present a series of different applications ranging from non-destructive testing, such as X-ray diffractometry (Section 1.1), via phase retrieval (Section 1.2) and parameter identification for partial differential equations (Section 1.3) to an inverse problem in finance (Section 1.4). All these applications are characterized by operator equations

$$F(x) = y\,,$$

where the so-called forward operator $F : \mathcal{D}(F) \subseteq X \to Y$ denotes a continuous linear or nonlinear mapping between Banach spaces X and Y. Since F is in general 'smoothing', i.e. it has a non-closed range in the linear case, these equations are *ill-posed* in the sense that even small perturbations in y severely corrupt the solution x. Stable approximate solutions are accomplished only by using *regularization methods*, which attenuate the discontinuity or unboundedness of the inverse mapping F^{-1}.

Sparsity means that the searched-for solution has only a few non-zero coefficients with respect to a specific, given basis. To take this into account, we aim to minimize a Tikhonov functional whose penalty term promotes sparsity. In their seminal article [55] Defrise, Daubechies and De Mol use ℓ^p-norms of the coefficients, but other penalty terms like BV- or Besov norms are also possible. We present a short introduction to this field in Section 1.5.

Hence this first part serves as a teaser, motivating the subsequent analysis. The basic material of Banach space and regularization theory is then provided in Part II, Tikhonov functionals with general convex and specific power type penalties Ω are the subject of Part III, in Part IV iterative regularization methods are presented, and Part V concludes the book with the *method of approximate inverse*.

Chapter 1

Applications with a Banach space setting

1.1 X-ray diffractometry

We summarize some results on X-ray diffractometry as outlined in the articles [219, 221]. X-ray diffractometry is a type of non-destructive testing and aims to recover the stress tensor $\sigma = (\sigma_{ij})$ of a specimen by measuring the Bragg angle of reflected X-rays. Mechanical stresses and the elastic strain tensor $\varepsilon = \varepsilon_{ij}$ in isotropic media are connected by Hooke's law

$$\varepsilon_{ij} = \frac{\nu+1}{E}\sigma_{ij} - \delta_{ij}\frac{\nu}{E}\sum_{k=1}^{3}\sigma_{kk}\,, \tag{1.1}$$

where ν is the Poisson number and E denotes the modulus of elasticity. In a laboratory system the probe under investigation is rotated by an angle φ about the x_3-axis and tilted by an angle ψ about the x_2-axis, transforming the strain tensor to

$$\varepsilon^L = U^t_{\varphi\psi}\varepsilon U_{\varphi\psi}\,, \tag{1.2}$$

where $U_{\varphi\psi} \in SO(3)$ is the orthogonal matrix that models the tilting and rotating by ψ and φ, respectively. In X-ray diffractometry only near-surface strains can be detected, that is

$$\varepsilon_{\varphi\psi} := \varepsilon^L_{33}\,.$$

Introducing (1.2) into Hooke's law (1.1), we obtain an explicit expression for the relation between $\varepsilon_{\varphi\psi}$ and the stress tensor σ,

$$\varepsilon_{\varphi\psi} = \sum_{i,j=1}^{3}\alpha_{ij}(\varphi,\psi)\sigma_{ij} \tag{1.3}$$

for particular $\alpha_{ij}(\varphi,\psi)$. Using Bragg's reflection model for a crystal lattice, an X-ray is reflected only if the beam hits the object under the so-called Bragg angle θ, which is determined by Bragg's condition

$$2d\sin\theta = n\lambda\,,$$

where n is an integer, d is the distance between the lattice planes and λ is the wavelength of the applied X-rays. Bragg's reflection model is shown in Figure 1.1. Bragg's condition implies $\mathrm{d}d = -d\cot\theta\,\mathrm{d}\theta$ and thus

$$\varepsilon_{\varphi\psi} = \frac{\mathrm{d}d}{d} \approx -\cot\theta_0(\theta_{\varphi\psi} - \theta_0)\,, \tag{1.4}$$

where $\theta_{\varphi\psi}$ denotes the maximum peak position of the reflected X-ray and θ_0 is the Bragg angle of the unstressed specimen. Stresses cause a shift of the maximum peak position $\theta_{\varphi\psi}$ away from θ_0 and by (1.4) we can relate this shift $\theta_{\varphi\psi} - \theta_0$ to the strain $\varepsilon_{\varphi\psi}$ and thus to σ by (1.3). Taking into account that the intensity $I(z)$ of the X-ray

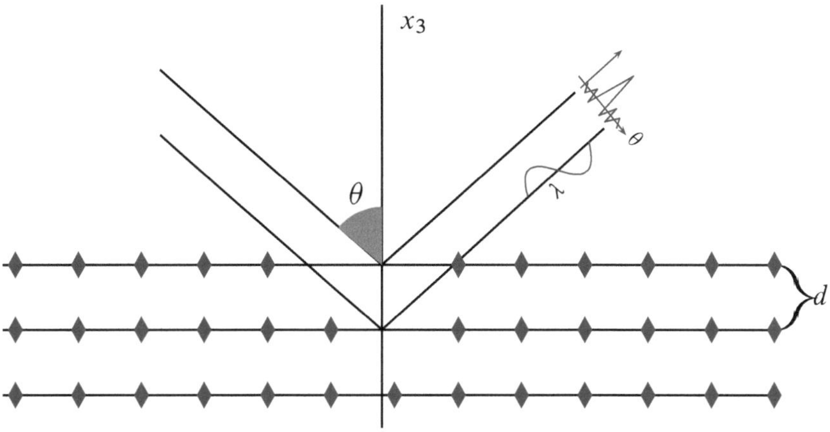

Figure 1.1. Bragg's reflection model on a crystal lattice. The x_3-axis is the axis of rotation.

beam is attenuated within the specimen according to Lambert-Beer's law

$$I(z) = I_0 \mathrm{e}^{-\mu z} ,$$

where z denotes the penetration depth, I_0 is the initial intensity of the emitted X-rays and μ is the material's specific attenuation coefficient, we finally find the fundamental relation

$$-\cot\theta_0(\theta_{\varphi\psi} - \theta_0) = \sum_{i,j=1}^{3} \alpha_{ij}(\varphi, \psi)\check{\sigma}_{ij}(\tau_\psi) , \tag{1.5}$$

where $\tau_\psi = \cos\psi \sin\theta_0/(2\mu)$ is the maximum penetration depth of the X-ray, depending on the tilt angle ψ and

$$\check{\sigma}_{ij}(\tau) = \frac{1}{\tau} \int_0^\infty \sigma_{ij}(z)\mathrm{e}^{-z/\tau}\,\mathrm{d}z$$

denotes the reciprocal Laplace transform of the stress σ_{ij}. The problem of X-ray diffractometry now consists of reconstructing the stress tensor σ from measurements of the Bragg angle shifts $\theta_{\varphi\psi} - \theta_0$ for different rotation and tilt angles φ and ψ. Thus, X-ray diffractometry involves the problem of inverting the Laplace transform

$$Lf(\tau) = \int_0^\infty f(z)\mathrm{e}^{-\tau z}\,\mathrm{d}z$$

where the penetration depth τ is only given for finitely many - say m - discrete points $\tau = \tau_j$, $j = 1, \ldots, m$. The stresses σ_{ij} can be assumed to be smooth, or at least continuous, to have compact support, and to be close to the surface, meaning that we may consider L as a mapping between the Banach spaces $\mathcal{C}([\omega_1, \omega_2])$ and $\mathcal{C}([\tau_{\min}, \tau_{\max}])$, where

$$\tau_{\min} := \min\{\tau_j : j = 1, \ldots, m\}, \quad \tau_{\max} := \max\{\tau_j : j = 1, \ldots, m\}$$

and $\omega_2 > \omega_1 > 0$.

1.2 Two phase retrieval problems

Many applications in optics, such as electron microscopy, analysis of neutron reflective data, and astronomy lead to the inverse problem of identifying the phase of a function from measurements of the amplitude of its Fourier transform, see, e.g., [18, 19, 44, 58, 70, 104, 116, 117, 134, 135].

Here, we will consider two different versions of the problem of phase retrieval:

(1) Given the intensity $r : \mathbb{R} \to \mathbb{R}^+$ of the Fourier transform $\mathcal{F}$ of a real-valued function f, reconstruct $f : \mathbb{R} \to \mathbb{R}$, i.e., find f such that $|\mathcal{F} f| = r$.

(2) Given the magnitude $f : \mathbb{R}^m \to \mathbb{R}^+$ ($m \in \mathbb{N}$) of a complex valued function and the intensity $r : \mathbb{R}^m \to \mathbb{R}^+$ of its Fourier transform, reconstruct its phase, i.e., find $\phi : \mathbb{R}^m \to \mathbb{R}$ such that $|\mathcal{F}(f \cdot e^{\iota\phi})| = r$.

Both versions lead to ill-posed problems, cf. [18, 58], hence regularization has to be applied. While in previous publications (except for [104, 116]) these problems have been considered in Hilbert spaces, we will here study possible formulations in Banach space settings.

Reconstruction of $f : \mathbb{R} \to \mathbb{R}$ from $|\mathcal{F} f| : \mathbb{R} \to \mathbb{R}^+$

Formulation as a nonlinear operator equation

$$F(f) = r$$

leads to the definition

$$F(f)(s) := |\mathcal{F} f|(s) \quad s \in \mathbb{R}$$

of the forward operator $F : X \to Y$. In view of the Hausdorff–Young Theorem, which states that $\mathcal{F} : L^Q(\mathbb{R}^n) \to L^{Q^*}(\mathbb{R}^n)$ is a linear and bounded operator for all $Q \in [1,2]$ and conjugate exponents $Q^* = Q/(Q-1)$, a natural choice of the preimage and image space of F is

$$X = L^Q_{\mathbb{R}}(\mathbb{R})\,, \quad Y = L^{Q^*}_{\mathbb{R}}(\mathbb{R}) \quad \text{with } Q \in [1,2]\,,\ Q^* = \frac{Q}{Q-1}\,,$$

so that F is a bounded operator. Here, the subscript $\mathbb{R}$ indicates values in $\mathbb{R}$. The a priori knowledge of f being real valued implies a certain symmetry of its Fourier transform

$$f \in L^Q_{\mathbb{R}}(\mathbb{R}) \Rightarrow (\mathcal{F} f)(-s) = \overline{(\mathcal{F} f)(s)}\,, \quad s \in \mathbb{R}\,.$$

The one-sided directional derivative, according to the definition

$$\lim_{t \to 0+} \frac{1}{t}\,(F(f+th) - F(f)) = F'(f;h) \in Y \tag{1.6}$$

is given by

$$F'(f;h)(s) = \begin{cases} \dfrac{\Re((\mathcal{F} f)(s)\overline{(\mathcal{F} h)(s)})}{|(\mathcal{F} f)(s)|} & \text{if } s \in \mathbb{R} \setminus \Sigma\,, \\ |(\mathcal{F} h)(s)| & \text{if } s \in \Sigma\,, \end{cases}$$

where

$$\Sigma = \{s \in \mathbb{R} : (\mathcal{F} f)(s) = 0\}\,.$$

Note that $F'(f;\cdot)$ does not necessarily need to be a linear operator. In both cases we have $|F'(f;h)(s)| \le |(\mathcal{F} h)(s)|$, hence $F'(f;h) \in Y = L^{Q^*}_{\mathbb{R}}(\mathbb{R})$ follows from $h \in X = L^Q_{\mathbb{R}}(\mathbb{R})$. However, due to $|\mathcal{F} f|$ appearing in the denominator, F cannot be expected to be Lipschitz continuously differentiable. By the Hausdorff-Young Theorem and the second triangle inequality, F obviously is continuous with respect to the norms in X, Y. However, it is not continuous with respect to the weak topologies in X, Y, as the counterexample given by $(\mathcal{F} f_n)(s) = \sqrt{2}\cos((2\pi n + \pi/2)s)\chi_{[-1,1]}(s)$ (cf. [104]) shows. This fact also illustrates the nonlinearity of the forward operator.

Reconstruction of $\phi : \mathbb{R}^m \to \mathbb{R}$ from $|\mathcal{F}(f \cdot e^{\imath\phi})| : \mathbb{R}^m \to \mathbb{R}^+$

With the forward operator $F : X \to Y$ defined as

$$F(\phi)(\xi) = |\mathcal{F}(f \cdot e^{\imath\phi})|(\xi)$$

this problem can be written as a nonlinear operator equation

$$F(\phi) = r\,.$$

The one-sided directional derivative according to (1.6) is given by

$$F'(\phi;h)(\xi) = \begin{cases} \dfrac{\Re((\mathcal{F}(f\cdot e^{\imath\phi}))(\xi)\overline{(\mathcal{F}(f\cdot e^{\imath\phi}\imath h))(\xi)}}{|(\mathcal{F}(f\cdot e^{\imath\phi}))(\xi)|} & \text{if } \xi\in\mathbb{R}^m\setminus\Sigma\,, \\ |(\mathcal{F}(f\cdot e^{\imath\phi}\imath h))(\xi)| & \text{if } \xi\in\Sigma\,, \end{cases} \tag{1.7}$$

where

$$\Sigma = \{\xi\in\mathbb{R} : (\mathcal{F}(f\cdot e^{\imath\phi}))(\xi) = 0\}\,.$$

Here, we will also consider F as a mapping between Lebesgue spaces

$$X = L^Q(\mathbb{R}^m)\,,\quad Y = L^P(\mathbb{R}^m)\,,$$

and explore possible choices of the exponents Q, P.

Proposition 1.1. *Let* $X = L^Q_{\mathbb{R}}(\mathbb{R}^m)$, $Y = L^P_{\mathbb{R}}(\mathbb{R}^m)$.

(a) *For any*

$$Q\in[1,\infty]\,,\ P\in[2,\infty]\,,\ f\in L^{P^*}(\mathbb{R}^m)\,,$$

the operator $F : X\to Y$, $F(\phi)(\xi) = |\mathcal{F}(f\cdot e^{\imath\phi})|$, *is well-defined and continuous.*

(b) *For any*

(i) $Q\in(1,\infty)\,,\ P\in[\max\{2,Q^*\},\infty)\,,\ f\in L_{\mathbb{R}}^{\frac{QP}{QP-Q-P}}(\mathbb{R}^m)\,,$ *or*

(ii) $Q=1\,,\ P=\infty\,,\ f\in L^\infty_{\mathbb{R}}(\mathbb{R}^m)\,,$ *or*

(iii) $Q=P=\infty\,,\ f\in L^1_{\mathbb{R}}(\mathbb{R}^m)\,,$ *or*

(iv) $Q\in(1,\infty)\,,\ P=\infty\,,\ f\in L^{Q^*}_{\mathbb{R}}(\mathbb{R}^m)\,,$ *or*

(v) $Q=\infty\,,\ P\in[2,\infty)\,,\ f\in L^{P^*}_{\mathbb{R}}(\mathbb{R}^m)\,,$

and $\phi\in X$, *the operator* $F'(\phi;\cdot) : X\to Y$, *defined in* (1.7), *is well-defined and bounded.*

Proof. Considering the Nemytskii operator $N^{\exp} : \phi\mapsto e^{\imath\phi} = \cos(\phi)+\imath\sin(\phi)$, we can invoke standard results, e.g., Theorem 2.2 in [6]. A general Nemytskii operator $N : v\mapsto k(\cdot, v(\cdot))$, defined by a kernel function $k : \Omega\times\mathbb{R}\to\mathbb{R}^j$, is a continuous mapping from $L^p(\Omega)$ to $L^q(\Omega)$ if k satisfies a Caratheodory condition, i.e. measurability with respect to the first and continuity with respect to the second argument, as well as a growth condition $|k(\xi,\eta)|\le a(\xi)+b\eta^{\frac{q}{p}}$ with $a\in L^q$, $a,b\ge 0$. Applying this with $q=\infty$, $\Omega=\mathbb{R}^m$, $a\equiv 1$, $b=0$, we can conclude that $N^{\exp}$ is a continuous mapping from $L^Q_{\mathbb{R}}(\mathbb{R}^m)$ (with $Q\in[1,\infty]$ arbitrary) to $L^\infty_{\mathbb{C}}(\mathbb{R}^m)$. Hence

$F : L^Q_{\mathbb{R}}(\mathbb{R}^m) \to L^P_{\mathbb{R}}(\mathbb{R}^m)$ is continuous, being the concatenation of continuous operators

$$F: \; L^Q_{\mathbb{R}}(\mathbb{R}^m) \underset{N^{\exp}}{\to} L^\infty_{\mathbb{C}}(\mathbb{R}^m) \underset{\mathrm{mult}_f}{\to} L^{P^*}_{\mathbb{C}}(\mathbb{R}^m)$$

$$\underset{\mathcal{F}}{\to} L^P_{\mathbb{C}}(\mathbb{R}^m) \underset{\mathrm{abs}}{\to} L^P_{\mathbb{R}}(\mathbb{R}^m),$$

where abs denotes the pointwise absolute value and mult_f the pointwise multiplication with the function f.

Boundedness of $F'(\phi, \cdot) : L^Q_{\mathbb{R}}(\mathbb{R}^m) \to L^P_{\mathbb{R}}(\mathbb{R}^m)$ follows by using the Hausdorff-Young Theorem together with Hölder's Inequality and $Q \geq P^*$

$$\begin{aligned}
&\left\| F'(\phi, h) \right\|_{L^P_{\mathbb{R}}(\mathbb{R}^m)} \\
&= \left(\int_{\mathbb{R}^m \setminus \Sigma} \left| \frac{\Re(\mathcal{F}(f \cdot e^{\imath\phi}) \overline{\mathcal{F}(f \cdot e^{\imath\phi} \imath h)})}{|\mathcal{F}(f \cdot e^{\imath\phi})|} \right|^P d\xi + \int_\Sigma |\mathcal{F}(f \cdot e^{\imath\phi} \imath h)|^P \, d\xi \right)^{\frac{1}{P}} \\
&\leq \left(\int_{\mathbb{R}^m} |\mathcal{F}(f \cdot e^{\imath\phi} \imath h)|^P \, d\xi \right)^{\frac{1}{P}} \leq \left\| f \cdot e^{\imath\phi} \imath h \right\|_{L^{P^*}} \\
&\leq \|f\|_{L^{\frac{QP}{QP-Q-P}}} \|h\|_{L^Q}
\end{aligned}$$

and inspection of the single cases listed as (i)-(v) in (b). □

1.3 A parameter identification problem for an elliptic partial differential equation

The problem of identifying coefficients or source terms in partial differential equations (PDEs) from data obtained from the PDE solutions arises in a variety of applications ranging from medical imaging, via nondestructive testing, to material characterization as well as model calibration.

Here, we consider a model problem that has been studied repeatedly in the literature (see, e.g., [50, 68, 87, 88, 127, 198]) to illustrate theoretical conditions and numerically test for convergence of regularization methods.

In doing so, we will especially point out the crucial enhancement of freedom in the formulation of the problem when moving from a Hilbert space setting, as has been used in the papers mentioned above, to a general Banach space framework.

Consider the identification of the space-dependent coefficient c in the elliptic boundary value problem

$$-\Delta u + cu = f \quad \text{in } \Omega \tag{1.8}$$

$$u = 0 \quad \text{on } \partial\Omega \tag{1.9}$$

from measurements of u in Ω. Here $\Omega \subseteq \mathbb{R}^m$, $m \in \mathbb{N}$ is assumed to be a smooth, bounded domain. The forward operator

$$F : \mathcal{D}(F) \subseteq X \to Y \,, \tag{1.10}$$

where the spaces X and Y are to be specified below, and its derivative can formally be written as

$$F(c) = A(c)^{-1} f \,, \quad F'(c)h = -A(c)^{-1}(h \cdot F(c)) \,,$$

with

$$\begin{array}{rcl} A(c) : H^2(\Omega) \cap H_0^1(\Omega) & \to & L^2(\Omega) \\ u & \to & -\Delta u + cu \,. \end{array}$$

In order to preserve ellipticity, a straightforward choice of the domain $\mathcal{D}(F)$ is

$$\mathcal{D}(F) = \{c \in X \mid c \geq 0 \text{ a.e. } \|c\|_X \leq \rho\} \tag{1.11}$$

for some sufficiently small $\rho > 0$. For those situations in which the theory requires a nonempty interior of $\mathcal{D}(F)$ in X, the choice

$$\mathcal{D}(F) = \{c \in X \mid \exists \hat{c} \in L^\infty(\Omega), \hat{c} \geq 0 \text{ a.e. } : \; \|c - \hat{c}\|_X \leq \rho\} \,, \tag{1.12}$$

for some sufficiently small $\rho > 0$, has been devised in [88].

So far, the preimage and image spaces X and Y have both been set to $L^2(\Omega)$, in order to fit into the Hilbert space theory described in the existing literature. However, the natural topology for measured data is typically $Y = L^\infty(\Omega)$ or a weighted version thereof. By extending to $Y = L^2(\Omega)$ information is lost, more precisely the problem is made more ill-posed than strictly necessary. On the other hand, $Y = L^1(\Omega)$ can also be of interest, since in the practically relevant situation of impulsive noise, L^1-data fitting provides a more robust option than the choice $Y = L^2(\Omega)$, cf. [49]. Also in preimage space, one often aims at actually reconstructing a uniformly bounded coefficient, corresponding to $X = L^\infty(\Omega)$, or of a coefficient that is sparse in some sense, suggesting the use of the $L^1(\Omega)$-norm in preimage space, cf. [120]. This motivates us to study the possible use of

$$X = L^Q(\Omega) \,, \quad Y = L^P(\Omega) \,,$$

with general exponents $Q, P \in [1, \infty]$, within the context of this example. To adequately limit the exposition, we will only consider the choice (1.11) of the domain here.

Proposition 1.2. *Let* $X = L^Q(\Omega)$, $Y = L^P(\Omega)$.

(a) *For any*

$$Q, P \in [1, \infty] \,, \; f \in L^{\max\{1, \frac{m}{2}\}}(\Omega) \,,$$

the operator $F : \mathcal{D}(F) \subseteq X \to Y$, $F(c) = A(c)^{-1} f$, *is well defined and bounded on* $\mathcal{D}(F)$ *as in* (1.11) *with sufficiently small* $\rho > 0$.

(b) *For any*

(i) $Q, P \in [1,\infty]$, $f \in L^1(\Omega)$, $m \in \{1,2\}$ *or*

(ii) $Q \in (\frac{m}{2},\infty]$, $Q \geq 1$, $P \in [1,\infty]$, $f \in L^{\frac{m}{2}+\epsilon}(\Omega)$, $\epsilon > 0$

and $c \in \mathfrak{D}(F)$, *the operator* $F'(c) : X \to Y$, $F(c) = -A(c)^{-1}(h \cdot F(c))$, *is well-defined and bounded.*

We will not dwell on the question whether, and if so in which sense the operator formally denoted by $F'(c)$ is indeed a derivative of F here but only mention that this point can be treated analogously to the Hilbert space case, cf. [50].

The proof of Proposition 1.2 is based on the following Lemma, whose last item is an enhanced version of Lemma 2 in [126]. Here, with a slight abuse of notation we denote for $S \in [1,\infty]$ by $(W^{2,S} \cap H_0^1)(\Omega)$ the closure of the space $C_0^\infty(\Omega)$ of infinitely differentiable functions with compact support, with respect to the norm

$$\|v\|_{W^{2,S}\cap H_0^1} := \|\Delta v\|_{L^S} + \|\nabla v\|_{L^2}$$

invoking Friedrichs' Inequality.

Lemma 1.3. *Let* $Q, S, P \in [1,\infty]$.

(a) *Let*

(i) $S \geq \frac{m}{2}$ *and* $\{S = 1$ *or* $S > \max\{1, \frac{m}{2}\}$ *or* $P < \infty\}$ *or*

(ii) $S < \frac{m}{2}$ *and* $P \leq \frac{mS}{m-2S}$

Then $(W^{2,S} \cap H_0^1)(\Omega) \subseteq L^P(\Omega)$ *and there exists* $C_{(a)}^P > 0$ *such that*

$$\forall v \in (W^{2,S} \cap H_0^1)(\Omega) : \|v\|_{L^P} \leq C_{(a)}^P \|v\|_{W^{2,S}\cap H_0^1} .$$

(b) *Let* $Q \geq S$ *and*

(i) $S \geq \frac{m}{2}$ *and* $\{S = 1$ *or* $S > \max\{1, \frac{m}{2}\}$ *or* $Q > S\}$ *or*

(ii) $S < \frac{m}{2}$ *and* $Q \geq \frac{m}{2}$

Then $\{u \cdot v : u \in L^Q(\Omega), v \in (W^{2,S} \cap H_0^1)(\Omega)\} \subseteq L^S(\Omega)$ *and there exists* $C_{(b)}^S > 0$ *such that*

$$\forall u \in L^Q(\Omega),\ v \in (W^{2,S} \cap H_0^1)(\Omega) :$$

$$\|u \cdot v\|_{L^S} \leq C_{(b)}^S \|v\|_{L^Q} \|v\|_{W^{2,S}\cap H_0^1} .$$

(c) *Let*

(i) $S \geq \frac{m}{2}$ *or*

(ii) $\frac{2m}{m+2} \leq S < \frac{m}{2}$ *and* $m > 2$.

Then $(W^{2,S} \cap H_0^1)(\Omega) \subseteq L^{S^*}(\Omega)$ *and there exists* $C_{(c)}^S > 0$ *such that*

$$\forall v \in (W^{2,S} \cap H_0^1)(\Omega) : \|v\|_{L^{S^*}} \leq C_{(c)}^S \|v\|_{W^{2,S} \cap H_0^1} .$$

(d) *Let* $c \in \mathfrak{D}(F)$ *with sufficiently small* $\rho > 0$, *and let*

$$1 = S \geq \frac{m}{2} \text{ or } S > \max\left\{1, \frac{m}{2}\right\} \tag{1.13}$$

be satisfied. Then $A(c)^{-1} : L^S(\Omega) \to (W^{2,S} \cap H_0^1)(\Omega)$ *is well-defined and bounded by some constant* $C_{(d)}^S$.

Proof. Assertions (a)–(c) directly follow from the Sobolev Imbedding Theorem (see, e.g., Theorem 4.12 in [2]) with the use of Hölder's inequality for (b):

$$\|u \cdot v\|_{L^S} \leq \|u\|_{L^Q} \|v\|_{L^{\frac{QS}{Q-S}}} \leq C_{(a)}^{\frac{QS}{Q-S}} \|u\|_{L^Q} \|v\|_{W^{2,S} \cap H_0^1}$$

Assumption (1.13) implies that the conditions of (a) with $P = \infty$ and of (c) are satisfied. Multiplying (1.8) with u, integrating by parts, and using $c \geq 0$, as well as Hölder's inequality, we obtain

$$\|\nabla u\|_{L^2(\Omega)}^2 \leq \|f\|_{L^S} \|u\|_{L^{S^*}} \leq C_{(c)}^S \|f\|_{L^S} \left(\|\Delta u\|_{L^S} + \|\nabla u\|_{L^2}\right),$$

hence,

$$\begin{aligned}
\|\nabla u\|_{L^2(\Omega)} &\leq \frac{C_{(c)}^S}{2} \|f\|_{L^S} + \sqrt{\frac{\left(C_{(c)}^S\right)^2}{4} \|f\|_{L^S}^2 + C_{(c)}^S \|f\|_{L^S} \|\Delta v\|_{L^S}} \\
&\leq C_{(c)}^S \|f\|_{L^S} + \|\Delta u\|_{L^S} ,
\end{aligned} \tag{1.14}$$

where we used $\frac{a}{2} + \sqrt{\frac{a^2}{4} + ab} \leq a + b$ for $a, b, \geq 0$. We easily obtain an estimate of $\|\Delta u\|_{L^S}$ by using the triangle inequality in (1.8), as well as inserting (1.14):

$$\begin{aligned}
\|\Delta u\|_{L^S(\Omega)} &\leq \|f\|_{L^S} + \|c\|_{L^S} \|u\|_{L^\infty} \\
&\leq \|f\|_{L^S} + C_{(a)}^\infty \|c\|_{L^S} \left(\|\Delta u\|_{L^S} + \|\nabla u\|_{L^2}\right) \\
&\leq \|f\|_{L^S} + C_{(a)}^\infty \|c\|_{L^S} \left(2 \|\Delta u\|_{L^S} + C_{(c)}^S \|f\|_{L^S}\right).
\end{aligned}$$

Hence for $\|c\|_{L^S} \le \rho < \frac{1}{2C^{\infty}_{(a)}}$

$$\|\Delta u\|_{L^S(\Omega)} \le \frac{1}{1-2C^{\infty}_{(a)}\rho}\left(1 + C^{\infty}_{(a)}\,\|c\|_{L^S}\, C^{S}_{(c)}\right)\|f\|_{L^S}\Big).$$

Re-inserting this into (1.14) gives the desired overall $(W^{2,S} \cap H^1_0)(\Omega)$ of u. □

Now we are in the position to show the proof of Proposition 1.2.

Proof. (Proposition 1.2) To see (a), readily note that, under the assumptions we made, we can set $S > \max\{1, \frac{m}{2}\}$ in (a), (c), (d) of Lemma 1.3 to obtain for

$$\left\|A(c)^{-1}f\right\|_{L^P} \le C^P_{(a)}\left\|A(c)^{-1}f\right\|_{W^{2,S}\cap H^1_0} \le C^P_{(a)}C^S_{(d)}\,\|f\|_{L^S}\,.$$

Similarly, in order to prove (b) we set $S = 1$ in (i), $S = \min\{Q\,,\,\frac{m}{2} + \epsilon\}$ in (ii), and estimate, using Lemma 1.3 (a)–(d):

$$\begin{aligned}\left\|A(c)^{-1}(h \cdot F(c)\right\|_{L^P} &\le C^P_{(a)}C^S_{(d)}\,\|h \cdot F(c)\|_{L^S}\\ &\le C^P_{(a)}C^S_{(b)}C^S_{(d)}\,\|h\|_{L^Q}\,\|F(c)\|_{W^{2,S}\cap H^1_0}\\ &\le C^P_{(a)}C^S_{(b)}\left(C^S_{(d)}\right)^2\,\|h\|_{L^Q}\,\|f\|_{L^S}\,.\end{aligned}$$

□

1.4 An inverse problem from finance

Inverse problems arising in financial markets, preferably aimed at finding volatility functions from option prices, have found increasing interest in the years around the change of the millennium, see, e.g., [25,53,56,62,140]. Our focus here is on a specific nonlinear inverse problem from this field mentioned in [104, Section 6], namely, the problem of calibrating purely time-dependent volatility functions based on maturity-dependent prices of European vanilla call options with fixed strike. See [96, 105] for details. It certainly is only a toy problem in mathematical finance, but due to its simple and explicit structure it serves as a benchmark problem for case studies in mathematical finance. However, following the decoupling approach suggested in [63], variants of this problem also occur as serious subproblems for the recovery of local volatility surfaces. Such surfaces are of considerable practical importance in finance.

More precisely, we consider a family of European vanilla call options written on an asset with actual asset price $S > 0$ for varying maturities $t \in [0, 1]$, but fixed strike price $K > 0$, and a fixed risk-free interest rate $r \ge 0$. We denote the associated function of option prices, observed at an arbitrage-free financial market, by $y(t)$ $(0 \le t \le 1)$. From that function we will determine the unknown volatility term-structure. Furthermore, we denote the square of the volatilities at time t by $x(t)$ $(0 \le t \le 1)$ and neglect a possible dependence of the volatilities on the asset price. Using a generalized Black–Scholes formula we obtain as the fair price function for the family of options

$$[F(x)](t) = U_{\mathrm{BS}}(S, K, r, t, [J\,x](t)) \quad (0 \le t \le 1) \tag{1.15}$$

exploiting the simple integration operator

$$[J\,h]\,(s) \;=\; \int_0^s h(t)\,dt \quad (0 \le s \le 1)\,,$$

the Black–Scholes function U_{BS} defined as

$$U_{\mathrm{BS}}(S, K, r, \tau, s) := \begin{cases} S\Phi(d_1) - Ke^{-r\tau}\Phi(d_2) & (s > 0) \\ \max\,(S - Ke^{-r\tau}, 0) & (s = 0) \end{cases}$$

with

$$d_1 = \frac{\ln\frac{S}{K} + r\tau + \frac{s}{2}}{\sqrt{s}}, \quad d_2 = d_1 - \sqrt{s}\,,$$

and the cumulative density function of the standard normal distribution

$$\Phi(\zeta) \;=\; \frac{1}{\sqrt{2\pi}} \int_{-\infty}^{\zeta} e^{-\frac{\xi^2}{2}}\,d\xi.$$

Hence, the inverse problem appears as a nonlinear operator equation (see Section 3.1.2 below). More precisely, the solution $x \in L^2(0,1)$ to equation

$$[F(x)](t) = y(t) \quad (0 \le t \le 1)$$

is to be determined from a given data function $y \in L^2(0,1)$, where the mapping $F : x \mapsto y$ from (1.15) acts as the nonlinear forward operator

$$F : \mathcal{D}(F) \subseteq X \to Y$$

with Hilbert spaces

$$X = Y = L^2(0,1)$$

and a convex domain

$$\mathcal{D}(F) \;=\; \{x \in X : x(t) \ge c_0 > 0 \;\text{ a.e. on } [0,1]\}\,.$$

This seems to be an appropriate model for the kind of inverse option pricing problem under investigation.

Obviously, the forward operator F is a composition

$$F = N \circ J$$

of the linear integral operator J with the nonlinear Nemytskii operator

$$[N(z)](t) = k(t, z(t)) = U_{\mathrm{BS}}(S, K, r, t, z(t)) \quad (0 \le t \le 1)$$

possessing a smooth generator function k. Furthermore, for all $x \in \mathcal{D}(F)$ the Gâteaux derivative $F'(x)$ of the forward operator exists and attains the form

$$F'(x) = G(x) \circ J$$

with the linear multiplication operator

$$[G(x)\,h](t) = m(x,t)\,h(t) \quad (0 \le t \le 1,\ h \in L^2(0,1))$$

determined by a nonnegative multiplier function

$$m(x,0) = 0, \quad m(x,t) = \frac{\partial U_{\mathrm{BS}}(S, K, r, t, [J\,x](t))}{\partial s} > 0 \quad (0 < t \le 1)$$

cf. [96, Lemma 2.1], for which, for every $0 < t \le 1$, we can derive the formula

$$m(x,t) = \frac{X}{2\sqrt{2\pi[J\,x](t)}} \exp\left(-\frac{(\kappa + rt)^2}{2[J\,x](t)} - \frac{(\kappa + rt)}{2} - \frac{[J\,x](t)}{8}\right) > 0.$$

This formula contains the logmoneyness

$$\kappa = \ln\left(\frac{S}{K}\right).$$

Note that due to $c_0 > 0$ we have

$$\underline{c}\,t \le [J\,x](t) \le \overline{c}\,\sqrt{t} \quad (0 \le t \le 1)$$

with $\underline{c} = c_0 > 0$ and $\overline{c} = \|x\|_{L^2(0,1)}$. Hence, we may estimate for all $x \in \mathcal{D}(F)$

$$\underline{C}\,\frac{\exp\left(-\frac{\kappa^2}{2\underline{c}\,t}\right)}{\sqrt[4]{t}} \le m(x,t) \le \overline{C}\,\frac{\exp\left(-\frac{\kappa^2}{2\overline{c}\,\sqrt{t}}\right)}{\sqrt{t}} \quad (0 < t \le 1) \tag{1.16}$$

with some positive constants $\underline{C}$ and $\overline{C}$.

For in-the-money options and out-of-the-money options, i.e. for $S \ne K$ or equivalently $\kappa \ne 0$, the functions $m(x,t)$ are continuous at t and have an essential zero at $t = 0$. In the neighborhood of this zero the multiplier function goes to zero exponentially, i.e. faster than any power of t, whenever the moneyness κ does not vanish, see formula (1.16). From [96] we can make the following assertions: The multiplier functions $m(x,\cdot)$ all belong to $L^\infty(0,1)$ and hence $G(x)$ is a bounded multiplication operator in $L^2(0,1)$. Then $F'(x)$ is a compact linear operator mapping in $L^2(0,1)$.

The nonlinear operator F, however, is injective, continuous, compact, weakly continuous and hence weakly closed. Moreover, for all $x \in \mathcal{D}(F)$, $F'(x)$ is even a Fréchet derivative, since it satisfies the condition

$$\|F(x_2) - F(x_1) - F'(x_1)(x_2 - x_1)\|_Y \le \gamma \|x_2 - x_1\|_X^2 \quad \text{for all } x_1, x_2 \in \mathcal{D}(F) \tag{1.17}$$

with

$$\gamma = \frac{1}{2} \sup_{(t,s)\in[0,1]^2:\, 0\le ct\le s} \left| \frac{\partial^2 U_{\mathrm{BS}}(S,K,r,t,s)}{\partial s^2} \right| < \infty .$$

Note that γ, which can be interpreted as a Lipschitz constant for $F'(x)$ for varying x, follows from the uniform boundedness of the second partial derivative of the Black–Scholes function U_{BS} with respect to the last variable, whereas the multiplier function $m(x,\cdot)$ defining $G(x)$ is due to the corresponding first partial derivative of U_{BS}.

As a consequence of the structure of F mentioned above, the inverse operator $F^{-1} : \mathrm{Range}(F) \subset Y \to X$ exists, but cannot be continuous, and the corresponding operator equation is ill-posed and requires regularization methods (see Chapter 3) for its stable approximate solution when instead of y only noisy option data y^δ with $\|y^\delta - y\|_Y \le \delta$ and noise level $\delta > 0$ are available. For the nonlinear Tikhonov regularization as the most prominent method (for a general approach see Chapter 4), where the regularized solutions x_α^δ in the *Hilbert space setting* are minimizers of the penalized least squares problem

$$T_\alpha(x) := \frac{1}{2}\|Ax - y^\delta\|_Y^2 + \alpha \|x - \overline{x}\|_X^2 \to \min, \quad \text{subject to } x \in X ,$$

with reference element $\overline{x} \in L^2(0,1)$ and regularization parameter $\alpha > 0$, the well-known theory from [67, Chap. 10] is applicable for $\kappa \ne 0$ because of (1.17) and yields for the convergence rates to the true volatility function $x^\dagger$ (see [96, Section 5])

$$\|x_\alpha^\delta - x^\dagger\|_{L^2(0,1)} = O(\sqrt{\delta}) \quad \text{as} \quad \delta \to 0. \tag{1.18}$$

Proposition 1.4. *Provided that $\kappa \ne 0$ we have a convergence rate* (1.18), *whenever the regularization parameter is chosen as $\alpha(\delta) \sim \delta$ and the true volatility $x^\dagger \in \mathcal{D}(F)$ fulfils a source condition*

$$x^\dagger - \overline{x} = F'(x^\dagger)^* w \tag{1.19}$$

for some $w \in L^2(0,1)$ satisfying the smallness condition

$$\gamma \|w\|_{L^2(0,1)} < 1 .$$

Note that the condition (1.19) in Proposition 1.4 implies $x^\dagger(1) = \overline{x}(1)$ and $x^\dagger - \overline{x} \in W^{1,2}(0,1)$. Vice versa, for measurable $(x^\dagger - \overline{x})'$ the function

$$w(t) = (x^\dagger(t) - \overline{x}(t))'/m(x^\dagger, t),$$

$0 < t \le 1$, is in $L^2(0,1)$ and satisfies (1.19).

However for at-the-money options, i.e. for $S = K$ or equivalently $\kappa = 0$, the classical Hilbert space approach fails, because the corresponding Fréchet derivative of F becomes degenerate. Nevertheless based on the Banach space theory from [104] one can find the same rate result (1.18) by exploiting the well-known explicit structure of the forward operator when working with *auxiliary Banach spaces* $L^{2-\varepsilon}(0,1)$, $0 < \varepsilon < 1$, complementing the Hilbert space $L^2(0,1)$.

For $S = K$ one can also write $F'(x) = G(u) \circ J$ for the Gâteaux derivative of the forward operator, but now for the first partial derivative of the Black–Scholes function the explicit expression is

$$\frac{\partial U_{\mathrm{BS}}(S,K,r,t,s)}{\partial s} = \frac{S}{2\sqrt{2\pi\, s}} \exp\left(-\frac{r^2 t^2}{2s} - \frac{r\,t}{2} - \frac{s}{8}\right) > 0,$$

and for the second partial derivative

$$\begin{aligned}&\frac{\partial^2 U_{\mathrm{BS}}(S,K,r,t,s)}{\partial s^2}\\ &= -\frac{S}{4\sqrt{2\pi\, s}} \left(-\frac{r^2 t^2}{s^2} + \frac{1}{4} + \frac{1}{s}\right) \exp\left(-\frac{r^2 t^2}{2s} - \frac{r\,t}{2} - \frac{s}{8}\right),\end{aligned} \tag{1.20}$$

see ([105]). However, on inspection of formula (1.20) we see that

$$\sup_{(t,s)\in[0,1]^2:\, 0\le \underline{c}t\le s} \left| \frac{\partial^2 U_{\mathrm{BS}}(X,K,r,t,s)}{\partial s^2} \right| = \infty.$$

Hence, for vanishing logmoneyness $\kappa = 0$ an inequality (1.17) cannot be shown in this way, since the constant γ explodes. But (1.17) was required to prove Proposition 1.4 in [96]. On the other hand, for $\kappa = 0$ the forward operator F from (1.15) is less smoothing than for $\kappa \neq 0$, because the multiplier function $m(x^\dagger, t)$ has a pole at $t = 0$ for at-the-money options instead of a zero in all other cases. Hence, the ill-posedness of the problem is less severe in the singular case $\kappa = 0$ than in all other cases $\kappa \neq 0$. However, to obtain rates, a more sophisticated approach is necessary, that allows us to compensate the degeneration of essential properties of the derivative $F'(x^\dagger)$.

More precisely, we consider in addition to the Hilbert spaces $X = Y = L^2(0,1)$ the Banach space

$$\tilde{Y} = L^{2-\varepsilon}(0,1) \supset Y \quad (0 < \varepsilon < 1)$$

with dual space

$$\tilde{Y}^* \text{ isometrically isomorph to } L^{\frac{2-\varepsilon}{1-\varepsilon}}(0,1)\,.$$

We assign $\varepsilon > 0$ a small value $\nu := \frac{\varepsilon}{1-\varepsilon} > 0$, where evidently we have $2 - \varepsilon = \frac{2+\nu}{1+\nu}$ and $\frac{2-\varepsilon}{1-\varepsilon} = 2 + \nu$. Then it can be shown that for $\kappa = 0$ inequalities of the form

$$\begin{aligned}\|F(x_2) - F(x_1) - F'(x_1)\,(x_2 - x_1)\|_{\tilde{Y}} &\le \tilde{\gamma}\, \|x_2 - x_1\|_X^2\\ &\text{for all } x_1, x_2 \in \mathcal{D}(F)\end{aligned} \tag{1.21}$$

hold taking the place of (1.17) (see the discussion in Section 3.2.4), where $\tilde{\gamma}$ is of the form $\tilde{\gamma} = C\, \| t\, \frac{\partial^2 U_{\mathrm{BS}}(S,K,r,t,\tilde{s})}{\partial s^2} \|_{L^{2-\varepsilon}(0,1)}$. Note that the norm of the Taylor remainder on the left-hand side of (1.21) is a norm in a Banach space $\tilde{Y}$, whereas in (1.17) the corresponding Taylor remainder is measured in the norm of a Hilbert space Y. In order to establish the consequences of this use of auxiliary Banach spaces we refer to [104, Section 6] for the following proposition.

Proposition 1.5. *Provided that* $\kappa = 0$ *we have a convergence rate* (1.18), *whenever the regularization parameter is chosen as* $\alpha(\delta) \sim \delta$ *and the true volatility* $x^\dagger \in \mathcal{D}(F)$ *fulfils the conditions that* $(x^\dagger - \overline{x})'$ *is measurable with* $x^\dagger(1) = \overline{x}(1)$ *and that there is an arbitrarily small* $\nu > 0$ *such that the function*

$$\tilde{w}(t) = \frac{x^\dagger - \overline{x}}{m(x^\dagger, t)}, \qquad 0 < t < 1,$$

satisfies the condition

$$\tilde{w} \in L^{2+\nu}(0,1) \tag{1.22}$$

and the smallness condition

$$\tilde{\gamma}\, \|\tilde{w}\|_{L^{2+\nu}(0,1)} < 1\,.$$

Note that condition (1.22) implies that $x^\dagger - \overline{x} \in W^{1,1}(0,1)$ and that the occurring conditions now refer to Banach spaces. The result of Proposition 1.5 for the convergence rate could only be obtained by leaving the Hilbert space setting. It is a nice example for the advantage of extending components of the regularization theory from Hilbert spaces to Banach spaces.

1.5 Sparsity constraints

In many applications one can assume that the solution of the underlying ill-posed problem is a linear combination of only a few elements of a specific, given system of functions. For example in image processing (e.g. deblurring) such a system may be given by a (bi-)orthogonal wavelet basis (cf. e.g. [190]). In signal processing it is often the peak system or the bell functions system (cf. e.g. [146]). In parameter identification problems for PDEs it may be a spline basis, like the basis of hat-functions or the basis of box-functions (cf. [120]). Such a priori knowledge is called a *sparsity constraint* of the solution. In particular, a function x is called *sparse* with respect to the system $\{\psi_n\}$, if only finitely many expansion coefficients $\langle x, \psi_n \rangle$ are non-zero.

There are several ways to incorporate sparsity constrains into a regularization scheme. Here, we focus on the so called Tikhonov-type regularization. To this end let A be a linear operator with non-closed range and y^δ a noisy version of the true data

$y = Ax$. Then, as was shown by Daubechies, M. Defrise and C. De Mol in [55], the minimizers of the Tikhonov functional

$$\frac{1}{2}\|Ax - y^\delta\|^2 + \alpha \frac{1}{q} \sum_n w_n |\langle x, \psi_n \rangle|^q, \quad 1 \le q \le 2 \tag{1.23}$$

can be used to regularize the inverse of A.

To see this, consider first the minimization with respect to the first summand, the so called fidelity term, only. Due to ill-posedness, the range of the operator A is not closed and therefore, in all practical cases, the minimizers will tend to be unbounded. If, on the other hand, we consider the minimizer with respect to the second summand, the so called penalty term, we get only the trivial solution. The minimization with respect to the Tikhonov functional balances the fidelity term and the penalty term, where the weighting between these two terms is given by the regularization parameter α. We note that a thorough introduction to Tikhonov-type regularization will be given later in Chapter 3.

One can easily see that for $q = 2$ the functional in (1.23) is just the classical smooth Tikhonov functional, cf. [68]. However, here we are especially interested in the case where q tends to one. Surprisingly, for $q = 1$, the minimizers of the Tikhonov functional are sparse, even if the true solution is not, which can be seen in several ways.

First, we justify the sparsity of the minimizers by a Lagrangian approach. In fact, one might consider the Tikhonov functional

$$\frac{1}{2}\|Ax - y^\delta\|^2 + \alpha \sum_n w_n |\langle x, \psi_n \rangle|$$

as the Lagrangian functional of a restricted minimization problem, either

$$\frac{1}{2}\|Ax - y^\delta\|^2 \to \min, \quad \text{subject to} \quad \sum_n w_n |\langle x, \psi_n \rangle| \le c$$

or

$$\sum_n w_n |\langle x, \psi_n \rangle| \to \min, \quad \text{subject to} \quad \frac{1}{2}\|Ax - y^\delta\|^2 \le c.$$

Hence, in the optimal point the level sets of the fidelity functional $\frac{1}{2}\|Ax - y^\delta\|^2$ and the penalty $\sum_n w_n |\langle x, \psi_n \rangle|$ are tangential. Therefore, as can be seen in Figure 1.2, for small α the reconstructions are large with respect to the ℓ^1-penalty. However, as α grows, the reconstructions will approach one of the vertices of the ℓ^1-ball, i.e. they will become sparse. Therefore, for appropriately chosen α we might hope to get a sparse reconstruction of the original data. Finally, if α is chosen too large, the reconstructions will be sparse but no longer meaningful as approximate solutions to $Ax = y$.

One can also justify the sparsity of the minimizers purely by regarding the penalty for $q = 1$ and $q = 2$, i.e. $\alpha \sum_n w_n |\langle x, \psi_n \rangle|$ and $\alpha \frac{1}{2} \sum_n w_n |\langle x, \psi_n \rangle|^2$. As can be

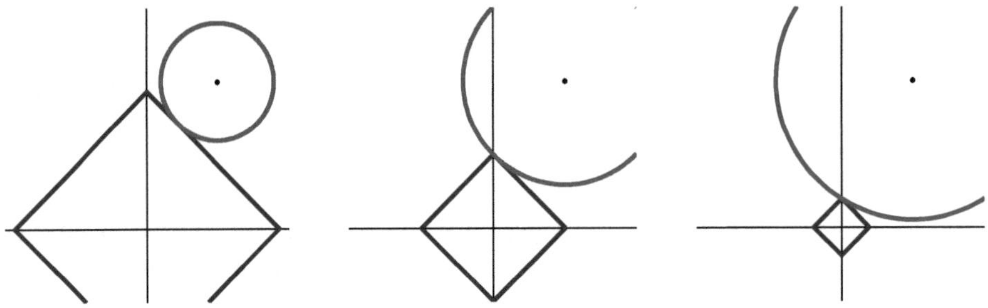

(a) Small α **(b)** Medium-sized α **(c)** Large α

Figure 1.2. Lagrangian view of sparsity promoting Tikhonov reconstruction in two dimensions. Let the black dot be the noisy data y^δ and the function system be given by the canonical basis of $\mathbb{R}^2$. As α grows, the reconstruction point, i.e. the tangent point of the ℓ^1-ball around the origin and the ℓ^2-ball around the noisy data, will approach one of the vertices of the ℓ^1-ball, i.e. it will become sparse (cf. [29]).

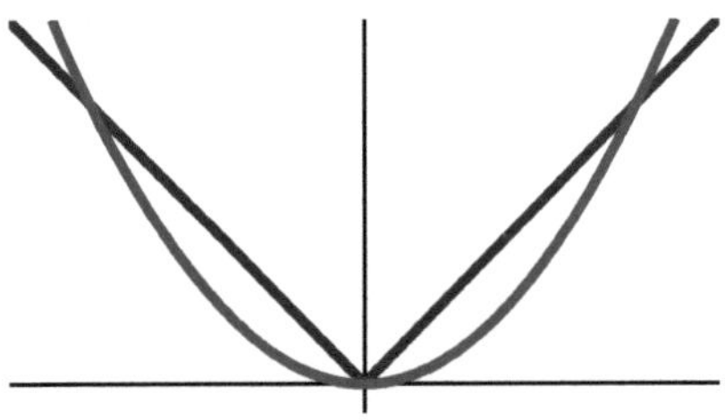

Figure 1.3. Penalty based view of sparsity promoting Tikhonov reconstruction. For small values a quadratic function is smaller than the absolute value function. Therefore, for $q = 2$ the penalty of the Tikhonov functional (1.23) favors elements with many small expansion coefficients, whereas for $q = 1$ elements with few relatively large coefficients are favored.

seen in Figure 1.3 the ℓ^1-functional penalizes small coefficients more severely than the quadratic ℓ^2-penalty. Hence, the reconstructions will tend to have expansion vectors with few large coefficients, rather than vectors with many small coefficients.

In order to arrive at a numerical realization of the Tikhonov-type regularization, one needs a minimization scheme for the functional (1.23). Several such schemes are available cf. [149] and the references therein.

The main drawback of a strict sparsity assumption, as introduced above, is that the system $\{\psi_n\}$ has to be perfectly adapted to the original data x, in the sense that x is only allowed to have finitely many non-zero expansion coefficients with respect to $\{\psi_n\}$. While in some application the system $\{\psi_n\}$ arises naturally, in other applications this might not be the case. For example if x is a natural image e.g. of a human face, which system $\{\psi_n\}$ should be chosen to ensure sparsity?

One remedy to this problem might be to choose a more general system than an orthogonal basis, say a frame or a dictionary. However, the heart of the problem remains unchanged, even in these cases: the function is still allowed to have only finitely many non-zero expansion coefficients. Furthermore, one must be very cautious when dealing with function systems which are non-orthonormal, since a badly chosen system might worsen the stability of the reconstruction and hence render it meaningless.

Fortunately, in most cases one has an idea which system should be chosen, in order to ensure *almost sparsity*. By this we mean that only a small number of expansion coefficients $\langle x, \psi_n \rangle$ carry almost all information about x.

Returning to our example of face images, it is well-known that the Fourier-basis or the Daubechies-basis are good choices, since images are almost sparse with respect to these systems, cf. [55]. Image compression algorithms like the JPEG algorithm or the JPEG2000 algorithm employ this idea in order to store images in an efficient way.

With the same arguments as in the case $q = 1$, one might deduce that for almost sparse functions penalty terms with q slightly larger than 1 might be better suited than $q = 1$.

In the last part of this section we will illustrate the latter claim by means of a simple numerical example, the retrieval of the velocity from GPS data, which can be mathematically modeled as numerical differentiation. We start with the following question: Given the GPS coordinates of a vehicle, is it possible to recover its velocity? We know that for given velocity v the position of the vehicle may be computed via

$$s(T) = \int_0^T v(t)\,dt + s(0).$$

We can choose the coordinates such that $s(0) = 0$. Altogether we have the setting $Av = s^\delta$, $\|s - s^\delta\| \le \delta$, where A is the integral operator $(Av)(T) = \int_0^T v(t)\,dt$. The system of functions with respect to which we expand the function is that of box-splines.

First, we consider a smooth velocity, as depicted in Figure 1.4. The resulting exact position and the simulated GPS position can also be found in Figure 1.4. We see that the noise clutter is negligible compared to the value of the position. As expected, the naïve approach, where v is reconstructed by v^δ, the minimizers of (1.23) for $\alpha = 0$, does not result in a reasonable reconstruction, as can be seen in Figure 1.4 (d). On the other hand, by choosing the classical case $q = 2$ and α appropriately, we improve the quality of the reconstruction significantly, as can be seen in Figure 1.4 (e).

However, the situation changes drastically if we consider the velocity profile depicted in Figure 1.5, which consists of three sharp peaks and is much more realistic for a vehicle. We remark that the peaks have steep slopes but are not discontinuous due to the physical setting of our problem. As we can see in Figure 1.5 (b) and (c), the quadratic Tikhonov reconstructions, i.e. reconstructions for $q = 2$, are not able to recover simultaneously the height and the support of the peaks.

Since the fact that the true signal consists of peaks can be considered as a sparsity information, we next consider the minimizers of (1.23) for $q = 1$. As is demonstrated

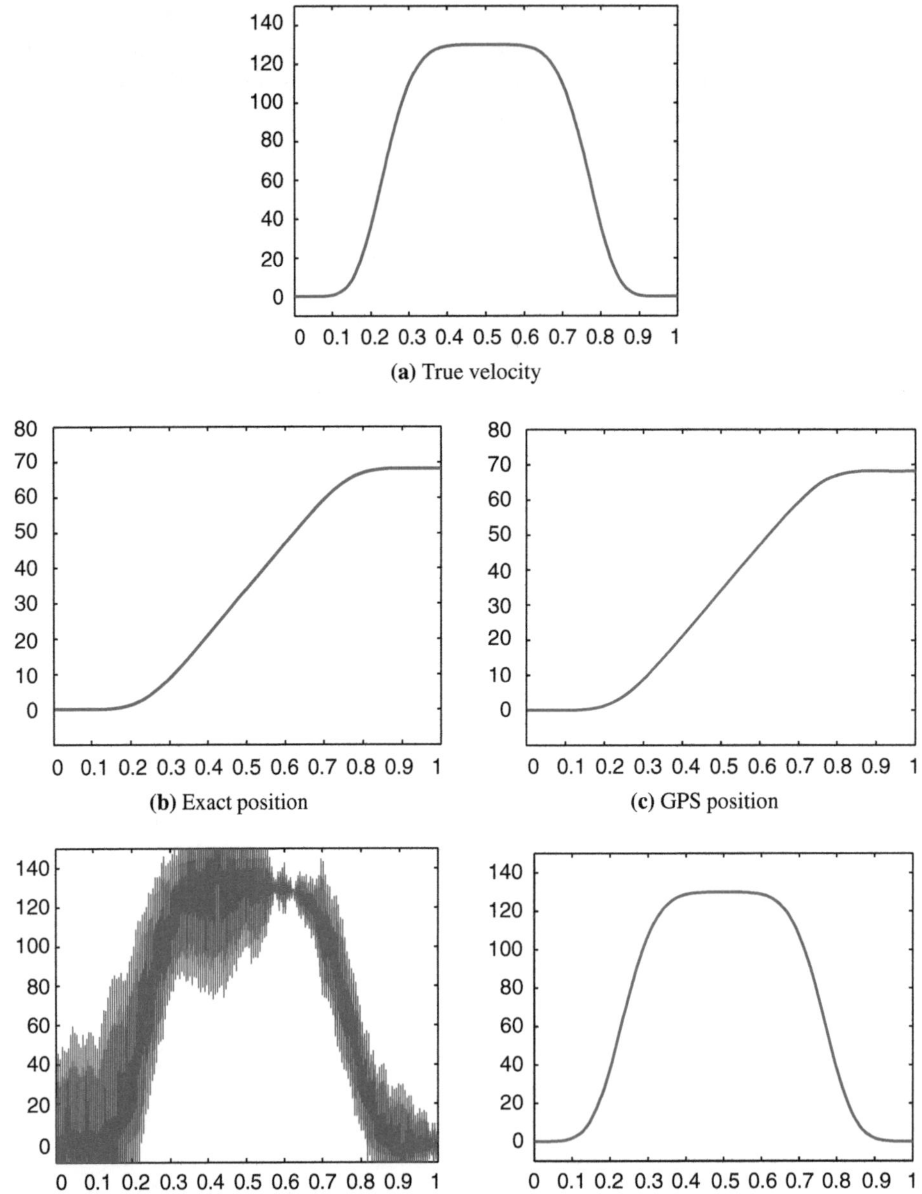

(**a**) True velocity

(**b**) Exact position

(**c**) GPS position

(**d**) Naïve reconstruction

(**e**) Tikhonov reconstruction

Figure 1.4. Naïve and Tikhonov reconstruction of the velocity from GPS coordinates (x-axis: time in hours; y-axis: velocity in km/h respectively position in km).

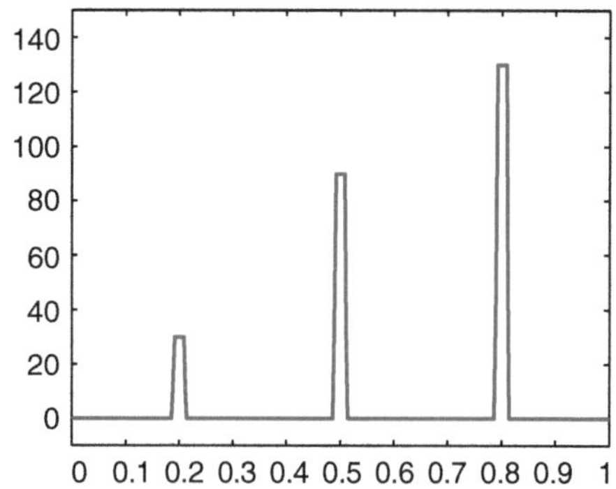

(a) True velocity

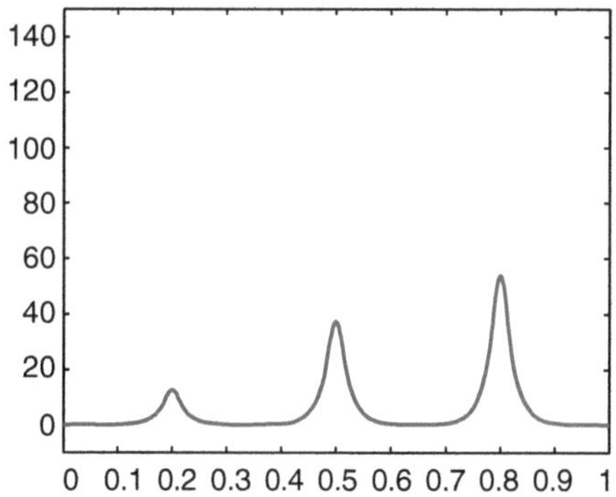

(b) Oversmoothed ℓ^2-reconstruction

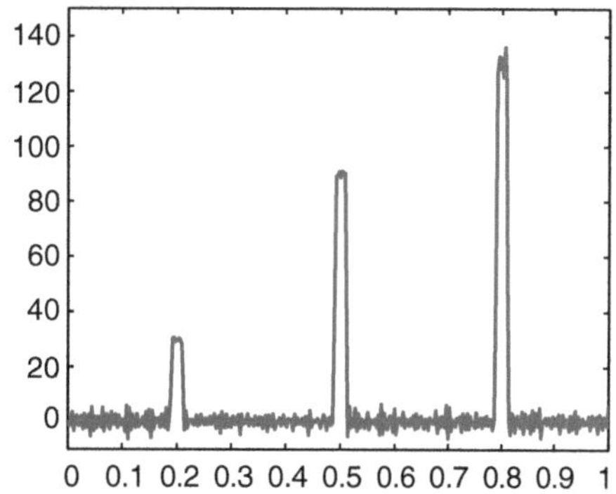

(c) ℓ^2-reconstruction with noise clutter

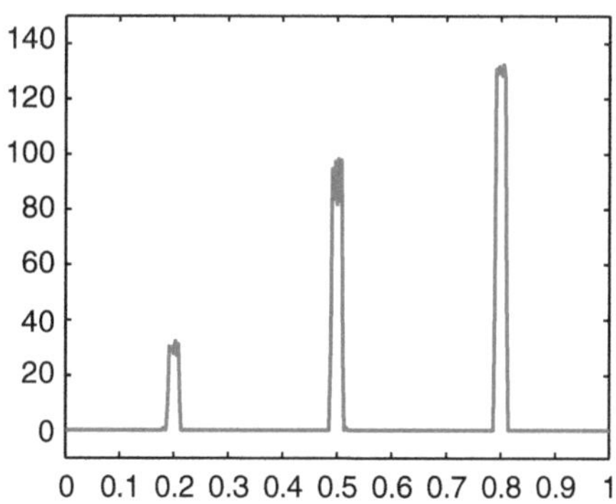

(d) ℓ^1-reconstruction

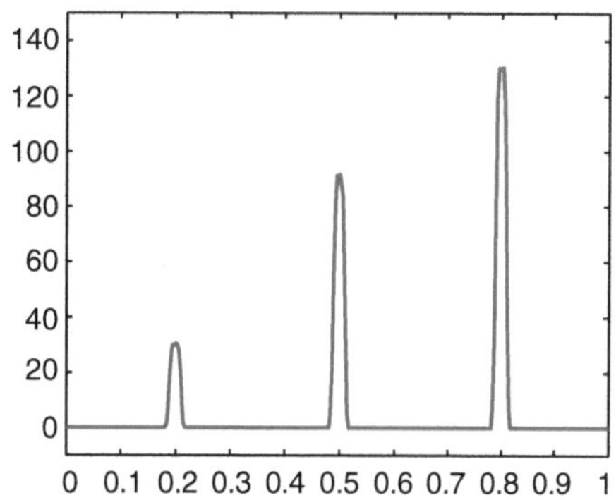

(e) $\ell^{1.1}$-reconstruction

Figure 1.5. True velocity and Tikhonov-type reconstructions from GPS coordinates for different penalties based on Hilbert and Banach space norms (x-axis: time in hours; y-axis: velocity in km/h).

in Figure 1.5 (d), this improves the reconstruction. But as can be seen in Figure 1.5 (e), the choice of a q slightly larger than 1 gives even better results than $q = 1$.

We conclude that the use of non-quadratic penalties and in particular penalties based on Banach space norms in certain cases, like the sparsity constraint described here, improves the quality of the reconstructions.

Part II

Geometry and mathematical tools of Banach spaces

The aim of the second part of this book is to introduce elementary results from convex analysis and Banach space theory (cf., e.g., [15, 48, 69, 164]) in Chapter 2, as well as from regularization theory in Chapter 3 and thereby to provide the basis for the analysis of regularization methods in Banach spaces.

We begin Chapter 2 with an introduction to the basic mathematical notation. Then we will introduce three important tools of convex analysis: the subgradient, the duality mapping and the Bregman distance. On the one hand, the subgradient is a generalization of the derivative concept for convex functions. On the other hand, the duality mapping is one of the most important tools in our analysis and it will turn out to be the subgradient of a power of the norm. The Bregman distance is a common way of measuring the deviation of elements in Banach spaces with respect to some convex functional, thus complementing the norm concept. In the last part of Chapter 2 we present the definitions and properties of several important classes of Banach spaces like uniformly smooth spaces, uniformly convex spaces, spaces smooth of power type (p-smooth spaces) and spaces convex of power type (p-convex spaces). We also discuss the connection between the properties of Banach spaces and the properties of the related duality mappings, as well as Bregman distances.

Chapter 3 presents some fundamental elements of regularization theory applied to finding the stable approximate solution of ill-posed linear and nonlinear operator equations formulated in Banach spaces. First, in Section 3.1 we explain the ill-posedness phenomenon and its consequences. Then in Section 3.2 we introduce the regularization approach in general and Tikhonov-type regularization in particular. In order to obtain convergence rates for the regularized solutions, additional conditions on the solution smoothness and, for nonlinear problems, on the structure of nonlinearity have to be imposed. In this context, we outline the concepts of source conditions and approximate source conditions, and emphasize the distinguished role of a specific type of variational inequality, which expresses smoothness and nonlinearity conditions in a unified and concise manner. The chapter will be completed by presenting a comprehensive discussion of nonlinearity conditions and of the differences between the linear and the nonlinear case.

Chapter 2

Preliminaries and basic definitions

The main aim in the first section of this chapter is to introduce the basic notations and to provide some definitions of quintessential terms used in this book.

2.1 Basic mathematical tools

We start with some standard definitions, inequalities and nomenclature. A *generic constant*

$$C > 0$$

can take a different value every time it is used. We exploit this notation in some proofs to avoid unnecessary enumeration of constants.

Definition 2.1 (conjugate exponents). For $p > 1$ we denote by $p^* > 1$ the *conjugate exponent* of p, satisfying the equation

$$\frac{1}{p} + \frac{1}{p^*} = 1.$$

Definition 2.2 (Banach space). A normed linear space $(X, \|\cdot\|_X)$ is called a *Banach space* if it is complete, i.e., if every Cauchy sequence is convergent with respect to the norm $\|\cdot\|_X$. We recall that a sequence $\{x_n\}$ is a Cauchy sequence if, for every $\varepsilon > 0$, there exists an integer $N(\varepsilon)$ such that $\|x_n - x_m\| < \varepsilon$ for all $n, m \geq N(\varepsilon)$.

Definition 2.3 (dual space). We denote by X^* the *dual space* of a Banach space X, which is the Banach space of all bounded (continuous) linear functionals $x^* : X \to \mathbb{R}$ equipped with the norm

$$\|x^*\|_{X^*} := \sup_{\|x\|=1} |x^*(x)|.$$

Definition 2.4 (dual pairing). For $x^* \in X^*$ and $x \in X$ we denote by $\langle x^*, x\rangle_{X^* \times X}$ and $\langle x, x^*\rangle_{X \times X^*}$ the *duality pairing (duality product* or *dual composition)* defined as

$$\langle x^*, x\rangle_{X^* \times X} := \langle x, x^*\rangle_{X \times X^*} := x^*(x).$$

In norms and dual pairings, when clear from the context, we will omit the indices indicating the spaces. Let us mention that the duality pairing is not to be confused with the duality mapping, which will be introduced in Definition 2.27.

Definition 2.5 (annihilator). Let X be a Banach space and let $M \subseteq X$, $N \subseteq X^*$ be subspaces of X, X^*, respectively. The *annihilators* $M^\perp \subseteq X^*$ and ${}^\perp N \subseteq X$ are defined as

$$M^\perp = \{x^* \in X^* : \langle x^*, x\rangle_{X^* \times X} = 0 \text{ for all } x \in M\}$$
$${}^\perp N = \{x \in X : \langle x, x^*\rangle_{X \times X^*} = 0 \text{ for all } x^* \in N\}.$$

Theorem 2.6 (adjoint operator). *Let X, Y be Banach spaces and $A : X \to Y$ be a bounded (continuous) linear operator. Then the bounded (continuous) linear operator $A^* : Y^* \to X^*$, defined as*

$$\langle A^* y^*, x\rangle_{X^* \times X} = \langle y^*, Ax\rangle_{Y^* \times Y} \quad \forall x \in X, y^* \in Y^*$$

is called the adjoint operator *of A (for more details see* [164, Section 3.1]*).*

Definition 2.7 (null-space). By $\mathcal{N}(A)$, we denote the *null-space*

$$\mathcal{N}(A) := \{x \,:\, Ax = 0\}$$

of the linear operator $A : X \to Y$ and by $\mathcal{R}(A)$ the *range*

$$\mathcal{R}(A) := \{y \,:\, \exists x \,:\, y = Ax\} = A(X)$$

of A. The space consisting of all bounded (continuous) linear operators $A : X \to Y$ between two Banach spaces X and Y is denoted by $\mathcal{L}(X, Y)$. We write $\mathcal{L}(X)$ instead of $\mathcal{L}(X, Y)$ if X and Y coincide.

Theorem 2.8 (Cauchy's inequality). *Let X denote a Banach space and let $x \in X$ and $x^* \in X^*$. Then we have*

$$|\langle x^*, x\rangle_{X^* \times X}| \le \|x^*\|_{X^*} \cdot \|x\|_X \,.$$

Theorem 2.9 (Hölder's inequality). *For sequences $\{x_k\} \in \ell^p$ and $\{y_k\} \in \ell^{p^*}$ of real numbers and conjugate exponents $p, p^* > 1$ we have*

$$\sum_k x_k y_k \le \left(\sum_k |x_k|^p\right)^{1/p} \cdot \left(\sum_k |y_k|^{p^*}\right)^{1/p^*} .$$

For functions $f \in L^p(\Omega)$, $g \in L^{p^*}(\Omega)$, Hölder's inequality accordingly reads as

$$\left|\int_\Omega f(x) g(x)\, dx\right| \le \left(\int_\Omega |f(x)|^p \, dx\right)^{1/p} \left(\int_\Omega |g(x)|^{p^*} \, dx\right)^{1/p^*} .$$

Definition 2.10. For real numbers $a, b \in \mathbb{R}$ we write

$$a \vee b := \max\{a, b\}, \quad a \wedge b := \min\{a, b\}.$$

Later we will use the shortcut symbols of Definition 2.10 instead of max and min whenever it renders the expressions more transparent.

Theorem 2.11 (Young's inequality). *Let a and b denote real numbers and $p, p^* > 1$ conjugate exponents. Then we have*

$$a \cdot b \le \tfrac{1}{p}|a|^p + \tfrac{1}{p^*}|b|^{p^*}.$$

Definition 2.12 (equivalent quantities). We call two positive quantities a and b *equivalent* if there exist constants $0 < c \le C < \infty$, such that

$$c \cdot b \le a \le C \cdot b$$

and write in shorthand

$$a \sim b.$$

In particular for two norms $\|\cdot\|_A : X \to \mathbb{R}$ and $\|\cdot\|_B : X \to \mathbb{R}$ we mean by

$$\|\cdot\|_A \sim \|\cdot\|_B$$

that

$$c\|x\|_B \le \|x\|_A \le C\|x\|_B \quad \forall x \in X$$

with some constants $0 < c \le C < \infty$ that do not depend on x. If $\alpha > 0$ is a function of $\delta > 0$ then we mean by

$$\alpha \sim \delta^\kappa$$

for $\kappa \ge 0$ that

$$c \cdot \delta^\kappa \le \alpha(\delta) \le C \cdot \delta^\kappa$$

with some constants $0 < c \le C < \infty$ that do not depend on δ.

Definition 2.13 (Gâteaux differentiability). Let $f : \mathcal{D}(f) \subseteq X \to Y$ be a mapping between normed spaces X and Y, on the domain $\mathcal{D}(f)$. We call f *Gâteaux differentiable* in $x \in \mathcal{D}(f)$ if there exists a continuous linear mapping $A_x : X \to Y$ such that for all $h \in X$ we have

$$\lim_{t \to 0} \frac{f(x + th) - f(x)}{t} = A_x h.$$

Definition 2.14 (Fréchet differentiability). Let $f : \mathcal{D}(f) \subseteq X \to Y$ be a mapping between normed spaces X and Y with domain $\mathcal{D}(f)$. We call f *Fréchet differentiable* in $x \in \mathcal{D}$ if there exists a continuous linear mapping $A_x : X \to Y$ such that for $z \in X$ we have

$$\lim_{\|z\| \to 0} \frac{\|f(x + z) - f(x) - A_x z\|}{\|z\|} = 0.$$

Remark 2.15. If f is Fréchet differentiable at x, then it is also Gâteaux differentiable in x. This can be seen by setting $z := th$ for fixed $h \in X$ in Definition 2.14 where $\|z\| \to 0$ can be written as $t \to 0$. Note that both definitions require the existence of elements $x + th$ for arbitrary directions $h \in X$ and sufficiently small $t > 0$, i.e., x has to belong to the core of $\mathcal{D}(f)$ which mostly coincides with the interior of $\mathcal{D}(f)$. Consequently, f cannot be Gâteaux differentiable or Fréchet differentiable if x is a boundary point of $\mathcal{D}(f)$.

2.2 Convex analysis

2.2.1 The subgradient of convex functionals

Definition 2.16 (convex functional). A functional $f : X \to \mathbb{R} \cup \{\infty\}$, defined on a linear space X, is *convex* if

$$f(\lambda x + (1-\lambda)y) \leq \lambda f(x) + (1-\lambda) f(y) \quad \forall x, y \in X,\ \forall \lambda \in [0,1].$$

Definition 2.17 (effective domain, proper functions). Let $f : X \to \mathbb{R} \cup \{\infty\}$ be convex. The *effective domain* $\mathcal{D}(f)$ of the functional f is defined as

$$\mathcal{D}(f) := \{x \in X : f(x) < \infty\}.$$

The functional f is called *proper* if

$$\mathcal{D}(f) \neq \emptyset.$$

Example 2.18 (Tikhonov functional). Let $A : X \to Y$ be a bounded (continuous) linear operator mapping between Banach spaces X and Y, $y^\delta \in Y$, $p \geq 1$ and $\Omega : X \to \mathbb{R} \cup \{\infty\}$ convex. Then for $\alpha > 0$ the functional $T_\alpha : X \to Y$, defined as

$$T_\alpha(x) := \frac{1}{p}\|Ax - y^\delta\|_Y^p + \alpha \cdot \Omega(x) \tag{2.1}$$

is called *Tikhonov functional.* Tikhonov functionals are convex. If $\mathcal{D}(\Omega) = X$ then $\mathcal{D}(T_\alpha) = X$. In particular, this is true if Ω is a monomial of the norm in X, i.e. $\Omega(x) := \frac{1}{q}\|x\|_X^q$ for some $q \geq 1$. In Chapter 3 we will extend the concept of Tikhonov functionals to include the case of nonlinear operators.

Remark 2.19. Some authors use a symbol which emphasizes the dependence of the Tikhonov functional on the noisy data y^δ, say T_{α,y^δ} or ${T_\alpha}^{y^\delta}$. To keep the notation more concise we will preferrably use the notation introduced in Example 2.18.

Definition 2.20 (subgradient of convex functionals). Let $f : X \to \mathbb{R} \cup \{\infty\}$ be a convex functional. Then, $x^* \in X^*$ is a *subgradient* of f in x if

$$f(y) \geq f(x) + \langle x^*, y - x\rangle \quad \forall y \in X.$$

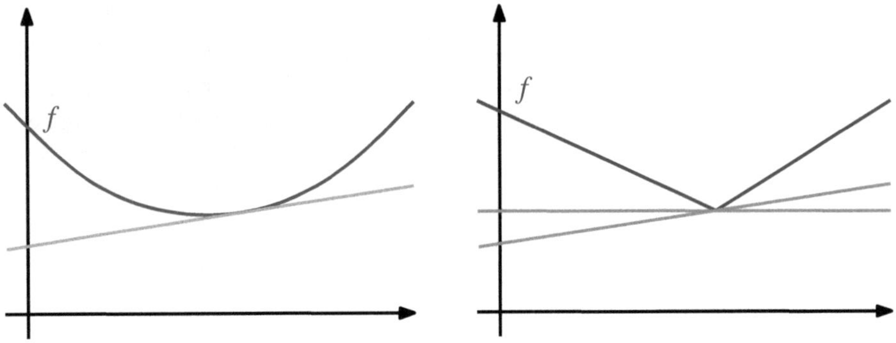

Figure 2.1. The subgradient of a function at x as the slope of the hyperplane supporting the function in x. For smooth functions the subgradient is single-valued, whereas for non-smooth functions in general it is not.

The set $\partial f(x)$ of all subgradients of f in x is called the *subdifferential*. We will call both the subgradient and the subdifferential as subgradient if it is clear from the context which one is meant. Especially for single-valued subdifferentials we will identify the whole set with its single element.

All affine linear functionals $g : X \to \mathbb{R}$ which are tangential to the functional f at x, meaning that $g(y) \leq f(y)$ for all y and $g(x) = f(x)$, can be written in the form

$$g(y) = \langle x^*, y - x\rangle + f(x)$$

with some element $x^* \in X^*$. Notice that the graph of g is a hyperplane. Therefore, geometrically, the subgradient is the slope (or set of slopes) of a hyperplane, which supports f (cf. Figure 2.1).

Theorem 2.21 (optimality conditions). *Let $f : X \to \mathbb{R} \cup \{\infty\}$ be convex and $z \in \mathcal{D}(f)$ then*

$$f(z) = \min_{x \in X} f(x) \quad \Leftrightarrow \quad 0 \leq f(x) - f(z) \quad \forall x \in X \quad \Leftrightarrow \quad 0 \in \partial f(z)$$

(cf. [211, 4.1.2]*).*

Therefore, $0 \in \partial f(z)$ is the generalization of the classical optimality condition $f'(z) = 0$. The subgradient also is the generalization of the gradient in the sense that if f is Gâteaux-differentiable, then $\partial f(x) = \{\nabla f(x)\}$.

In general the subgradient of a function may also be the empty set. However, if the functional is Lipschitz-continuous, its subgradient is not empty, cf. [226, Chap. II.7.5].

Let f and g be two convex functionals. Then

$$\partial(f+g)(x) \supseteq \partial f(x) + \partial g(x),$$

since for any $x^* \in \partial f(x)$ and $z^* \in \partial g(x)$ we have

$$f(x) + \langle x^*, y-x\rangle + g(x) + \langle z^*, y-x\rangle \leq f(y) + g(y).$$

The next theorem provides a sufficient condition for f and g to satisfy $\partial(f+g)(x) \subseteq \partial f(x) + \partial g(x)$, too.

Theorem 2.22 (subgradient of sums). *Let X be a Banach space and $f, g : X \to \mathbb{R} \cup \{\infty\}$ be convex functionals. If there is a point in $\mathcal{D}(f) \cap \mathcal{D}(g)$ such that f is continuous, then for all $x \in X$ the equation*

$$\partial(f+g)(x) = \partial f(x) + \partial g(x)$$

holds (cf. [243, Theorem 47.B]*).*

Theorem 2.23 (subgradient of compositions). *Let $f : Y \to \mathbb{R} \cup \{\infty\}$ be a convex functional and $A : X \to Y$ a bounded (continuous) linear operator. If f is continuous at some point of the range of A then for all $x \in X$*

$$\partial(f \circ A)(x) = A^*(\partial f(Ax))$$

(cf. [226, II.7.8]*).*

Theorem 2.24 (subgradient of translation). *Let $f : X \to \mathbb{R} \cup \{\infty\}$ be convex. Then*

$$\partial(f(\cdot - y))(x) = (\partial f)(x-y).$$

Proof. Let $x^* \in \partial(f(\cdot - y))(x)$, then

$$f(z-y) \geq f(x-y) + \langle x^*, z-x\rangle \quad \forall z \in X.$$

Hence with $a = z - y$

$$f(a) \geq f(x-y) + \langle x^*, a-(x-y)\rangle \quad \forall a \in X.$$

Therefore, we get $\partial(f(\cdot - y))(x) \subset (\partial f)(x-y)$. In the same way one can prove that $\partial(f(\cdot - y))(x) \supset (\partial f)(x-y)$. □

As a consequence of the last three theorems, we are now able to formulate the optimality conditions for the Tikhonov functional (2.1).

Corollary 2.25. *We have*

$$\partial(\tfrac{1}{p}\|A\cdot - y^\delta\|_Y^p)(x) = (A^*\partial\tfrac{1}{p}\|\cdot - y^\delta\|_Y^p)(Ax) = (A^*\partial\tfrac{1}{p}\|\cdot\|_Y^p)(Ax - y^\delta).$$

Hence, if x_α^δ minimizes the functional T_α, then

$$0 \in A^*(\partial\tfrac{1}{p}\|\cdot\|_Y^p)(Ax_\alpha^\delta - y^\delta) + \alpha\cdot\partial\Omega(x_\alpha^\delta).$$

In particular for $\Omega(x) := \frac{1}{q}\|x\|^q$, with $q \geq 1$, we get

$$0 \in A^*(\partial\tfrac{1}{p}\|\cdot\|_Y^p)(Ax_\alpha^\delta - y^\delta) + \alpha(\partial\tfrac{1}{q}\|\cdot\|_X^q)(x_\alpha^\delta). \tag{2.2}$$

As a final fact emerging from general convex analysis we mention that the notion of the subgradient also generalizes the concept of monotonicity. We recall that, if $f : \mathbb{R} \to \mathbb{R}$ is differentiable and convex, the derivative f' is monotone, i.e.

$$(f'(x) - f'(y))\cdot(x - y) \geq 0 \quad \forall x, y.$$

This is also true for all convex mappings if the derivative is replaced by the subgradient.

Theorem 2.26. *The subgradient of a proper convex functional $f : X \to \mathbb{R}$ defined on a Banach space X is monotone, i.e.*

$$\langle x^* - y^*, x - y\rangle \geq 0 \quad \forall x^* \in \partial f(x), y^* \in \partial f(y),\ x, y \in X$$

(cf. [203, Theorem 12.17]*).*

2.2.2 Duality mappings

In Corollary 2.25 we derived the optimality condition for the Tikhonov functional (2.1). This condition uses the mappings

$$\partial(\tfrac{1}{p}\|\cdot\|_Y^p) \quad \text{and} \quad \partial(\tfrac{1}{q}\|\cdot\|_X^q).$$

To make effective use of the optimality condition, we need more information about the structure of the above subgradients of powers of Banach space norms. We start by introducing the so-called duality mapping J_p^X.

Definition 2.27 (duality mapping). The (set-valued) mapping $J_p^X : X \rightrightarrows X^*$ with $p \geq 1$ defined by

$$J_p^X(x) := \{x^* \in X^* : \langle x^*, x\rangle = \|x\|\|x^*\|, \|x^*\| = \|x\|^{p-1}\} \tag{2.3}$$

is called the *duality mapping* of X with gauge function $t \mapsto t^{p-1}$.

By j_p^X we will denote a single-valued selection of J_p^X, i.e. $j_p^X : X \to X^*$ is a mapping with $j_p^X(x) \in J_p^X(x)$ for all $x \in X$. We will sometimes abbreviate this to $j_p^X \in J_p^X$.

However, for single-valued duality mappings J_p^X we will use the symbols J_p^X and j_p^X interchangeably, i.e. formally $j_p^X \equiv J_p^X$.

Again we will omit the index indicating the space, where it is clear from the context which one is meant.

The duality mapping is the desired subgradient of powers of Banach norms.

Theorem 2.28 (Asplund Theorem). *Let X be a normed space and $p \geq 1$. Then*

$$J_p^X = \partial\left(\frac{1}{p}\|\cdot\|_X^p\right)$$

(cf. [10], [211, Section 4.6]*).*

As a corollary of Theorem 2.26 we get

Theorem 2.29. *The duality mapping J_p^X is monotone, i.e.*

$$\langle x^* - y^*, x - y\rangle \geq 0 \quad \textit{for all } x, y \in X, x^* \in J_p^X(x), y^* \in J_p^X(y).$$

With the definition of the duality mapping J_p^X and its characterization by the Theorem of Asplund (Theorem 2.28) we already have some tools to describe the subgradients of powers of norms at our disposal. However, what we still lack is the exact form of J_p^X. In the next example we provide J_p^X in an explicit form for the case of Hilbert spaces and sequence spaces.

Example 2.30. In Hilbert spaces we have $J_2^X(x) = x$ and therefore

$$J_p^X(x) = \|x\|^{p-2} \cdot x.$$

Next, we consider the family of ℓ^r-spaces with exponents $1 < r < \infty$ and infinite sequences $x = \{x_i\}_{i=1}^\infty$ of real numbers as elements. These are Banach spaces with norms defined as

$$\|x\|_r := \left(\sum_{i=1}^{\infty} |x_i|^r\right)^{1/r}$$

equipped with the duality pairing $\langle y, x\rangle = \sum_{i=1}^\infty x_i y_i$, where $y = \{y_i\}_{i=1}^\infty \in \ell^{r^*}$. Here, we use r as the exponent of the sequence spaces, since we need the letter p otherwise. We can easily compute

$$(J_p^{\ell^r}(x))_i := ((\nabla \tfrac{1}{p}\|\cdot\|_r^p)(x))_i = \|x\|_r^{p-r}|x_i|^{r-1}\operatorname{sign}(x_i).$$

We can write ∇ instead of ∂ since the functional is smooth, cf. [143, 213, 239].

For $r = 1$ we have a dual space $(\ell^1)^* = \ell^\infty$ and

$$(J_p^{\ell^1}(x))_i := (\partial \tfrac{1}{p}\|\cdot\|_1^p)_i = \|x\|_1^{p-1} \cdot \mathrm{Sign}(x_i),$$

where Sign is the set-valued sign function, defined in the same way as the regular sign function except that $\mathrm{Sign}(0) = [-1, 1]$ (cf. [144, Corollary 3.3]).

We do not consider the case $r = \infty$ since it already becomes difficult to identify the dual space of ℓ^∞ and the subgradient of its norm in closed form. The situation is, however, much simpler if we only consider the finite dimensional spaces $\mathbb{R}^N$ of N-dimensional vectors with real components, and equipped with the norm $\|x\|_\infty := \max_{1\le i\le N} |x_i|$. Then, the dual space is the finite dimensional version of ℓ^1 with the norm $\|x\|_1 := \sum_{i=1}^N |x_i|$ and

$$(J_p(x))_i = \|x\|_\infty^{p-1} \cdot y_i\,,$$

where $y_i = 0$ for i such that $|x_i| \neq \|x\|_\infty$, $\mathrm{sign}(y_i) = \mathrm{sign}(x_i)$ for all other i and $\sum_{i=1}^N |y_i| = 1$. One of the elements of the subgradient is given by $y_i = \mathrm{sign}(x_i)$, for exactly one i such that $|x_i| = \|x\|_\infty$ and $y_i = 0$ for all other i. For more details we refer to [213, 216].

Remark 2.31. The above example shows that at least for the finite dimensional case the numerical complexity (i.e. the number of floating point operations) for the evaluation of the norm is comparable to the evaluation of the duality mapping and is of order $\mathcal{O}(N)$, where N is the dimension of the vector space under consideration. Furthermore, the evaluation of the duality mapping is not slower than the evaluation of a usual operator mapping from $\mathbb{R}^N$ to $\mathbb{R}^N$, which we assume to be of order $\mathcal{O}(N)$ or $\mathcal{O}(N \cdot \log N)$ floating point operations for the so-called fast operators and $\mathcal{O}(N^2)$ for all other operators.

In all algorithms presented in this work we will make the silent assumption that the evaluation of the norm and the duality mapping is computationally not more expensive than the evaluation of the operator.

To present the most important properties of the duality mapping we need some notions from the geometry of Banach spaces. Therefore, we will first introduce the basic notions in the subsequent section and then state the properties of the duality mapping in Theorem 2.53.

2.3 Geometry of Banach space norms

Is there any way to say that a Banach space has ‘good’ geometric properties? We know that the norm of a Banach space is convex, hence also the monomials of norms are convex supposed that the exponent is not smaller than one. So far, the only geometrical notion connected to convex functions is the subgradient. Therefore, it is natural to introduce 'niceness' of a Banach space by extending (or enhancing) the subgradient definition. Further we keep in mind that the subgradient of a power of the norm is the duality mapping.

From Definition 2.20 and Theorem 2.28 we know that for all $x, z \in X$ we have

$$\tfrac{1}{p}\|z\|^p - \tfrac{1}{p}\|x\|^p - \langle J_p^X(x), z - x\rangle \geq 0,$$

and setting $y = -(z - x)$ yields

$$\tfrac{1}{p}\|x - y\|^p - \tfrac{1}{p}\|x\|^p + \langle J_p^X(x), y\rangle \geq 0.$$

We are interested in the upper and lower bounds of the left-hand side of the above inequality, in terms of the norm of y, say

$$\tfrac{1}{p}\|x - y\|^p - \tfrac{1}{p}\|x\|^p + \langle J_p^X(x), y\rangle \geq \tfrac{c_p}{p}\|y\|^p$$

or

$$\tfrac{G_p}{p}\|y\|^p \geq \tfrac{1}{p}\|y - x\|^p - \tfrac{1}{p}\|x\|^p + \langle J_p^X(x), y\rangle$$

for some $G_p, c_p > 0$.

2.3.1 Convexity and smoothness

We start with the definitions of p-convexity and p-smoothness of a Banach space.

Definition 2.32 (p-convex). We call a Banach space X *convex of power type* p or p*-convex* if there exists a constant $c_p > 0$ such that

$$\tfrac{1}{p}\|x - y\|^p \geq \tfrac{1}{p}\|x\|^p - \langle j_p^X(x), y\rangle + \tfrac{c_p}{p}\|y\|^p$$

for all $x, y \in X$ and all $j_p^X \in J_p^X$.

Definition 2.33 (p-smooth). We call a Banach space X *smooth of power type* p or p*-smooth* if there exists a constant $G_p > 0$ such that

$$\tfrac{1}{p}\|x - y\|^p \leq \tfrac{1}{p}\|x\|^p - \langle j_p^X(x), y\rangle + \tfrac{G_p}{p}\|y\|^p \tag{2.4}$$

for all $x, y \in X$ and all $j_p^X \in J_p^X$.

In some cases we will consider more general Banach spaces.

Definition 2.34 (strictly convex). We call X *strictly convex* if $\left\|\tfrac{1}{2}(x + y)\right\| < 1$ for all x, y from the unit sphere of X satisfying the condition $x \neq y$.

Definition 2.35 (uniformly convex). We say X to be *uniformly convex* if, for the *modulus of convexity* $\delta_X : [0,2] \to [0,1]$, defined via

$$\delta_X(\varepsilon) := \inf\left\{1 - \left\|\frac{1}{2}(x+y)\right\| : \|x\| = \|y\| = 1, \|x-y\| \geq \varepsilon.\right\}$$

we have

$$\delta_X(\varepsilon) > 0 \quad \forall\, 0 < \varepsilon \leq 2.$$

Definition 2.36 (smooth). We call X *smooth* if for every $x \in X$ with $x \neq 0$ there is a unique $x^* \in X^*$ such that $\|x^*\| = 1$ and $\langle x^*, x\rangle = \|x\|$.

Definition 2.37 (uniformly smooth). We say X to be *uniformly smooth* if for the *modulus of smoothness* $\rho_X : [0,\infty) \to [0,\infty)$ defined via

$$\rho_X(\tau) := \frac{1}{2}\sup\{\|x+y\| + \|x-y\| - 2 : \|x\| = 1, \|y\| \leq \tau\}$$

we have the limit condition

$$\lim_{\tau\to 0} \frac{\rho_X(\tau)}{\tau} = 0.$$

Remark 2.38. Notice that if a Banach space is p-smooth for some $p > 1$, then the p-th power of norm is Fréchet differentiable, hence Gâteaux differentiable and therefore $J_p^X(x)$ is single valued for all x. By Theorem 2.53, we will see that if $J_p^X(x)$ is single-valued for all x and *some* $p > 1$ then $J_p^X(x)$ is single-valued for all x and *all* $p > 1$.

Xu and Roach [239] have proved the following inequalities, which are very useful in a series of proofs below.

Theorem 2.39 (Xu–Roach inequalities I). *Let X be uniformly convex and $1 < p < \infty$. Then for every $j_p^X \in J_p^X$ and for every pair of elements $x, y \in X$ we have*

$$\|x-y\|^p \geq \|x\|^p - p\langle j_p^X(x), y\rangle + \sigma_p(x,y) \tag{2.5}$$

with

$$\sigma_p(x,y) = pK_p \int_0^1 \frac{(\|x-ty\| \vee \|x\|)^p}{t}\,\delta_X\left(\frac{t\|y\|}{2(\|x-ty\| \vee \|x\|)}\right) dt\,, \tag{2.6}$$

where

$$\begin{aligned} K_p = 4(2+\sqrt{3}) \min\Big\{&\frac{1}{2}p(p-1)\wedge 1\,,\ \left(\frac{1}{2}p\wedge 1\right)(p-1)\,,\\ &(p-1)\left(1-(\sqrt{3}-1)^q\right),\ 1-\left(1+(2-\sqrt{3})q\right)^{1-p}\Big\}. \end{aligned} \tag{2.7}$$

In the special case that X is convex of power type we can deduce interesting variants of these inequalities.

Theorem 2.40 (Xu–Roach inequalities II). *The following statements are equivalent:*

(a) *X is s-convex.*

(b) *For some p with $1 < p < \infty$, some $j_p^X \in J_p^X$ and for every $x, y \in X$ we have*

$$\langle j_p^X(x) - j_p^X(y), x - y\rangle \geq C \max\{\|x\|, \|y\|\}^{p-s} \|x - y\|^s.$$

(c) *The statement (b) holds for all p with $1 < p < \infty$, all $j_p^X \in J_p^X$ and all $x, y \in X$.*

(d) *For some p with $1 < p < \infty$ and some $j_p^X \in J_p^X$ the inequality* (2.5) *holds for all $x, y \in X$. Moreover, for $\sigma_p(x, y)$ as defined in* (2.6) *we have*

$$\sigma_p(x, y) \geq C \int_0^1 t^{s-1} \max\{\|x - ty\|, \|x\|\}^{p-s} \|y\|^s dt.$$

(e) *The statement (d) holds for all p such that $1 < p < \infty$, all $j_p^X \in J_p^X$ and all $x, y \in X$.*

The generic constant $C > 0$ can be chosen independently of x and y.

For uniformly smooth Banach spaces we also have corresponding Xu–Roach inequalities.

Theorem 2.41 (Xu–Roach inequalities III). *Let X be uniformly smooth, $1 < p < \infty$, and $j_p^X \in J_p^X$. Then, for all $x, y \in X$, we have*

$$\|x - y\|^p \leq \|x\|^p - p\langle j_p^X(x), y\rangle + \tilde{\sigma}_p(x, y) \tag{2.8}$$

with

$$\tilde{\sigma}_p(x, y) = pG_p \int_0^1 \frac{(\|x - ty\| \vee \|x\|)^p}{t} \rho_X\left(\frac{t\|y\|}{\|x - ty\| \vee \|x\|}\right) dt\,, \tag{2.9}$$

where $G_p = 8 \vee 64cK_p^{-1}$ with K_p defined according to (2.7) *and*

$$c = 4\frac{\tau_0}{\sqrt{1 + \tau_0^2} - 1} \prod_{j=1}^{\infty}\left(1 + \frac{15}{2^{j+2}}\tau_0\right) \quad \textit{with} \quad \tau_0 = \frac{\sqrt{339} - 18}{30}.$$

If X is smooth of power type then the following characterizations of these estimates hold.

Theorem 2.42 (Xu–Roach inequalities IV). *The following statements are equivalent:*

(a) *X is s-smooth.*

(b) *For some p with $1 < p < \infty$ the duality mapping J_p^X is single-valued and for all $x, y \in X$ we have*

$$\|J_p^X(x) - J_p^X(y)\| \le C \max\{\|x\|, \|y\|\}^{p-s} \|x - y\|^{s-1}.$$

(c) *The statement (b) holds for all p such that $1 < p < \infty$.*

(d) *For some p with $1 < p < \infty$, some $j_p^X \in J_p^X$, and for all $x, y \in X$ the inequality* (2.8) *holds. Moreover, for $\tilde{\sigma}_p(x, y)$ as defined in* (2.9) *we have*

$$\tilde{\sigma}_p(x, y) \le C \int_0^1 t^{s-1} \max\{\|x - ty\|, \|x\|\}^{p-s} \|y\|^s \, dt.$$

(e) *The statement (d) holds for all p such that $1 < p < \infty$ and all $j_p^X \in J_p^X$.*

The generic constant $C > 0$ can be chosen independently of x and y.

Corollary 2.43. *Let X be p-smooth. Then for all q such that $1 < q < p$ the space X is also q-smooth. If on the other hand X is p-convex then for all q such that $p < q < \infty$ the space X is also q-convex.*

Proof. By the Xu–Roach characterization of smoothness of power type in Theorem 2.42 we have for a p-smooth space and $q > 1$ that

$$\|J_q(x) - J_q(y)\| \le C \max\{\|x\|, \|y\|\}^{q-p} \|x - y\|^{p-1}.$$

Due to $q < p$ we get

$$\max\{\|x\|, \|y\|\}^{q-p} \|x - y\|^{p-1} \le C \|x - y\|^{q-1}$$

and therefore

$$\|J_q(x) - J_q(y)\| \le C \|x - y\|^{q-1}.$$

Hence, again by the Xu–Roach characterization of smoothness of power type, the space X is q-smooth. The proof for convexity of power type is analogous. □

Corollary 2.44. *Let X be s-smooth and $p > 1$. Then the duality mapping J_p^X is $\min\{p - 1, s - 1\}$-Hölder continuous on bounded sets and single-valued.*

Proof. If $p \le s$ then, by Corollary 2.43, we know that X is p-smooth too. Therefore, by Theorem 2.42 the duality mapping J_p^X is $(p - 1)$-Hölder continuous. If $s \le p$, then by Theorem 2.42 we have

$$\|J_p^X(x) - J_p^X(y)\| \le C \max\{\|x\|, \|y\|\}^{p-s} \|x - y\|^{s-1} \le C \|x - y\|^{s-1}$$

on bounded sets, which proves the claim. □

Remark 2.45. The definitions of convexity of power type in Definition 2.32 and smoothness of power type in Definition 2.33 can be regarded as special cases of Theorems 2.40 and 2.42 where the index p of the duality mapping has the same value as the index s denoting the type of power of the space.

Next, we present prominent examples of spaces convex of power type and smooth of power type.

Example 2.46. The polarization identity

$$\tfrac{1}{2}\|x-y\|^2 = \tfrac{1}{2}\|x\|^2 - \langle x, y\rangle + \tfrac{1}{2}\|y\|^2$$

ensures that Hilbert spaces are 2-convex and 2-smooth.

Example 2.47. Let $\mathcal{G} \subset \mathbb{R}^n$ be a domain. It is known [130, 143, 239] that

- the spaces ℓ^r of infinite real sequences,
- the Lebesgue spaces $L^r(\mathcal{G})$, and
- the Sobolev spaces $W^{m,r}(\mathcal{G})$

equipped with the usual norms and $1 < r < \infty$ are

$$\max\{2, r\} - \text{convex}$$

and

$$\min\{2, r\} - \text{smooth}.$$

The Besov spaces $B^s_{p,q}(\mathbb{R})$ are

$$\max\{2, p, q\} - \text{convex}$$

and

$$\min\{2, p, q\} - \text{smooth}.$$

For ℓ^1 one can show with $x = (1, 0, \ldots)$, $J_p^{\ell^1}(x) = (1, 0, \ldots)$, $y = (0, 0, \ldots)$, and $j_p^{\ell^1}(y) = (1, 0, \ldots)$, that the space cannot be p-convex or p-smooth for any p.

The examples above justify our view that convexity and smoothness of power type is quite common. Moreover, we see that the powers in the convexity and smoothness of power type interpolate between the space ℓ^2 which, as a Hilbert space, may be regarded as a geometrically well-behaved space, and the spaces like ℓ^1 or ℓ^∞, which are regarded as geometrically 'not so nice', because no direct generalizations of the polarization identity can be established in these spaces.

In Examples 2.46 and 2.47 we have seen that most of the usual spaces are smooth and convex of some power type. Further, one can even show that spaces uniformly convex *or* uniformly smooth are (up to some equivalent renorming) also smooth *and* convex of power type.

Theorem 2.48. *Let X be uniformly convex* or *uniformly smooth. Then, there exists an equivalent norm such that X equipped with this norm is smooth of power type* and *convex of power type. Furthermore, every uniformly convex and every uniformly smooth space is reflexive (cf.* [57, Theorem 5.2]*).*

Remark 2.49. We note that although the proof of this theorem is constructive, it relies heavily on Banach space geometry, especially notions of superreflexivity and Banach spaces of Rademacher-type.

In the light of the above facts we will now go into more detail about spaces that are smooth and convex of power type. These spaces share many very interesting properties, which we summarize in the following theorems.

Theorem 2.50. *If X is p-convex, then we have the following assertions:*

(a) $p \geq 2$,

(b) *X is uniformly convex and the modulus of convexity satisfies $\delta_X(\varepsilon) \geq C\varepsilon^p$,*

(c) *X is strictly convex,*

(d) *X is reflexive*

(cf. [239, p. 193], [48, Chapter II]*).*

Theorem 2.51. *If X is p-smooth, then we have:*

(a) $p \leq 2$,

(b) *X is uniformly smooth and the modulus of smoothness of X satisfies $\rho_X(\tau) \leq C\tau^p$,*

(c) *X is smooth,*

(d) *X is reflexive*

(cf. [239, p. 193], [48, Chapter II]*).*

The next theorem allows us to connect properties of the primal space to the properties of the dual space.

Theorem 2.52 (duality of convexity and smoothness). *We have the assertions:*

(a) *X is p-smooth if and only if X^* is p^*-convex.*

(b) *X is p-convex if and only if X^* is p^*-smooth.*

(c) *X is uniformly convex (respectively uniformly smooth) if and only if X^* is uniformly smooth (respectively uniformly convex).*

(d) *X is uniformly smooth if and only if X is uniformly convex*

(cf. [143, Vol II 1.e]*).*

The following theorem represents a list of the most important properties of the duality mapping.

Theorem 2.53. *We have the following assertions:*

(a) *For every $x \in X$ the set $J_p^X(x)$ is non-empty and convex.*

(b) $J_p^X(-x) = -J_p^X(x)$ *and* $J_p^X(\lambda x) = \lambda^{p-1} J_p^X(x)$ *for all $x \in X$ and all $\lambda > 0$.*

(c) *If X is uniformly convex then X is reflexive and strictly convex.*

(d) *If X is uniformly smooth then X is reflexive and smooth.*

(e) *X is smooth if and only if every duality mapping J_p^X is single-valued.*

(f) *If X is uniformly smooth then J_p^X is single-valued and uniformly continuous on bounded sets.*

(g) *Let X be reflexive. Then, X is strictly convex (respectively smooth) if and only if X^* is smooth (respectively strictly convex).*

(h) *X is strictly convex if and only if every duality mapping J_p^X is strictly monotone, i.e. $\langle x^* - y^*, x - y\rangle > 0$ for all $x, y \in X$ with $x \neq y$ and $x^* \in J_p^X(x)$, $y^* \in J_p^X(y)$.*

(i) *For $p, q > 1$ we have*

$$\|x\|^{q-1} J_p^X(x) = \|x\|^{p-1} J_q^X(x).$$

(j) *If X is convex of power type and smooth, then J_p^X is single valued, norm-to-weak continuous, bijective, and the duality mapping $J_{p^*}^{X^*}$ is single-valued with*

$$J_{p^*}^{X^*}(J_p^X(x)) = x.$$

(k) *Let $M \neq \emptyset$ be a closed convex subset of X. If X is uniformly convex, there exists a unique $x \in M$ such that*

$$\|x\| = \inf_{z \in M} \|z\|.$$

If in addition X is smooth then $\langle J_p^X(x), x\rangle \leq \langle J_p^X(x), z\rangle$ for all $z \in M$

(cf. [213, Lemma 2.3 and 2.5], [48, Proposition I.4.7.f and II.3.6] *and* [239]*).*

Remark 2.54. Statement (a) is a consequence of the Hahn–Banach theorem, assertions (b) and (c) follow by a straightforward application of the definition of the duality mapping. Property (j) holds even under the weaker condition that X is smooth, strictly convex and reflexive. The importance of (j) can hardly be overestimated, since it states that e.g. for spaces being smooth of power type and convex of power type the duality mappings on the primal space and the dual space can be used to transport *all* elements from the primal to dual space and vice versa. We have

$$J_{p^*}^{X^*}(J_p^X(x)) = x \quad \text{and} \quad J_p^X(J_{p^*}^{X^*}(x^*)) = x^* \quad \forall x \in X \quad \forall x^* \in X^*.$$

Lemma 2.55. *Suppose that X, Y are Banach spaces and $A \in \mathcal{L}(X,Y)$. Then*

$$\mathcal{N}(A^*) = \mathcal{R}(A)^{\perp} \quad \textit{and} \quad \mathcal{N}(A) = {}^{\perp}\mathcal{R}(A^*).$$

If additionally X is uniformly convex, then we have

$${}^{\perp}\mathcal{N}(A^*) = \overline{\mathcal{R}(A)} \quad \textit{and} \quad \mathcal{N}(A)^{\perp} = \overline{\mathcal{R}(A^*)},$$

where the closures are always to be understood as the norm-closures in the corresponding spaces.

Proof. The first part is Theorem 4.12 from [205]. Theorem 4.7 in [205] says that ${}^{\perp}(M^{\perp})$ is the norm-closure of M in X and $({}^{\perp}N)^{\perp}$ is the weak*-closure of N in X^*. Since X is uniformly convex and hence reflexive, the weak*-closure coincides with the norm-closure in X^*. □

2.3.2 Bregman distance

Due to geometrical properties of Banach spaces it is often more appropriate to exploit the *Bregman distance* instead of functionals like $\|x-y\|_X^p$ or $\|j_p^X(x) - j_p^X(y)\|_{X^*}^p$ to prove the convergence of algorithms. The main idea of the Bregman distance is to use the gap between a functional and its linearization instead of the functional itself in order to measure distances.

Definition 2.56 (Bregman distance – special version). Let $j_p : X \to X^*$ be a single-valued selection of the duality mapping J_p. Then, the functional

$$D_{j_p}(x,y) := \tfrac{1}{p}\|x\|^p - \tfrac{1}{p}\|y\|^p - \langle j_p(y), x-y\rangle, \quad x,y \in X$$

is called the Bregman distance (with respect to the functional $\frac{1}{p}\|\cdot\|^p$).

If X is a smooth Banach space, then j_p is single-valued according to Theorem 2.53 and the Bregman distance can also be calculated as

$$\begin{aligned} D_{j_p}(x,y) &= \frac{1}{p^*}\|y\|^p + \frac{1}{p}\|x\|^p - \langle J_p(y), x\rangle \\ &= \frac{1}{p^*}(\|y\|^p - \|x\|^p) + \langle J_p(x) - J_p(y), x\rangle. \end{aligned} \tag{2.10}$$

Remark 2.57. To the best of the authors' knowledge the concept of Bregman distances was introduced by Bregman in [28]. Moreover, we note that there are several possible ways to define the Bregman distance of a convex functional. However, the main idea is always to measure the gap between the functional and its linearization. We visualize this concept in Figure 2.2.

The definition of the Bregman distance in this work is the same as in [32, 104]. We point out that in some papers (cf., e.g., [213,216]) the Bregman distance is defined via

$$\Delta_p(x, y) = \frac{1}{p}\|y\|^p - \frac{1}{p}\|x\|^p - \langle j_p(x), y - x\rangle,$$

which is slightly different from the Definition 2.56, because the arguments are interchanged.

For general convex penalty functionals we will also use the following, more general version of the Bregman distance concept.

Definition 2.58 (Bregman distance – general version). Let the functional $\Omega : X \to \mathbb{R} \cup \{\infty\}$ be convex and let $x, x^\dagger \in X$ as well as $\xi^\dagger \in \partial\Omega(x^\dagger)$. Then the Bregman distance in $x^\dagger$ and $\xi^\dagger$ with respect to Ω is defined as

$$D^{\Omega}_{\xi^\dagger}(x, x^\dagger) := \Omega(x) - \Omega(x^\dagger) - \langle \xi^\dagger, x - x^\dagger\rangle.$$

Setting $\Omega(x) = \|x\|^p/p$ and $x^\dagger = y$ we have $D^{\Omega}_{\xi^\dagger} = D_{j_p}$ with D_{j_p} from Definition 2.56.

Remark 2.59. In all cases where we make use of $D^{\Omega}_{\xi^\dagger}(x, x^\dagger)$ we will measure the distance between a variable point x and a fixed point $x^\dagger$. In this context, our focus is also on another fixed element $\xi^\dagger$ from $\partial\Omega(x^\dagger)$. We mention that, in this case, the Bregman distance at some point is only defined if the subdifferential at this point is not empty, and we refer to the concept of *Bregman domain*, which will be introduced in Assumption 3.26 (a). For $\Omega(x) = \frac{1}{q}\|x\|^q$ the subdifferential is obviously never an empty set (cf. Theorem 2.53).

Theorem 2.60. *Let X be a Banach space and $j_p \in J_p$ a fixed single-valued selection of the duality mapping J_p. Then the following properties are valid:*

(a) $D_{j_p}(x, y) \geq 0$.

(b) $D_{j_p}(x, y) = 0$ *if and only if* $j_p(y) \in J_p(x)$.

(c) *If X is smooth and uniformly convex, then a sequence $\{x_n\} \subset X$ remains bounded in X if $\{D_{j_p}(y, x_n)\}$ is bounded in $\mathbb{R}$. In particular, this assertion is true if X is convex of power type.*

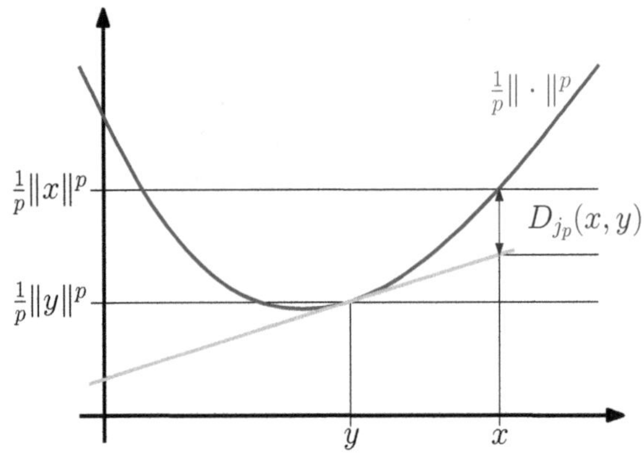

Figure 2.2. Geometrical interpretation of the Bregman distance as the gap between the function and its linearization.

(d) *$D_{j_p}(x,y)$ is continuous in its first argument. If X is smooth and uniformly convex, then J_p is continuous on bounded subsets and $D_{j_p}(x,y)$ is continuous in its second argument, too. In particular, this is true if X is convex of power type.*

(e) *If X is smooth and uniformly convex, then the following statements are equivalent:*

   i. $\lim_{n\to\infty}\|x_n - x\| = 0$,

   ii. $\lim_{n\to\infty}\|x_n\| = \|x\|$ *and* $\lim_{n\to\infty}\langle J_p(x_n), x\rangle = \langle J_p(x), x\rangle$,

   iii. $\lim_{n\to\infty} D_{j_p}(x, x_n) = 0$.

   In particular, this assertion is true if X is convex of power type.

(f) *The sequence $\{x_n\}$ is a Cauchy sequence in X if it is bounded and for all $\varepsilon > 0$ there is an $N(\varepsilon) \in \mathbb{N}$, such that $D_{j_p}(x_k, x_l) < \varepsilon$ for all $k, l \geq N(\varepsilon)$.*

(g) *X is p-convex if and only if $D_{j_p}(x,y) \geq C\|x-y\|^p$,*

(h) *X is p-smooth if and only if $D_{j_p}(x,y) \leq C\|x-y\|^p$*

(cf. [213, Theorem 2.12] *and* [239]*).*

Next, we formulate corollaries of the Xu–Roach inequalities (cf. Theorem 2.40) regarding the Bregman distances.

Corollary 2.61. *Let X be s-convex. Then, with the function σ_p from the Xu–Roach characterization of the convexity of power type in Theorem* 2.40 *we have:*

(a) *If* $1 < p \leq s$ *then*

$$D_{j_p}(y,x) \geq C \cdot \sigma_p(x, x-y) \geq C \cdot (\|x\| + \|y\|)^{p-s} \|x-y\|^s.$$

(b) *If* $s \leq p < \infty$ *then*

$$D_{j_p}(y,x) \geq C \cdot \sigma_p(x, x-y) \geq C \cdot \|x-y\|^p.$$

The generic constant $C > 0$ *can always be chosen independently of* x *and* y.

Proof. By the Xu–Roach characterization of convexity of power type in Theorem 2.40 we have

$$\begin{aligned} D_{j_p}(y,x) &= \tfrac{1}{p}\|y\|^p - \tfrac{1}{p}\|x\| - \langle j_p(x), y - x\rangle \\ &\geq \sigma_p(x, x-y) \\ &\geq C \int_0^1 t^{s-1} (\max\{\|x - t(x-y)\|, \|x\|\}^{p-s} \|x-y\|^s \, dt\,. \end{aligned}$$

Then the claims follow from the fact that

$$\|x\| + \|y\| \geq \max\{\|x - t(x-y)\|, \|x\|\} \geq \tfrac{t}{2}\|x-y\|$$

for $0 \leq t \leq 1$. □

We again return to the properties of the Bregman distances. The following identity for Bregman distances is known as the *three-point identity*.

Lemma 2.62. *Let* j_p *be a single-valued selection of the duality mapping* J_p^X. *Then we have*

$$D_{j_p}(x,y) = D_{j_p}(x,z) + D_{j_p}(z,y) + \langle j_p(z) - j_p(y), x - z\rangle.$$

Proof. We have

$$\begin{aligned} D_{k_p}(x,z) &+ D_{j_p}(z,y) + \langle j_p(z) - j_p(y), x - z\rangle \\ &= \tfrac{1}{p}\|x\|^p - \tfrac{1}{p}\|z\|^p - \langle j_p(z), x - z\rangle \\ &\quad + \tfrac{1}{p}\|z\|^p - \tfrac{1}{p}\|y\|^p - \langle j_p(y), z - y\rangle \\ &\quad + \langle j_p(z), x - z\rangle - \langle j_p(y), x - z\rangle \\ &= \tfrac{1}{p}\|x\|^p - \tfrac{1}{p}\|y\|^p - \langle j_p(y), x - y\rangle \\ &= D_{j_p}(x,y). \end{aligned}$$

□

We note that there is a close connection between the primal Bregman distances $D_{j_p}^X$ and the related dual Bregman distances $D_{j_{p^*}}^{X^*}$.

Lemma 2.63. *Let j_p^X be a single-valued selection of the duality mapping J_p^X. If there exists a single-valued selection $j_{p^*}^{X^*}$ of $J_{p^*}^{X^*}$ such that for some fixed $y \in X$ we have $j_{p^*}^{X^*}(j_p^X(y)) = y$, then*

$$D_{j_p}^X(y,x) = D_{j_{p^*}}^{X^*}(j_p^X(x), j_p^X(y))$$

for all $x \in X$.

Proof. We omit the indices indicating the spaces. From straightforward computations we obtain

$$\begin{aligned} D_{j_p}(y,x) &= \tfrac{1}{p}\|y\|^p - \tfrac{1}{p}\|x\|^p - \langle j_p(x), y - x\rangle \\ &= \tfrac{1}{p^*}\|x\|^p - \tfrac{1}{p^*}\|y\|^p - \langle y, j_p(x) - j_p(y)\rangle. \end{aligned}$$

For all $x \in X$ we have $\|x\|^p = \|x\|^{(p-1)\cdot p^*} = \|j_p(x)\|^{p^*}$. Therefore, we have

$$D_{j_p}(y,x) = \tfrac{1}{p^*}\|j_p(x)\|^{p^*} - \tfrac{1}{p^*}\|j_p(y)\|^{p^*} - \langle y, j_p(x) - j_p(y)\rangle.$$

The final right-hand side is equal to $D_{j_{p^*}}(j_p(x), j_p(y))$, since by assumption $y = j_{p^*}(j_p(y))$. □

From the above lemmas follows an alternative form of the three-point identity.

Theorem 2.64. *Let X be convex of power type and smooth of power type. Furthermore let $y = j_{p^*}(y^*)$ and $z = j_{p^*}(z^*)$, where j_{p^*} is the unique single-valued selection of $J_{p^*}^{X^*}$. Then we have for the unique single-valued selection j_p of J_p^X the equation*

$$D_{j_p}(x,y) = D_{j_p}(x,z) + D_{j_{p^*}}(y^*, z^*) + \langle z^* - y^*, x - z\rangle.$$

Proof. Since X is convex of power type and smooth of power type, by Theorem 2.52 the dual X^* is convex of power type and smooth of power type, too. Therefore, by Theorem 2.42, the mappings J_p^X and $J_{p^*}^{X^*}$ have only one single-valued selection, say j_p and j_{p^*} and by Theorem 2.53 we have $j_p(j_{p^*}(x^*)) = x^*$ for all $x^* \in X^*$ and $j_{p^*}(j_p(x)) = x$ for all $x \in X$. Hence, $j_p(y) = j_p(j_{p^*}(y^*)) = y^*$ and $j_p(z) = j_p(j_{p^*}(z^*)) = z^*$ as well as $D_{j_p}(z,y) = D_{j_{p^*}}(j_p(y), j_p(z)) = D_{j_{p^*}}(y^*, z^*)$, due to Lemma 2.63. The claim is then a consequence of the three-point identity of Lemma 2.62. □

Chapter 3

Ill-posed operator equations and regularization

3.1 Operator equations and the ill-posedness phenomenon

In this chapter, we will introduce the mathematical expression of inverse problems in terms of operator equations, i.e., abstract equations in Banach spaces. Depending on whether the associated forward operator is linear or nonlinear we have to distinguish between linear and nonlinear operator equations.

We should note that of course not all operator equations have an associated inverse problem. More precisely, it is an intrinsic property of inverse problems that the forward operator is typically *smoothing*, since clearly distinct causes may be transformed to nearby effects by the forward operator. Hence the inverse transformation is typically *roughening*. This relationship is known as the so-called *ill-posedness phenomenon* characterized by the disadvantage that smoothing mappings reduce information. In this context we refer to Hadamard's classic concept (cf. [85]) of well-posedness and ill-posedness, which we can reformulate as follows for an operator equation in abstract spaces with an observable right-hand side and a non-observable solution:

Definition 3.1 (Hadamard's definition). An operator equation is called *well-posed* in the sense of Hadamard if (a) it is solvable for all right-hand sides, if (b) the solutions are uniquely determined and if (c) the solution is stable in the sense that small perturbations in the right-hand side only lead to small perturbations in the solution. If at least one of requirements (a), (b) or (c) is violated, then the operator equation is called *ill-posed* in the sense of Hadamard.

The most serious difficulty is the violation of the stability requirement (c) nearly always occurring in the context of inverse problems. In applied mathematics one tries to solve inverse problems approximately, which in mathematical terms corresponds to the approximate solution of operator equations, and the main goal is to do so in a stable manner. Also non-uniqueness, violating (b), is a serious difficulty of ill-posed operator equations if the solution represents a well-defined quantity from natural sciences, engineering or finance to be recovered, whereas non-existence, violating (a), often can be overcome by appropriate modeling concepts.

3.1.1 Linear problems

Let X and Y be two Banach spaces with generic norms for which we will use the uniform symbol $\|\cdot\|$ with respect to both spaces. Only if we use norms of elements from a third Banach space or if we temporarily need some alternative norm in X or Y we characterize this by indices, as for example $\|\cdot\|_Z$. The typical mathematical expression of linear inverse problems consists in the solution of *linear operator equations*

$$Ax = y, \qquad x \in X,\ y \in \mathcal{R}(A) \subset Y, \tag{3.1}$$

where $A : X \to Y$ is a bounded (continuous) linear operator and $\mathcal{R}(A)$ its range. If A fails to be injective, i.e., the null-space $\mathcal{N}(A)$ of A is non-trivial, then the problem (3.1) has more than one solution and is ill-posed in the sense of Definition 3.1 since requirement (b) is violated. In such case one is usually interested in the solution which has the smallest norm. To overcome the possible non-uniqueness we define:

Definition 3.2 (minimum norm solution). We call $x^\dagger \in X$ a *minimum norm solution* of the operator equation (3.1) if

$$Ax^\dagger = y \qquad \text{and} \qquad \|x^\dagger\| = \inf\{\|\tilde{x}\| \ :\ \tilde{x} \in X,\ A\tilde{x} = y\}.$$

The following lemma (cf. [213, Lemma 2.10]) gives us an important characterization of the minimum norm solution:

Lemma 3.3. *Let X be a smooth and uniformly convex Banach space. Moreover, let Y be an arbitrary Banach space. Then the minimum norm solution $x^\dagger$ of* (3.1) *exists and is unique. Furthermore, it satisfies the condition $J_p^X(x^\dagger) \in \overline{\mathcal{R}(A^*)}$ for $1 < p < \infty$. If additionally there is some $x \in X$ such that $J_p^X(x) \in \overline{\mathcal{R}(A^*)}$ and $x - x^\dagger \in \mathcal{N}(A)$ then we have $x = x^\dagger$.*

Proof. The set $M := \{z \in X \ :\ Az = y\}$ is a nonempty closed convex subset of X since $y \in \mathcal{R}(A)$ and A is a continuous linear operator. Item (k) of Theorem 2.53 guarantees the existence and uniqueness of the minimum-norm solution $x^\dagger$ of (3.1). Now let z be an arbitrary element of $\mathcal{N}(A)$. Then $x^\dagger \pm z \in M$ and again by Theorem 2.53 (k) we have $\langle J_p^X(x^\dagger), x^\dagger\rangle_{X^*\times X} \le \langle J_p^X(x^\dagger), x^\dagger \pm z\rangle_{X^*\times X} = \langle J_p^X(x^\dagger), x^\dagger\rangle_{X^*\times X} \pm \langle J_p^X(x^\dagger), z\rangle_{X^*\times X}$. It follows that $\langle J_p^X(x^\dagger), z\rangle_{X^*\times X} = 0$ and hence $J_p^X(x^\dagger) \in (\mathcal{N}(A))^\perp$. This, however, implies that $J_p^X(x^\dagger) \in \overline{\mathcal{R}(A^*)}$ because we have $(\mathcal{N}(A))^\perp = \overline{\mathcal{R}(A^*)}$ for uniformly convex Banach spaces, see Lemma 2.55. If $J_p^X(x) \in \overline{\mathcal{R}(A^*)}$ for $x \in X$, then we can find a sequence $u_n \in Y^*$ such that $J_p^X(x^\dagger) - J_p^X(x) = \lim_{n\to\infty} A^*u_n$. This shows

$$\begin{aligned}\langle J_p^X(x^\dagger) - J_p^X(x), x^\dagger - x\rangle_{X^*\times X} &= \lim_{n\to\infty} \langle A^*u_n, x^\dagger - x\rangle_{X^*\times X}\\ &= \lim_{n\to\infty} \langle u_n, A(x^\dagger - x)\rangle_{Y^*\times Y} = 0\end{aligned}$$

since $x^\dagger - x \in \mathcal{N}(A)$. Using Theorem 2.53 (h) we conclude that $x = x^\dagger$. □

Remark 3.4. Even if the mapping $y \in \mathcal{R}(A) \mapsto x^\dagger \in X$ to the minimum norm solution is uniquely determined, this mapping needs not be linear in general Banach spaces. It is homogeneous but not necessarily additive. If, however, A is injective then $A^{-1} : \mathcal{R}(A) \subset Y \to X$ is linear and we have $x^\dagger = A^{-1}y$. For non-injective A but Hilbert spaces X and Y we have $x^\dagger = A^\dagger y$ with a linear operator $A^\dagger : \mathcal{R}(A) \oplus \mathcal{R}(A)^\perp \subset Y \to X$ called the Moore-Penrose inverse. Notice that the term minimum norm solution for Hilbert spaces is sometimes used to describe the solution of the normal equations $A^*Ax = A^*y$ with minimum norm. However, this is *not* the sense in which we use this term here.

The stability requirement (c) in Definition 3.1 translates to a range condition for linear operator equations, and we here refer to the concept of Nashed (cf. [171]):

Definition 3.5 (Nashed's definition). A linear operator equation (3.1) is called *well-posed* in the sense of Nashed if the range of A is closed, i.e., $\mathcal{R}(A) = \overline{\mathcal{R}(A)}$, and *ill-posed* in the sense of Nashed if the range of A is not closed, i.e., $\mathcal{R}(A) \neq \overline{\mathcal{R}(A)}$. Moreover, for nonclosed range $\mathcal{R}(A)$ we call the equation *ill-posed of type I* if $\mathcal{R}(A)$ contains a closed infinite dimensional subspace $\tilde{Y}$ and the nullspace $\mathcal{N}(A)$ is topologically complemented in the subspace $A^{-1}(\tilde{Y}) := \{x \in X \ : \ Ax \in \tilde{Y}\}$. If not, then *ill-posed of type II*.

Remark 3.6. Note that $\mathcal{R}(A) \neq \overline{\mathcal{R}(A)}$ cannot occur if $\dim(\mathcal{R}(A)) < \infty$, because all finite dimensional subspaces are closed. Hence for $\dim(X) < \infty$ and for degenerating A such that $\dim(X) = \infty$, but $\dim(\mathcal{R}(A)) < \infty$, the equation (3.1) is always well-posed in the sense of Nashed. Thus any finite dimensional *discretization*

$$\tilde{A}\tilde{x} = \tilde{y}, \quad \tilde{A} \in \mathbb{R}^{m \times n},\ \tilde{x} \in \mathbb{R}^n,\ \tilde{y} \in \mathbb{R}^m \tag{3.2}$$

of the operator equation (3.1) is well-posed. Such discretizations occur in all numerical approaches as approximations of (3.1). However, if $\tilde{A}$ approximates A sufficiently well, the system of linear equations (3.2) is in general *ill-conditioned.* This means that for sufficiently fine discretization the condition number $\mathrm{cond}(\tilde{A})$ of the matrix $\tilde{A}$ is large and tends to infinity as $n \to \infty$.

Remark 3.7. Provided that X and Y are infinite dimensional Hilbert spaces, ill-posedness of type II occurs if and only if A is compact with $\dim(\mathcal{R}(A)) = \infty$. Otherwise, in the case of Hilbert spaces, we have ill-posedness of type I if and only if the range $\mathcal{R}(A)$ is not closed, but contains a closed infinite dimensional subspace.

Based on Lemma 3.8 the subsequent proposition explains the essential connection between the ill-posedness concepts by Hadamard and Nashed for injective linear operators A (cf. also [170]).

Lemma 3.8. [171, Proposition 2.1]. *If the range $\mathcal{R}(A)$ of a bounded linear operator $A : X \to Y$ mapping between two Banach spaces X and Y is not closed, then A*

has no bounded inner inverse, i.e., no bounded linear operator $B : \mathcal{R}(A) \subset Y \to X$ *satisfying the equation* $ABA = A$.

Proposition 3.9. *If the bounded linear operator* $A : X \to Y$ *mapping between Banach spaces* X *and* Y *is injective, then the operator equation* (3.1) *is ill-posed in the sense of Nashed if and only if the inverse operator* $A^{-1} : \mathcal{R}(A) \subset Y \to X$ *is unbounded, i.e.,* A^{-1} *fails to be continuous and the stability requirement* (c) *of Definition* 3.1 *is violated.*

Proof. If equation (3.1) is ill-posed in the sense of Nashed, then A^{-1} is an inner inverse of A and, by Lemma 3.8, cannot be bounded. Vice versa if (3.1) is well-posed in the sense of Nashed we have $\mathcal{R}(A) = \overline{\mathcal{R}(A)}$ and the closed subspace $\mathcal{R}(A)$ of Y endowed with the same norm as Y is a Banach space. In this case Banach's bounded inverse theorem applies and ensures that A^{-1} is a bounded and hence continuous operator. □

3.1.2 Nonlinear problems

The typical mathematical expression of nonlinear inverse problems consists in the solution of *nonlinear operator equations*

$$F(x) = y, \qquad x \in \mathcal{D}(F) \subseteq X, \;\; y \in F(\mathcal{D}(F)) \subseteq Y, \tag{3.3}$$

where $F : \mathcal{D}(F) \subseteq X \to Y$ is a nonlinear operator with domain $\mathcal{D}(F)$ and range $F(\mathcal{D}(F)) = \{\tilde{y} \in Y : F(\tilde{x}) = \tilde{y} \;\; \text{for some } \tilde{x} \in \mathcal{D}(F)\}$.

Unfortunately, the majority of inverse problems cannot be treated by means of the mathematically simpler linear operator equations (3.1) since either the forward operator F is frequently far from being a linear mapping, or additional constraints coming from a priori information about the sought solution require the handling of a domain $\mathcal{D}(F)$, which also destroys the simple structure of the unconstrained linear equation (3.1). For some inverse problems in partial differential equations, aimed at finding unknown parameter functions in the differential equation, or in boundary conditions, e.g., in the context of heat transfer or wave propagation, the forward operator F is even nonlinear if the differential equation is linear.

According to the local character of solutions in nonlinear equations we have to focus on some neighborhood of a reference element $\overline{x} \in X$ which can be interpreted as an initial guess for the solution to be determined. Consequently, we shift our coordinate system from zero to $\overline{x}$ and search for $\overline{x}$-minimum norm solutions:

Definition 3.10 ($\overline{x}$-minimum norm solution). We say that $x^\dagger \in \mathcal{D}(F) \subseteq X$ is an *$\overline{x}$-minimum norm solution* of the operator equation (3.3) if

$$F(x^\dagger) = y \qquad \text{and} \qquad \|x^\dagger - \overline{x}\| = \inf\{\|\tilde{x} - \overline{x}\| : \tilde{x} \in \mathcal{D}(F),\; F(\tilde{x}) = y\}.$$

In order to ensure that $\overline{x}$-minimum norm solutions to the nonlinear equation (3.3) exist, we make some assumptions about the Banach spaces X and Y, as well as about the operator F and its domain $\mathcal{D}(F)$, which are of later importance. In this context, for the Banach spaces X and Y we distinguish between norm convergence, abbreviated by $\to$ and weak convergence abbreviated by $\rightharpoonup$, respectively. We note that uniqueness of $\overline{x}$-minimum norm solutions, in general, cannot be guaranteed in the case of nonlinear operator equations.

Assumption 3.11.

(a) X and Y are infinite dimensional reflexive Banach spaces.

(b) $\mathcal{D}(F)$ is a convex and closed subset of X.

(c) $F : \mathcal{D}(F) \subseteq X \to Y$ is weak-to-weak sequentially continuous, i.e., $x_n \rightharpoonup x_0$ in X with $x_n \in \mathcal{D}(F),\ n \in \mathbb{N}$, and $x_0 \in \mathcal{D}(F)$ implies $F(x_n) \rightharpoonup F(x_0)$ in Y.

We mention the following lemma (cf., e.g., [164, p.251]) and its corollary to clarify the consequences of Assumption 3.11.

Lemma 3.12. *The closed unit ball of a Banach space X is weakly sequentially compact if and only if X is reflexive. Moreover, in reflexive Banach spaces a convex and closed subset is also weakly sequentially closed.*

Corollary 3.13. *Any closed ball in a reflexive Banach space X is weakly sequentially compact. Hence, for all $c > 0$ an infinite sequence $\{x_n\}_{n=1}^{\infty}$ belonging to the sublevel set*

$$\mathcal{M}_{\overline{x}}(c) := \{x \in X \ :\ \|x - \overline{x}\| \le c\}$$

has a weakly convergent subsequence $x_{n_k} \rightharpoonup x_0 \in \mathcal{M}_{\overline{x}}(c)$ as $k \to \infty$.

Under Assumption 3.11, by Lemma 3.12 $\mathcal{D}(F)$ is weakly sequentially closed and $x_n \rightharpoonup x_0$ in X with $x_n \in \mathcal{D}(F),\ n \in \mathbb{N}$ implies $x_0 \in \mathcal{D}(F)$. Furthermore, by Corollary 3.13 the intersection of $\mathcal{D}(F)$ with a closed ball in X is weakly sequentially compact.

Proposition 3.14. *Under Assumption* 3.11 *the nonlinear operator equation* (3.3) *possesses an $\overline{x}$-minimum norm solution.*

Proof. By definition there exists a sequence $\{x_n\}_{n=1}^{\infty}$ of elements from $\mathcal{D}(F)$ such that $F(x_n) = y$ for all $n \in \mathbb{N}$ and $\lim_{n\to\infty} \|x_n - \overline{x}\| = \inf_{\tilde{x}\in\mathcal{D}(F)\,:\,F(\tilde{x})=y} \|\tilde{x} - \overline{x}\| \ge 0$. Since any such sequence belongs to a closed ball in X by Corollary 3.13, there is a subsequence $x_{n_k} \rightharpoonup x_0 \in \mathcal{D}(F)$. Since the norm is weakly lower semicontinuous this yields $\|x_0 - \overline{x}\| = \inf_{\tilde{x}\in\mathcal{D}(F)\,:\,F(\tilde{x})=y} \|\tilde{x} - \overline{x}\|$ and by item (c) of Assumption 3.11 we have $F(x_0) = y$. Hence x_0 is an $\overline{x}$-minimum norm solution. □

The violation of the stability requirement (c) in Definition 3.1 translates to a local condition at a solution point $x_0 \in \mathcal{D}(F)$ of (3.3) for nonlinear operator equations. Here, we refer to the concept of *local ill-posedness* (cf. [110, 111] and [101]) acting in closed balls $\overline{B}_r(x_0)$ with center x_0 and radius $r > 0$:

Definition 3.15 (local ill-posedness). A nonlinear operator equation (3.3) is called *locally ill-posed* at the point $x_0 \in \mathcal{D}(F)$ satisfying $F(x_0) = y$ if for arbitrarily small radii $r > 0$ there is a sequence $\{x_n\}_{n=1}^{\infty} \subset \overline{B}_r(x_0) \cap \mathcal{D}(F)$ such that

$$F(x_n) \to F(x_0) \quad \text{in } Y, \quad \text{but} \quad x_n \not\to x_0 \quad \text{in } X, \quad \text{as } n \to \infty.$$

Otherwise the equation is called *locally well-posed* at x_0.

If an equation (3.3) is locally ill-posed, then a solution x_0 cannot be recovered sufficiently well, even if the perturbations on the right-hand side $y = F(x_0)$ are arbitrarily small. This means that the error norm $\|x_n - x_0\|$ of a sequence of approximate solutions $\{x_n\}_{n=1}^{\infty} \subset \mathcal{D}(F)$ satisfying $\lim_{n\to\infty} \|F(x_n) - F(x_0)\| = 0$ need not converge to zero. Local ill-posedness includes the case of local non-identifiability which occurs when x_0 is an accumulation point with respect to the norm topology of elements x_n, such that $F(x_n) = F(x_0)$. Analogous to the fact that compact operators A are responsible for an essential class of ill-posed linear operator equations in the sense of Nashed (cf. Remark 3.7), wide classes of locally ill-posed nonlinear operator equations are associated with compact nonlinear forward operators F where, as in the linear case, we call F compact if it transforms every subset of $\mathcal{D}(F)$, which is bounded in X, to a sequentially pre-compact subset of Y.

Proposition 3.16. *Under Assumption* 3.11 *the nonlinear operator equation* (3.3) *is locally ill-posed at a point $x_0 \in \mathcal{D}(F)$ satisfying $F(x_0) = y$ if the operator F is* compact *and for arbitrarily small radii $r > 0$ there is a sequence $\{x_n\}_{n=1}^{\infty} \subset \overline{B}_r(x_0) \cap \mathcal{D}(F)$, such that $F(x_n) \rightharpoonup F(x_0)$ in Y, but $x_n \not\to x_0$ in X as $n \to \infty$.*

Proof. Since F is compact and, by Assumption 3.11 (c), also weak-to-weak continuous, weakly convergent sequences $x_n \rightharpoonup x_0$ in X will be transformed to strongly convergent sequences $F(x_n) \to F(x_0)$ in Y. This yields the local ill-posedness at the solution point x_0. □

In the context of Hilbert spaces , the main ideas of Proposition 3.16 can already be found in Proposition A.3 of [68]. If, for an infinite dimensional separable Hilbert space X, the domain $\mathcal{D}(F)$ contains a ball $\overline{B}_{r_0}(x_0)$ for some $r_0 > 0$ then, under Assumption 3.11, by Proposition 3.16 we have local ill-posedness at x_0 whenever the operator F is compact: for an orthonormal basis $\{e_n\}_{n=1}^{\infty}$ in X and for all $0 < r < r_0$ it holds $x_n \rightharpoonup x_0$ for $x_n := x_0 + r e_n \rightharpoonup x_0$, but $\|x_n - x_0\| = r > 0$. As a consequence of the Josefson-Nissenzweig theorem (cf. [69, p.88]), which ensures the existence of elements $e_n \rightharpoonup 0$ with $\|e_n\| = 1$ in any infinite dimensional reflexive Banach space we can formulate the following corollary. (Since bounded linear operators are weak-to-weak continuous we can apply this result to linear problems as well.)

Corollary 3.17. *Under Assumption* 3.11 *the nonlinear operator equation* (3.3) *is locally ill-posed at a point* $x_0 \in \mathcal{D}(F)$ *satisfying* $F(x_0) = y$ *if the operator* F *is compact and* $\overline{B}_{r_0}(x_0) \subset \mathcal{D}(F)$ *for some* $r_0 > 0$. *Moreover, for infinite dimensional reflexive Banach spaces* X *and compact linear operators* A, *the linear operator equation* (3.1) *is always ill-posed in the sense of Nashed of type II.*

3.1.3 Conditional well-posedness

For linear (3.1), as well as for nonlinear (3.3) operator equations, in a Banach space setting ill-posedness can be overcome by restricting the domain of the forward operator to appropriate subsets $\mathcal{M}$ of X. Frequently, the problems are then called *conditionally well-posed* with respect to $\mathcal{M}$. Here, we explain this for nonlinear equations. As a preliminary, we introduce the concept of index functions.

Definition 3.18 (index function). A function $\varphi : [0,\infty) \to [0,\infty)$ is called an *index function*, if it is strictly increasing and continuous with $\varphi(0) = 0$.

Index functions play an important role for the error analysis of ill-posed problems. Note that the name *index function* refers to indices of variable Hilbert scales, see [90, 91] and more recently [92, 162, 163].

Firstly, conditional well-posedness can be enforced by *compactness* of the domain $\mathcal{D}(F)$, and we refer to *Tikhonov's theorem*, see [12, 234], formulated now as Proposition 3.19.

Proposition 3.19. *Let* $F : \mathcal{D}(F) \subseteq X \to Y$ *be a* continuous *operator mapping between Banach spaces* X *and* Y *and let* $\mathcal{D}(F) := \mathcal{M}$ *be a* compact *subset of* X. *Moreover, suppose that for given* $y \in Y$ *there exists a uniquely determined solution* $x_0 \in \mathcal{M}$ *of the operator equation* (3.3). *Then, for a sequence* $y_n = F(x_n)$ *with* $x_n \in \mathcal{M}$, *the convergence* $\lim_{n\to\infty} \|y_n - y\| = 0$ *with respect to the norm in* Y *implies the norm convergence* $\lim_{n\to\infty} \|x_n - x_0\| = 0$ *in* X.

As one can see, compactness of $\mathcal{M}$ implies local well-posedness (cf. Definition 3.15) at any point $x_0 \in \mathcal{M}$ for which the solution of (3.3) with $y = F(x_0)$ is uniquely determined in $\mathcal{M}$. Compact subsets $\mathcal{M}$ occur, for example, if the domain of F is further restricted by *objective a priori information* in the form of monotonicity, convexity or concavity properties, imposed on functions x from a bounded set in $L^p(a,b)$ with $1 < p < \infty$ (cf. [233]). Note that norm-to-norm continuity of the operator F in Proposition 3.19 can be replaced by a weak-to-weak continuity without changing the assertion.

Secondly, conditional well-posedness can also be stated if (3.3) is ill-posed on the domain $\mathcal{D}(F)$, but there is a (not necessarily compact) subset $\mathcal{M}$ of $\mathcal{D}(F)$, such that a *conditional stability estimate*

$$\|x_1 - x_2\| \le \varphi(\|F(x_1) - F(x_2)\|) \qquad \text{for all} \quad x \in \mathcal{M} \subset X \tag{3.4}$$

holds for some index function φ (cf., e.g., [111,208]). Then under condition (3.4) and for $\mathcal{D}(F) = \mathcal{M}$ the operator F is injective and a proposition analogous to Proposition 3.19 can be formulated which, however, is without the requirements of compactness for $\mathcal{M}$ and of continuity for F. Consequently, under (3.4) the equation (3.3) is locally well-posed at each point $x_0 \in \mathcal{M}$. The function φ in condition (3.4) can be considered as an upper bound of the modulus of continuity of the inverse operator F^{-1}, restricted to $F(M) \subset Y$. An example of (3.4) for nonlinear F is given in [138] in the context of inverse option pricing. For linear forward operators A conditional stability estimates of the form

$$\|x\| \le \varphi(\|Ax\|) \qquad \text{for all} \quad x \in \mathcal{M} \subset X$$

are preferred to be considered, and we refer to [108, 230] for details and examples of linear inverse problems.

Note that a frequently used version of (3.4) refers to radius-dependent families of sets

$$\mathcal{M} = \mathcal{M}_R := \{x \in \mathcal{D}(F) \cap Z : \|x\|_Z \le R\},$$

where Z is a Banach space with norm $\|\cdot\|_Z$. This space is assumed to be densely defined in, and continuously embedded into the original space X. Then, for obtaining conditional stability it is assumed that there exist constants $C(R) > 0$ for all $R > 0$ and an index function φ such that

$$\|x_1 - x_2\| \le C(R)\,\varphi(\|F(x_1) - F(x_2)\|) \qquad \text{for all} \quad x_1, x_2 \in \mathcal{M}_R.$$

We refer to the articles [47] and [113] for details. Moreover, results in the context of Tikhonov-type regularization will be given in Section 4.2.5 below.

3.2 Mathematical tools in regularization theory

In general, if one tackles an applied inverse problem by solving the corresponding operator equations (3.1) or (3.3), formulated in infinite dimensional Banach spaces X and Y, the ill-posedness phenomenon appears. It produces serious practical problems since, instead of the exact right-hand side y, only noisy data, i.e., elements $y^\delta \in Y$ satisfying the inequality

$$\|y^\delta - y\| \le \delta \tag{3.5}$$

with *noise level* $\delta > 0$, are available. When using such perturbed data the goal consists of the *stable approximate solution* of the operator equations. Since for ill-posedness reasons the optimization problems

$$\|Ax - y^\delta\| \to \min, \quad \text{subject to } x \in X,$$

in the linear case, as well as

$$\|F(x) - y^\delta\| \to \min, \quad \text{subject to } x \in \mathcal{D}(F) \subseteq X,$$

in the nonlinear case, both aimed at minimizing the norm discrepancy as a *fidelity term* or *misfit term* for the operator equations, need not be solvable, or, if they are solvable, the approximate solutions are unstable and wrong, *stable auxiliary problems* must be used in order to overcome, or at least reduce the destructive influence of the ill-posedness. This is the idea of *regularization*, which can be realized in diverse ways.

Approximate solutions for ill-posed problems, which are called *regularized solutions* if they result from the use of regularization methods (see, for example, [67, 100, 150, 199, 232]), should not try to minimize the norm discrepancy, because this leads to highly oscillating solutions, and does not make sense. In contrast, the discrepancy norm of regularized solutions must be brought into agreement with the noise level $\delta > 0$, and from all potential solution elements compatible with the data in this sense we select only those that are reliable with respect to additional subjective or objective a priori information. Such information can refer to solution smoothness, further constraints or closeness to some reference element. Mostly, a *regularization parameter* $\alpha > 0$ controls the neighborhood properties of the auxiliary problems. Larger values of α express higher stability of the approximate solutions, but the auxiliary problem is rather far from the original one. On the other hand, values of α near zero represent auxiliary problems close to the original one, but they tend to become more and more unstable as $\alpha \to 0$. Hence, smart regularization approaches must find an appropriate trade-off between the conflicting goals of stability and approximation.

Regularized solutions based on y^δ must be stable, i.e., their fluctuation should be small for small changes in y^δ. Moreover, *convergence* is required, i.e., the approximate solutions should converge to exact solutions of the ill-posed operator equation under consideration. If the solutions are not uniquely determined, the convergence targets solutions with preferred properties, for example minimum norm solutions (cf. Definition 3.2), $\overline{x}$-minimum norm solutions (cf. Definition 3.10), or Ω-minimizing solutions (cf. Definition 3.25 below). A mathematically very challenging aspect in regularization is to derive *convergence rates*, i.e., to detect conditions under which the error of the regularized solution, evaluated by norm or alternative error measures, converges with some rate of order $\varphi(\delta)$ as $\delta \to 0$, where φ is an index function (cf. Definition 3.18). In the next subsection we will describe regularization procedures for linear problems and briefly mention the analog for the nonlinear case.

3.2.1 Regularization approaches

Definition 3.20 (regularization). A mapping that transforms every pair $(y^\delta, \alpha) \in Y \times (0, \overline{\alpha}]$ with $0 < \overline{\alpha} \le +\infty$ to some well-defined element $x_\alpha^\delta \in X$ is called a *regularization* (procedure) for the linear operator equation (3.1), if there exists an appropriate choice $\alpha = \alpha(y^\delta, \delta)$ of the regularization parameter such that, for every sequence $\{y_n\}_{n=1}^{\infty} \subset Y$ with $\|y_n - y\| \le \delta_n$ and $\delta_n \to 0$ as $n \to \infty$, the corresponding regularized solutions $x_{\alpha(y_n,\delta_n)}^{\delta_n}$ converge in a well-defined sense to the solution $x^\dagger$ of equation (3.1). If the solution is not unique, regularized solutions have to converge to solutions of equation (3.1) possessing the desired properties, e.g., to minimum norm solutions. For nonlinear operator equations (3.3) with regularized solutions $x_\alpha^\delta \in \mathcal{D}(F)$, the definition is analogous. If the solution of equation (3.3) is not unique, convergence to solutions possessing desired properties, e.g., to $\overline{x}$-minimum norm solutions, is required. In case of non-uniqueness, different subsequences of regularized solutions can converge to different solutions of the operator equation, which all possess the same desired property.

Remark 3.21. We have to distinguish between *a priori parameter choices* $\alpha = \alpha(\delta)$ independent of the data element y^δ and *a posteriori parameter choices* $\alpha = \alpha(y^\delta, \delta)$, that take advantage of knowledge of the present data. From [13] we know that parameter choices $\alpha = \alpha(y^\delta)$ that avoid the explicit use of the noise level δ cannot be the basis of regularization procedures. However, since in practice knowledge of δ is not always an acceptable premise, *heuristic parameter choices* like the L-curve rule, the quasi-optimality rule and the approach of generalized cross validation (cf., e.g., [89]) with different modifications and realizations like the model function method (cf. [139] and [99, 157, 237, 238]) are also of some importance for the stable approximate solution of ill-posed inverse problems. Recently, it was shown that they can even yield regularization procedures if additional conditions imposed on noise and solutions are satisfied (cf. [133, 174]). Further details concerning the choice of regularization parameters will also be presented in Sections 4.1.2, 4.2 and 5.2.

In the case of linear operator equations in a Hilbert space setting, the construction of regularization procedures is in general based on the approximation of the Moore-Penrose inverse $A^\dagger$ by an α-dependent family of bounded operators with regularized solutions

$$x_\alpha^\delta = g_\alpha(A^*A)A^*y^\delta, \qquad \alpha > 0,$$

and appropriate generator functions g_α (cf. [67, 84, 106, 150, 160, 167, 235]). Unfortunately, in our Banach space setting neither $A^\dagger$ nor A^*A is available, since the adjoint operator $A^* : Y^* \to X^*$ maps between the dual spaces. In the case of nonlinear operator equations a comparable phenomenon occurs, because the adjoint operator

$$F'(x^\dagger)^* : Y^* \to X^*$$

of a bounded linear derivative operator

$$F'(x^\dagger) : X \to Y$$

of F at the solution point $x^\dagger \in \mathcal{D}(F)$ also maps between the dual spaces.

Nevertheless, two large and powerful classes of regularization procedures with prominent applications, for example in imaging (cf. [209]), were recently promoted: the class of *Tikhonov-type regularization* methods in Banach spaces, to be presented in Part III (see also [104, 210]), and the class of *iterative regularization* methods in Banach spaces, to be presented in Part IV of this work (see also [14, 128]). Tikhonov-type regularization and iterative regularization can be applied to linear as well as to nonlinear problems in Banach spaces in order to overcome the ill-posedness. In the past decade, substantial progress has been made with respect to both classes.

For the former class of Tikhonov-type regularization, which is also called *variational regularization*, regularized solutions x_α^δ in our work are minimizers of an extremal problem of the form

$$\mathrm{T}_\alpha(x) := \frac{1}{p}\|Ax - y^\delta\|^p + \alpha\,\Omega(x) \to \min, \quad \text{subject to } x \in X, \tag{3.6}$$

for linear operator equations (3.1), and of the form

$$\mathrm{T}_\alpha(x) := \frac{1}{p}\|F(x) - y^\delta\|^p + \alpha\,\Omega(x) \to \min, \quad \text{subject to } x \in \mathcal{D}(F) \subseteq X, \tag{3.7}$$

for nonlinear operator equations (3.3), respectively. We can consider the approaches exploiting (3.6) or (3.7) as a *penalized minimization* of a fidelity term which is constructed by applying the index function

$$\psi(t) := \frac{t^p}{p}, \qquad t > 0,$$

to the discrepancy norms $\|Ax - y^\delta\|$ or $\|F(x) - y^\delta\|$, respectively. The Tikhonov-type regularization admits the inclusion of *subjective a priori information* of wide comprehension varying the penalty functionals Ω. Indeed, for fixed $\alpha > 0$ the extremal problem (3.7) is equivalent to an approach called *residual method*

$$\Omega(x) \to \min, \quad \text{subject to } \|F(x) - y^\delta\| \le K, \;\; x \in \mathcal{D}(F), \tag{3.8}$$

for some constant $K = K(\alpha, \delta) > 0$ (cf. [118, 227] and [83]). If one considers Ω as a *functional of sympathy*, i.e., just elements $x \in X$ with small values $\Omega(x)$ possess required properties, variational regularization tries to pitch on the most likable approximate solutions among all those compatible with the data. However, the amount of work required to compute the regularized solutions x_α^δ is in general huge, because often non-quadratic optimization problems must be solved for any α, and a sizable number of parameters has to be checked for selecting a best possible one with respect to the data.

In the following, for all fidelity terms and penalty functionals, as ingredients of the Tikhonov functional T_α under consideration, the following Assumption 3.22 is assumed to hold. This assumption complements Assumption 3.11, introduced in Section 3.1.2.

Assumption 3.22.

(a) For the exponent in (3.6) and (3.7), we have $1 < p < \infty$.

(b) $\Omega : X \to [0, \infty]$ is a proper *convex* and lower semi-continuous functional, where proper denotes that the domain of Ω

$$\mathcal{D}(\Omega) := \{x \in X \;:\; \Omega(x) < \infty\}$$

is nonempty. Moreover, for nonlinear operator equations we also assume

$$\mathcal{D} := \mathcal{D}(F) \cap \mathcal{D}(\Omega) \neq \emptyset.$$

(c) Ω is assumed to be a *stabilizing functional* in the sense that the sublevel sets

$$\mathcal{M}_{\Omega}(c) := \{x \in X \;:\; \Omega(x) \leq c\}$$

are weakly sequentially pre-compact in X for all $c \geq 0$.

Remark 3.23. In this work, in order to measure the discrepancy between operator images $F(x)$ and data y^δ in variational regularization, we restrict our consideration to terms of the form

$$\psi(\|F(x) - y^\delta\|) := \frac{1}{p}\|F(x) - y^\delta\|^p \quad \text{with} \quad 1 < p < \infty,$$

or their analogs for linear problems. Note that the limiting case $p = 1$ must be handled specifically, because it produces strange situations, where the regularization parameter needs not tend to zero as $\delta \to 0$. So for $p = 1$ the results for convergence rates in Tikhonov-type regularization are proved when the regularization parameter $\alpha > 0$ is chosen arbitrarily fixed but sufficiently small, see [32] where this phenomenon is called *exact penalization.* More curious is the situation of $p = 1$ in [177], where optimal convergence rates require that $\alpha > 0$ must equal to a fixed value, depending on properties of the solution $x^\dagger$. On the other hand, motivated by stochastic noise models, for example Poisson-distributed data occurring in imaging problems, an extension of Tikhonov regularization theory in Banach spaces to more general fidelity or misfit terms $S(F(x), y^\delta)$, sometimes also called similarity terms and replacing $\psi(\|F(x) - y^\delta\|)$, was recently promoted by several authors, and we refer to [17, 72, 74, 75, 116, 121, 176, 185].

Remark 3.24. Frequently, *coercivity* of Ω is considered as an important property of the functional Ω, which means that $\|x\| \to \infty$ implies $\Omega(x) \to \infty$. However, this is only used to conclude that all sublevel sets $\mathcal{M}_{\Omega}(c)$ are bounded in X, and therefore weakly pre-compact, provided X is reflexive. In Assumption 3.22 (c), we do not take this detour, but directly postulate the essential property of weak pre-compactness.

According to Assumptions 3.11 and 3.22, regularized solutions $x_\alpha^\delta \in \mathcal{D}$ minimizing (3.7) exist for all $\alpha > 0$, as will be proved in Proposition 4.1 below. We note that this existence assertion is also valid for linear problems, where x_α^δ is a minimizer of (3.6). This is a consequence of the fact that any bounded linear operator $A : X \to Y$ is weak-to-weak continuous between Banach spaces X and Y.

By extending Definition 3.10 of $\overline{x}$-minimum norm solutions to the case of general penalty functionals Ω, we introduce the concept of Ω-minimizing solutions of an operator equation (cf. [104]). In Definition 3.25, we will formulate this for nonlinear problems. Analogously, Ω-minimizing solutions extend the concept of minimum norm solutions from Definition 3.2 for linear problems.

Definition 3.25 (Ω-minimizing solution). We say that $x^\dagger \in \mathcal{D}(F) \subseteq X$ is an Ω-*minimizing solution* of the operator equation (3.3) if

$$F(x^\dagger) = y \qquad \text{and} \qquad \Omega(x^\dagger) = \inf\{\Omega(\tilde{x}) \,:\, \tilde{x} \in \mathcal{D}(F),\ F(\tilde{x}) = y\}.$$

Following along the lines of the proof of Proposition 3.14, one can easily show that there exists at least one Ω-minimizing solution for convex stabilizing functionals Ω. Recently, non-convex penalty functionals Ω were also introduced to the field of Tikhonov-type regularization (cf. [78, 242]). However, a couple of additional analytical and numerical difficulties occur for such extensions, which we will not discuss here.

We will now define a third set of requirements in Assumption 3.26, that further complement the Assumptions 3.11 and 3.22, concerning derivatives of F at the solution $x^\dagger$.

Assumption 3.26.

(a) There exists an Ω-minimizing solution $x^\dagger$ of equation (3.3), which belongs to the so-called *Bregman domain*

$$\mathcal{D}_B(\Omega) := \{x \in \mathcal{D} \subseteq X \,:\, \partial\Omega(x) \neq \emptyset\},$$

where $\partial\Omega(x) \subseteq X^*$ denotes the subdifferential of Ω in the point x.

(b) There is a bounded linear operator $F'(x^\dagger) : X \to Y$ such that for the one-sided directional derivative at $x^\dagger$ and for every $x \in \mathcal{D}$ we have the limit condition

$$\lim_{t \to +0} \frac{1}{t}\Big(F(x^\dagger + t(x - x^\dagger)) - F(x^\dagger)\Big) = F'(x^\dagger)(x - x^\dagger). \tag{3.9}$$

Definition 3.27 (starlike). We call a set $\mathcal{M}$ *starlike* with respect to $x^\dagger$ if $x \in \mathcal{M}$ implies that $x^\dagger + \lambda(x - x^\dagger) \in \mathcal{M}$ holds for all $0 \le \lambda < 1$.

We mention that a convex set $\mathcal{M}$ with $x^\dagger \in \mathcal{M}$ is obviously starlike with respect to $x^\dagger$. Hence, $\mathcal{D}$, being a convex subset of X, possesses this property. Note that $\mathcal{D}$ is convex because it is the intersection of the convex set $\mathcal{D}(F)$ (cf. Assumption 3.11 (b)) with the set $\mathcal{D}(\Omega)$ which is convex due to the convexity of the functional Ω (cf. Assumption 3.22 (b)). For the starlike set $\mathcal{M}$ however, the limit condition (3.9) makes sense. The directional derivative $F'(x^\dagger)$ from Assumption 3.26 (b) has the properties of a Gâteaux derivative, but $x^\dagger$ need not be an interior point of $\mathcal{D}$, and there can be directions, going out from $x^\dagger$, in which no points of $\mathcal{D}$ appear. Note that, under our assumptions, interior points of $\mathcal{D}$ need not exist. We mention the well-known example of the 'half-space' $\mathcal{D}(F) = \{x \in L^p(a,b) : x \geq 0 \text{ a.e.}\}$, which does not possess any interior point in the Banach space $L^p(a,b)$, $1 \leq p < \infty$.

In this work, in particular in the case of linear ill-posed problems, our focus is on convex penalties of norm power-type

$$\Omega(x) := \frac{1}{q}\|x\|_X^q, \qquad 1 \leq q < \infty, \tag{3.10}$$

and hence on regularized solutions x_α^δ that are minimizers of the extremal problem

$$\mathrm{T}_\alpha(x) := \frac{1}{p}\|Ax - y^\delta\|_Y^p + \alpha\frac{1}{q}\|x\|_X^q \to \min, \quad \text{subject to } x \in X, \tag{3.11}$$

where $1 < p < \infty$ and $1 \leq q < \infty$. The character of the Banach spaces X from which the norm powers are derived can be very different. For example, ℓ^p-spaces, L^p-spaces, Sobolev spaces and Besov spaces (cf. [148]) are of interest in choosing X in different mathematical models. If the space exponent $p > 1$ in $Y = L^p$ and the residual norm exponent p coincide, the analysis and numerics can be simplified. The same is true of $X = L^q$ and the norm exponent q in (3.11). On the other hand, the appropriate choice of p and q in (3.11) is also of serious interest from the point of view of mathematical modelling. If we choose p very large, this model can simulate a form of penalized Chebyshev approximation, whereas p close to one seems to be useful if robustness is required. Also, the choice of the exponent q can be interpreted, such that for example q close to one is appropriate in the case where the solution is sparse.

Indeed because of Corollary 3.13, according to (3.10), Ω is a stabilizing functional whenever X is a reflexive Banach space. The exponents p, q and the regularization parameter $\alpha > 0$ are weights controlling the interplay and character of the fidelity term and of the penalty in the Tikhonov functional T_α. This allows us to find an acceptable compromise between stability and approximation by making use of appropriate a priori or a posteriori criteria for selecting the regularization parameter α. Note that for Ω as in (3.10) any minimum norm solution in the sense of Definition 3.2 is an Ω-minimizing solution in the sense of Definition 3.25 and vice versa.

For nonlinear problems, the focus can be on a neighborhood of some reference element $\overline{x} \in X$, which acts as the origin of the coordinate system. In this case, penalties of the form

$$\Omega(x) := \frac{1}{q}\|x - \overline{x}\|_X^q, \qquad 1 \le q < \infty, \tag{3.12}$$

and minimizers of the extremal problem

$$\mathrm{T}_\alpha(x) := \frac{1}{p}\|F(x) - y^\delta\|_Y^p + \alpha\,\frac{1}{q}\,\|x - \overline{x}\|_X^q \to \min, \quad \text{subject to } x \in \mathcal{D}(F) \subseteq X, \tag{3.13}$$

are frequently used. In particular, the setting $p = q = 2$ is the basic approach for Hilbert spaces X and Y (cf. [68, 225] and more recently [156, 158]). Note again that, according to (3.12), for Ω, any $\overline{x}$-minimum norm solution in the sense of Definition 3.10 is an Ω-minimizing solution in the sense of Definition 3.25 and vice versa.

For the class of iterative regularization methods, stable approximate solutions are found in the course of an iterative procedure, e.g., aimed at minimizing the discrepancy norm $\|F(x) - y^\delta\|$ or $\|Ax - y^\delta\|$, yielding a sequence of iterates x_n. By means of a *stopping rule* $N = N(\delta)$ (an a priori stopping rule) or better $N = N(y^\delta, \delta)$ (an a posteriori stopping rule) the iteration is terminated and, by setting $\alpha := 1/N$, the regularized solution appears as $x_\alpha^\delta := x_N$. If, due to ill-posedness, the iteration would not terminate, i.e., $\alpha \to 0$, the iteration elements x_n would tend to oscillate and to explode as $n \to \infty$. Appropriate stopping rules aim at selecting values N such that the residual $\|F(x_N) - y^\delta\|$ or $\|Ax_N - y^\delta\|$, respectively, corresponds to the right order of the noise level $\delta > 0$. Hence, the iteration need not be executed until it converges, but only as long as necessary. Often the amount of work required for carrying out iterative regularization methods is much smaller than the comparable amount for a Tikhonov-type regularization. A drawback of iterative regularization, however, is its reduced ability to control properties of the approximate solution.

We should mention that some classes of linear and nonlinear ill-posed operator equations like Volterra and Hammerstein integral equations allow us to use a specific and sophisticated regularization approach called *local regularization*, which complements the approach of Tikhonov-type regularization by selecting solutions based on the penalty functional, preferably in a global manner. We refer for example to the papers [30,31,54,141] on local regularization and references therein. Another alternative to the Tikhonov-type regularization in the case that the spaces X and Y coincide is the *Lavrentiev regularization* (cf., e.g., [119, 159, 168, 229]), where stable neighboring problems of the ill-posed operator equations are constructed by adding the α-multiple of the identity operator to the original forward operator.

3.2.2 Source conditions and distance functions

If regularized solutions x_α^δ, as minimizers of the Tikhonov functional (3.11), converge to a minimum norm solution $x^\dagger$ of the ill-posed linear operator equation (3.1), then this convergence can be *arbitrarily slow* (cf. [212]). To obtain *convergence rates*

$$E(x_\alpha^\delta, x^\dagger) = O\,(\varphi(\delta)) \quad \text{as } \delta \to 0 \tag{3.14}$$

for an error measure E (cf. Section 4.2 below) and an index function φ, some *smoothness* of the solution element $x^\dagger$ *with respect to* $A : X \to Y$ is required. In other words, the rates depend on the interplay of intrinsic smoothness of $x^\dagger$ and the smoothing properties of the operator A with nonclosed range.

A classical tool for expressing the smoothness of $x^\dagger$ is the concept of *source conditions*, where either $x^\dagger$ itself or an element $\xi^\dagger$ from the subdifferential $\partial\Omega(x^\dagger)$ of the convex functional Ω in $x^\dagger$ belongs to the range of a linear operator G that interacts with A in an appropriate manner. More precisely, source conditions are of the form

$$x^\dagger = G\,w \tag{3.15}$$

or

$$\xi^\dagger = H\,v\,, \tag{3.16}$$

where $x^\dagger$ or $\xi^\dagger$ emerge from a source element w or v, transferred by the bounded linear operators G or H with nonclosed ranges.

At the beginning we briefly characterized the situation for linear ill-posed problems (3.1) in Hilbert spaces. For more details, see also [163, 228]. In Hilbert spaces X and Y with

$$A : X \to Y \qquad \text{and} \qquad A^* : Y \to X$$

source conditions for obtaining convergence rates (3.14) for the error measure

$$E(x, x^\dagger) := \|x - x^\dagger\|$$

in full generality attain the form (3.15), where $G : X \to X$ can be considered as a self-adjoint bounded linear operator with nonclosed range, satisfying *link conditions* (cf. [106]) for connecting G and A. The simplest version is of the form

$$x^\dagger = \kappa(A^*A)\,w, \qquad w \in X, \tag{3.17}$$

with $G := \kappa(A^*A)$ and some index function κ, which is defined essentially on the spectrum of the positive semi-definite self-adjoint operator $A^*A : X \to X$, which is a subset of $[0, \|A\|^2]$. In general, one can say that high convergence rates, with $\varphi(\delta)$ in (3.14) rapidly decreasing to zero as $\delta \to 0$, are associated with a fast decay rate of $\kappa(t)$ as $t \to 0$ and vice versa. It has been proved in [109, 161] that for any $x^\dagger$ from the Hilbert space X there are an index function κ and a source element $w \in X$ satisfying (3.17). However, the decay rate $\kappa(t) \to 0$ as $t \to 0$ can be arbitrarily slow.

The most prominent representative of a source condition (3.17) expressing medium smoothness occurs when $\kappa(t) = \sqrt{t}$. Because of $\mathcal{R}((A^*A)^{1/2}) = \mathcal{R}(A^*)$ this condition is equivalent to

$$x^\dagger = A^*\,v, \qquad v \in Y, \tag{3.18}$$

and admits the convergence rate

$$\|x_\alpha^\delta - x^\dagger\| = O\left(\sqrt{\delta}\right) \quad \text{as } \delta \to 0. \tag{3.19}$$

Another representative, expressing a higher smoothness, is

$$x^\dagger = A^* A\, w, \qquad w \in X, \tag{3.20}$$

which arises for $\kappa(t) = t$. This admits the convergence rate

$$\|x_\alpha^\delta - x^\dagger\| = O\left(\delta^{2/3}\right) \quad \text{as } \delta \to 0, \tag{3.21}$$

which is the best possible rate, the so-called *saturation rate*, in Tikhonov regularization. On the other hand, if the solution smoothness is rather low, relative to the smoothing property of the forward operator, then *logarithmic source conditions* (cf. [115] and [24]) with

$$\kappa(t) = \frac{1}{\left(\log\left(\frac{1}{t}\right)\right)^{\mu}}, \qquad \mu > 0,$$

play a prominent role, leading to convergence rates of the same type.

If the transfer operator G fails to attain the form (3.17), then other link conditions can help, such as inequalities like

$$\|\varrho(G)\| \le C\, \|Ax\| \qquad \text{for all} \quad x \in X,$$

with some index function ϱ and some positive constant C, which are equivalent to *range inclusions*

$$\mathcal{R}(\varrho(G)) \subseteq \mathcal{R}((A^*A)^{1/2}),$$

as shown in [24, Prop. 2.1]. For detailed results we refer to [106, 112].

When we consider nonlinear operator equations (3.3) and regularized solutions x_α^δ, minimizing the Tikhonov functional (3.13), then in the Hilbert space setting we aim at convergence rates to $\overline{x}$-minimum norm solutions. Source conditions occur as described above, but A in formulas (3.17), (3.18) and (3.20) is replaced by a derivative operator $F'(x^\dagger)$ at the solution point. Moreover, $x^\dagger$ on the left-hand sides is replaced by $x^\dagger - \overline{x}$. For more details in the Hilbert space setting see [67, Chap.10-11].

If, however, we turn our attention to linear ill-posed problems in Banach spaces X and Y, we see that, because of

$$A : X \to Y \qquad \text{and} \qquad A^* : Y^* \to X^*,$$

no representative is left over from the wide field of source conditions (3.17). However, we can fit the members (3.18) and (3.20) to the Banach space situation as

$$\xi^\dagger = A^*\, v, \qquad v \in Y^*, \tag{3.22}$$

and

$$\xi^\dagger = A^* J_p^Y(A\,w), \quad w \in X,\ p > 1\,, \tag{3.23}$$

respectively, where $\xi^\dagger \in \partial\Omega(x^\dagger) \subset X^*$ is an element of the subdifferential of the convex functional Ω under consideration in Tikhonov-type regularization. Note that for simplicity we can set $p := 2$ in the duality mapping J_p^Y of (3.23). Otherwise, the source element w can be amended in an appropriate manner. For the penalty functional $\Omega(x) = \frac{1}{q}\|x\|_X^q$ with $q > 1$ we recall that, according to Theorem 2.53 (e), for smooth Banach spaces X, the subgradient is uniquely determined and attains the explicit form $\xi^\dagger = J_q^X(x^\dagger)$.

Based on the seminal paper [32], Bregman distances

$$D_{\xi^\dagger}^\Omega(x, x^\dagger) = \Omega(x) - \Omega(x^\dagger) - \langle \xi^\dagger, x - x^\dagger\rangle_{X^*\times X} \tag{3.24}$$

at $x^\dagger \in \mathcal{D}_B(\Omega) \subseteq X$ and $\xi^\dagger \in \partial\Omega(x^\dagger) \subseteq X^*$ for general convex functionals Ω using the dual pairing $\langle\cdot,\cdot\rangle_{X^*\times X}$ between the Banach space X and its dual X^* have become a frequently used error measure E in studying convergence rates (3.14) of regularized solutions, see for example [104, 196, 197] and Section 4.2.2 below.

The rate

$$D_{\xi^\dagger}^\Omega(x_\alpha^\delta, x^\dagger) = O(\delta) \quad \text{as } \delta \to 0 \tag{3.25}$$

occurs whenever source conditions (3.22) are satisfied, and it represents the analog to the medium rate (3.19) in the Hilbert space setting. The Bregman rates are limited by

$$D_{\xi^\dagger}^\Omega(x_\alpha^\delta, x^\dagger) = O(\delta^{4/3}) \quad \text{as } \delta \to 0\,, \tag{3.26}$$

which is the extension of the saturation rate (3.21) for Tikhonov-type regularization in Hilbert spaces to the Banach space situation.

Since we only have the two clearly expressed levels of solution smoothness (3.22) and (3.23) in Banach spaces, we can compensate for the missing intermediate stages of the smoothness scale by the technique of *approximate source conditions*, which was originally developed for the Hilbert space setting (cf. [61, 77, 102, 103, 107, 112]), but can be immediately applied to the Banach space setting (cf. [97] and [95]). Let us explain this idea for the more evident case where the subgradient $\xi^\dagger$ fails to satisfy (3.22), i.e., the element $\xi^\dagger$ is not smooth enough. In this case, (3.22) plays the role of a *benchmark source condition* for this element and we ask how far from that benchmark the smoothness of $\xi^\dagger$ is. This can be measured by the *distance function*

$$d_{\xi^\dagger}(R) := \inf\{\|\xi^\dagger - A^*v\|_{X^*} \,:\, v \in Y^*,\ \|v\|_{Y^*} \le R\}, \qquad R > 0. \tag{3.27}$$

Proposition 3.28 ([20, 74]). *Let $\xi^\dagger \notin \mathcal{R}(A^*)$. Then, the distance function $d_{\xi^\dagger}(R)$ is strictly positive, continuous, convex and decreasing for $0 < R < \infty$. It tends to zero as $R \to \infty$ whenever*

$$\xi^\dagger \in \overline{\mathcal{R}(A^*)}^{\|\cdot\|_{X^*}} \setminus \mathcal{R}(A^*). \tag{3.28}$$

Furthermore, for every $0 < R < \infty$ there is an element $v_R \in Y^*$ with $\|v_R\|_{Y^*} = R$ such that $d_{\xi^\dagger}(R) = \|\xi^\dagger - A^*v_R\|_{X^*}$.

Remark 3.29. A sufficient condition for (3.28) is the *injectivity* of $A : X \to Y$ whenever X and Y are *reflexive* Banach spaces. In non-reflexive spaces it is sufficient that the biadjoint operator $A^{**} : X^{**} \to Y^{**}$[1] is injective. If with (3.28) the distance function $d_{\xi^\dagger}(R)$ tends to zero for large R, then its *decay rate* is a measure for the extent to which $\xi^\dagger$ violates the benchmark source condition (3.22). If (3.28) holds true, then convergence rates of the form (3.14), with the Bregman distance as error measure E, can be evaluated for the Tikhonov regularization with appropriate choices for the regularization parameter, where the derived rate functions φ in (3.14) depend on the associated distance function $d_{\xi^\dagger}$. For details, we refer to Section 4.2.2 below and [97]. If the decay rate of the distance function is slow for some $\xi^\dagger$, e.g., logarithmic in the sense $d_{\xi^\dagger}(R) \sim \log(R)^{-\mu}$, for some $\mu > 0$, then the benchmark source condition is strongly violated and only low (mostly logarithmic) convergence rates can be expected. On the other hand, for a decay of power type $d_{\xi^\dagger}(R) \sim R^{-\nu}$, the chances of better rates (Hölder convergence rates) are increasing and the obtained Hölder exponent in the rate function will grow with $\nu > 0$.

For nonlinear operator equations (3.3) the operator A^* in source conditions again must be replaced with the adjoint $(F'(x^\dagger))^*$ of a derivative of F at the solution point $x^\dagger$. This derivative can be a Fréchet, Gâteaux or directional derivative and must have appropriate properties. Then for our benchmark source condition (3.22) the extension to the nonlinear case takes the form

$$\xi^\dagger = (F'(x^\dagger))^* \, v, \qquad v \in Y^*, \tag{3.29}$$

and the assertions of Proposition 3.28 remain true for the corresponding distance functions

$$d_{\xi^\dagger}(R) := \inf\{\|\xi^\dagger - (F'(x^\dagger))^* v\|_{X^*} \; : \; v \in Y^*, \; \|v\|_{Y^*} \le R\}, \qquad R > 0. \tag{3.30}$$

3.2.3 Variational inequalities

Since the Bregman distance (3.24) at $x^\dagger$ for a convex functional Ω contains the term $\langle \xi^\dagger, x - x^\dagger \rangle_{X^* \times X}$, many authors (cf. [104] and also [20,74,81,126,210]) pointed out, that *variational inequalities* , estimating the negative of this term from above, prove to be a powerful tool to obtain convergence rates in regularization. In the context of variational inequalities for convergence rates we restrict our attention to rates up to a maximum of the form (3.25). Some initial ideas for combining variational inequalities and enhanced rates up to the limiting rate (3.26) can be found in the recent paper [80]. For higher rates we refer to Section 4.2.4 below.

For linear operator equations the variational inequalities under consideration attain the form

$$\langle \xi^\dagger, x^\dagger - x \rangle_{X^* \times X} \le \beta_1 \, D^{\Omega}_{\xi^\dagger}(x, x^\dagger) + \beta_2 \, \varphi(\|A(x - x^\dagger)\|), \tag{3.31}$$

[1] X^{**} denotes the bidual of X consisting of all linear, continuous functionals $\varphi : X^* \to \mathbb{R}$, Y^{**} is defined accordingly.

for constants $0 \le \beta_1 < 1$, $\beta_2 \ge 0$ and a *concave* index function φ. In the case of nonlinear equations, the variational inequalities are written as

$$\langle \xi^\dagger, x^\dagger - x \rangle_{X^* \times X} \le \beta_1 \, D^{\Omega}_{\xi^\dagger}(x, x^\dagger) + \beta_2 \, \varphi(\|F(x) - F(x^\dagger)\|). \tag{3.32}$$

In both cases, the inequalities must hold for all $x \in \mathcal{M}$, with some set $\mathcal{M}$, which contains all regularized solutions x_α^δ of interest, for sufficiently small $\delta > 0$. In some papers that deal with Tikhonov regularization, e.g. [104], $\mathcal{M}$ refers to sublevel sets

$$\mathcal{M} := \mathcal{M}_{\mathrm{T}_\alpha}(c) := \{x \in \mathcal{D} : \mathrm{T}_\alpha(x) \le c\} \tag{3.33}$$

of the Tikhonov functional T_α with sufficiently large $c > 0$. In this work we will concentrate on sets

$$\mathcal{M} := \mathcal{M}_\Omega(c) \cap \mathcal{D}(F). \tag{3.34}$$

It is sometimes helpful to know that $\mathcal{M}$ is starlike with respect to $x^\dagger \in \mathcal{D}$ (cf. Definition 3.27). Indeed, any set $\mathcal{M}$ from (3.34) with $c \ge \Omega(x^\dagger)$ is starlike with respect to $x^\dagger$, because $\mathcal{D}(F)$ and all sublevel sets $\mathcal{M}_\Omega(c)$ of the stabilizing functional Ω are convex.

It was shown that non-concave, in particular strictly convex, index functions φ with $\lim_{t \to 0} \frac{\varphi(t)}{t} = 0$ do not make sense in the variational inequalities (3.31) and (3.32) (see Proposition 4.18 below). On the other hand, the concave index functions φ in (3.31) and (3.32) are immediately responsible for obtaining convergence rates

$$D^{\Omega}_{\xi^\dagger}(x_\alpha^\delta, x^\dagger) = O(\varphi(\delta)) \quad \text{as } \delta \to 0 \tag{3.35}$$

of the regularized solutions, as will be shown in Section 4.2. Such rates are limited by (3.25) for $\varphi(t) = t$, where the variational inequalities triggering those rates attain the form

$$\langle \xi^\dagger, x^\dagger - x \rangle_{X^* \times X} \le \beta_1 \, D^{\Omega}_{\xi^\dagger}(x, x^\dagger) + \beta_2 \, \|A(x - x^\dagger)\| \tag{3.36}$$

for linear problems and the form

$$\langle \xi^\dagger, x^\dagger - x \rangle_{X^* \times X} \le \beta_1 \, D^{\Omega}_{\xi^\dagger}(x, x^\dagger) + \beta_2 \, \|F(x) - F(x^\dagger)\| \tag{3.37}$$

for nonlinear problems. In this limiting case there is an equivalence between the benchmark source condition (3.22) and the variational inequality (3.36). Namely, from [210, §3.2] one easily derives the following proposition.

Proposition 3.30. *If, for a convex functional $\Omega : \mathcal{D}(\Omega) \subseteq X \to [0, \infty)$ and $x^\dagger \in \mathcal{D}(\Omega) \subseteq X$, there is some $\xi^\dagger \in \partial\Omega(x^\dagger)$ satisfying the source condition* (3.22), *then there exist constants $0 \le \beta_1 < 1$ and $\beta_2 \ge 0$, such that the variational inequality* (3.36) *holds for all x in an appropriate set $\mathcal{M}$, which is starlike with respect to $x^\dagger$. Vice versa, under the variational inequality* (3.36), *valid for such $\mathcal{M}$ we have the source condition* (3.22).

For nonlinear problems a similar connection between (3.29) and (3.37) is true, based on the following assertion, where we refer to [76, Lemma 6.1] for the proof.

Proposition 3.31. *Let F be Gâteaux-differentiable at the point $x^\dagger \in \mathcal{D}(F)$, with Gâteaux-derivative $F'(x^\dagger) : X \to Y$ and $\xi^\dagger \in \partial\Omega(x^\dagger)$. If a variational inequality* (3.37) *is valid for all elements x of a set $\mathcal{M}$ starlike with respect to $x^\dagger$, we also have*

$$\langle \xi^\dagger, x^\dagger - x\rangle_{X^*\times X} \le \beta_1\, D^{\Omega}_{\xi^\dagger}(x, x^\dagger) + \beta_2\, \|F'(x^\dagger)(x - x^\dagger)\| \tag{3.38}$$

for all $x \in \mathcal{M}$.

In Chapter 7 we will also consider variational inequalities of multiplicative form

$$\langle \xi^\dagger, x^\dagger - x\rangle_{X^*\times X} \le \left(D^{\Omega}_{\xi^\dagger}(x, x^\dagger)\right)^{1/2} \kappa\left(\frac{\|F'(x^\dagger)(x - x^\dagger)\|^2}{D^{\Omega}_{\xi^\dagger}(x, x^\dagger)}\right) \tag{3.39}$$

for $D^{\Omega}_{\xi^\dagger}(x, x^\dagger) \neq 0$, complementing the additive forms (3.32) and (3.38). In particular, for the power type case $\kappa(t) = t^{\nu/2}$, for $\beta > 0$ and $0 < \nu \le 1$, the inequality

$$\langle \xi^\dagger, x^\dagger - x\rangle_{X^*\times X} \le \beta \left(D^{\Omega}_{\xi^\dagger}(x, x^\dagger)\right)^{\frac{1-\nu}{2}} \|F'(x^\dagger)(x - x^\dagger)\|^{\nu} \tag{3.40}$$

immediately implies by Young's inequality an additive form

$$\langle \xi^\dagger, x^\dagger - x\rangle_{X^*\times X} \le \beta_1\, D^{\Omega}_{\xi^\dagger}(x, x^\dagger) + \beta_2\, \|F(x^\dagger)(x - x^\dagger)\|^{\frac{2\nu}{1+\nu}} \tag{3.41}$$

with $0 \le \beta_1 < 1,\ \beta_2 \ge 0$.

3.2.4 Differences between the linear and the nonlinear case

Both (approximate) source conditions and variational inequalities are tools expressing the solution smoothness of $x^\dagger$ and $\xi^\dagger$, with respect to A and F, respectively. For linear problems the cross connections between the two tools are much simpler than in the nonlinear case, because for nonlinear problems the *structure of nonlinearity* around $x^\dagger$ is an additional factor complicating the interplay. The two forms of variational inequalities (3.37) and (3.38), valid for all x in a set $\mathcal{M}$ starlike with respect to $x^\dagger$, coincide if $F = A : X \to Y$ is a bounded linear operator. For nonlinear F however, by Proposition 3.31, we see that (3.37) implies (3.38), but for the reverse direction, we need some structural conditions. If the so-called η-condition (cf. [67, p.279])

$$\|F(x) - F(x^\dagger) - F'(x^\dagger)(x - x^\dagger)\| \le \eta\, \|F(x) - F(x^\dagger)\| \tag{3.42}$$

holds for all $x \in \mathcal{M}$ and some constant $\eta > 0$, then, using the triangle inequality

$$\|F'(x^\dagger)(x - x^\dagger)\| \le \|F(x) - F(x^\dagger) - F'(x^\dagger)(x - x^\dagger)\| + \|F(x) - F(x^\dagger)\|$$

one obtains the variational inequality (3.37) with constant $\tilde{\beta}_2 = (\eta + 1)\beta_2$, instead of β_2 directly from (3.38). We note that (3.42) expresses a rather strong condition in the sense that $F'(x^\dagger)$ characterizes the nonlinear operator around $x^\dagger$ very well. This condition can be relaxed by requiring a concave index function σ with $\lim_{t\to 0} \frac{t}{\sigma(t)} = 0$, such that

$$\|F'(x^\dagger)(x - x^\dagger)\| \le K\,\sigma(\|F(x) - F(x^\dagger)\|) \qquad \text{for all} \quad x \in \mathcal{M} \tag{3.43}$$

holds, with some constant $K > 0$. In this case, the following Proposition 3.32 published as Theorem 5.1 in the joint paper [20] with R. I. Boţ clearly shows that variational inequalities (3.32) merge solution smoothness of $x^\dagger$, expressed by the source conditions, and structural conditions on the nonlinearity of F around $x^\dagger$. Hence, in contrast to linear ill-posed problems, where source conditions and variational inequalities provide comparable information in different ways, variational inequalities for nonlinear ill-posed problems cover much more information and deliver an outstanding tool for deriving rates also for non-smooth situations with respect to operator F and solution $x^\dagger$.

Proposition 3.32. *Let $x^\dagger \in \mathcal{M}$ and let the condition* (3.43) *hold for all $x \in \mathcal{M}$ and some concave index function σ. Then the benchmark source condition* (3.29) *implies a variational inequality of the form* (3.32) *for all $x \in \mathcal{M}$, with index function $\varphi = \sigma$.*

Proof. Due to (3.29) and (3.43) we can estimate for all $x \in \mathcal{M}$

$$\begin{aligned} \langle \xi^\dagger, x^\dagger - x\rangle_{X^*\times X} &= \langle (F'(x^\dagger))^* v, x^\dagger - x\rangle_{X^*\times X} = \langle v, F'(x^\dagger)(x^\dagger - x)\rangle_{Y^*\times Y} \\ &\le \|v\|_{Y^*}\|F'(x^\dagger)(x - x^\dagger)\| \le K\,\|v\|_{Y^*}\,\sigma(\|F(x) - F(x^\dagger)\|)\,. \end{aligned}$$

This immediately yields the variational inequality (3.32) with $\beta_1 = 0$, $\beta_2 = K\,\|v\|_{Y^*}$, and with $\varphi = \sigma$. So the proof is complete. □

Hence, under the benchmark source condition (3.29) the decay rate of the function σ in (3.43) controls the convergence rate which is always smaller than (3.25), whenever we assume $\lim_{t\to 0} \frac{t}{\sigma(t)} = 0$.

If (3.29) is violated, and only a distance function $d_{\xi^\dagger}(R)$ for $R \ge \underline{R} > 0$ contains the information about the solution smoothness then, under the nonlinearity condition (3.43), the validity of a variational inequality (3.32) cannot be concluded in general. To find an index function φ in (3.32), a link condition between the functional Ω and the occurring norm functionals is necessary. This even is the case if the problem is linear for $F := A$. One such helpful link condition is the *q-coercivity* of the Bregman distance

$$D^{\Omega}_{\xi^\dagger}(x, x^\dagger) \ge c_q\,\|x - x^\dagger\|^q \qquad \text{for all} \quad x \in \mathcal{M} \tag{3.44}$$

with some $q \ge 2$ and some positive constant $c_q > 0$, which is a canonical condition for

$$\Omega(x) = \frac{1}{q}\|x\|^q \quad \text{and} \quad q\text{-convex Banach spaces } X\,.$$

For details concerning the smoothness and convexity of Banach spaces and cross connections with properties of the Bregman distance we refer to Chapter 2.

However, supposing that (3.29) is violated, under condition (3.44), the method of approximate source conditions applies and also yields a variational inequality (3.32), where the index function φ is controlled by both the distance function $d_{\xi^\dagger}$ and the function σ, characterizing the nonlinearity structure. We reformulate Theorem 5.2 from [20] and its proof as follows:

Proposition 3.33. *Let $x^\dagger \in \mathcal{M}$ and let the conditions* (3.43) *and* (3.44), *with $q \geq 2$ hold for all $x \in \mathcal{M}$ and let σ be some concave index function. If $\xi^\dagger$ violates the benchmark source condition* (3.29), *but the distance function $d_{\xi^\dagger}(R)$ tends to zero as $R \to \infty$, then a variational inequality of the form* (3.32), *with constants $\beta_1 = \frac{1}{q} < 1$ and $\beta_2 > 0$, is valid for all $x \in \mathcal{M}$, where the index function φ is defined by means of the auxiliary function*

$$\Psi(R) := \frac{(d_{\xi^\dagger}(R))^{q^*}}{R}, \qquad R > 0, \qquad \frac{1}{q} + \frac{1}{q^*} = 1,$$

as

$$\varphi(0) = 0, \quad \varphi(t) = \left[d_{\xi^\dagger}\left(\Psi^{-1}(\sigma(t))\right)\right]^{q^*}, \quad t > 0. \tag{3.45}$$

Proof. When setting

$$\xi^\dagger = F'(x^\dagger)^* v_R + r_R \qquad \text{with} \qquad \|v_R\|_{Y^*} = R, \ \|r_R\|_{X^*} = d_{\xi^\dagger}(R)$$

we can estimate by (3.43) and for all $x \in \mathcal{M}$ as

$$\begin{aligned}\langle \xi^\dagger, x^\dagger - x\rangle_{X^* \times X} &= \langle (F'(x^\dagger))^* v_R + r_R, x^\dagger - x\rangle_{X^* \times X} \\ &= \langle v_R, F'(x^\dagger)(x^\dagger - x)\rangle_{Y^* \times Y} + \langle r_R, x^\dagger - x\rangle_{X^* \times X} \\ &\leq R\,\|F'(x^\dagger)(x - x^\dagger)\| + d_{\xi^\dagger}(R)\,\|x - x^\dagger\| \\ &\leq K\,R\,\sigma(\|F(x) - F(x^\dagger)\|) + d_{\xi^\dagger}(R)\,\|x - x^\dagger\|.\end{aligned}$$

This summarizes to

$$\langle \xi^\dagger, x^\dagger - x\rangle_{X^* \times X} \leq K\,R\,\sigma(\|F(x) - F(x^\dagger)\|) + c_q^{-1/q}\, d_{\xi^\dagger}(R)\left(D_{\xi^\dagger}^{\Omega}(x, x^\dagger)\right)^{1/q}.$$

Then, when setting $a := D_{\xi^\dagger}^{\Omega}(x, x^\dagger)$ and $b := c_q^{-1/q}\, d_{\xi^\dagger}(R)$, for conjugate exponents q and q^* with $\frac{1}{q} + \frac{1}{q^*} = 1$, Young's inequality

$$ab \leq \frac{a^q}{q} + \frac{b^{q^*}}{q^*}, \qquad a, b \geq 0,$$

provides us with the estimate

$$\langle \xi^\dagger, x^\dagger - x\rangle_{X^*\times X} \le K\, R\,\sigma(\|F(x)-F(x^\dagger)\|) + \frac{1}{q}\, D^{\Omega}_{\xi^\dagger}(x,x^\dagger) + \frac{c_q^{-q^*/q}}{q^*}\,\left(d_{\xi^\dagger}(R)\right)^{q^*}.$$

The continuity of the distance function carries over to the auxiliary function $\Psi : (0,\infty) \to (0,\infty)$, which is strictly decreasing and satisfies the limit conditions $\lim_{R\to 0}\Psi(R) = \infty$ and $\lim_{R\to\infty}\Psi(R) = 0$. Its inverse $\Psi^{-1} : (0,\infty) \to (0,\infty)$ is also continuous and strictly decreasing. Hence for all $t > 0$ the equation $\Psi(R) = \sigma(t)$ has a uniquely determined solution $R > 0$. Note that, in principle, only sufficiently small $t > 0$ influence the rate results. By setting $R := \Psi^{-1}(\sigma(\|F(x) - F(x^\dagger)\|))$ we get with some constant $\hat{K} > 0$

$$\langle \xi^\dagger, x^\dagger - x\rangle_{X^*\times X} \le \frac{1}{q}\, D^{\Omega}_{\xi^\dagger}(x,x^\dagger) + \hat{K}\left[d_{\xi^\dagger}\left(\Psi^{-1}(\sigma(\|F(x)-F(x^\dagger)\|))\right)\right]^{q^*}$$

for all $x \in \mathcal{M}$. This completes the proof since $\varphi = \zeta^{q^*}$ with $\zeta := d_{\xi^\dagger} \circ \Psi^{-1} \circ \sigma$ is an index function (cf. Proposition 3.28). □

Remark 3.34. One can see that the index function $\varphi = [d_{\xi^\dagger} \circ \Psi^{-1} \circ \sigma]^{q^*}(t),\ t > 0$, in Proposition 3.33 tends to zero as $t \to 0$ *slower* than the index function $\sigma(t)$ in Proposition 3.32, which is reasonable since the smoothness of $\xi^\dagger$ is reduced: taking into account the one-to-one correspondence between large $R > 0$ and small $t > 0$, via $\Psi(R) = \sigma(t)$ and $\Psi(R) = (d_{\xi^\dagger}(R))^{q^*}/R$, we have for the quotient function

$$\frac{\sigma(t)}{[d_{\xi^\dagger} \circ \Psi^{-1} \circ \sigma]^{q^*}(t)} = \frac{\Psi(R)}{\left(d_{\xi^\dagger}(R)\right)^{q^*}} = \frac{1}{R} \to 0 \qquad \text{as} \quad R \to \infty,\ t \to 0.$$

If the structural condition (3.43) cannot be satisfied for any concave index function σ, but weaker nonlinearity conditions of the form

$$\|F(x) - F(x^\dagger) - F'(x^\dagger)(x - x^\dagger)\| \le \tilde{\eta}\, D^{\Omega}_{\xi^\dagger}(x,x^\dagger) \qquad \text{for all} \quad x \in \mathcal{M} \tag{3.46}$$

hold for some constant $\tilde{\eta} > 0$, this expresses the fact that $F'(x^\dagger)$ fails to characterize the nonlinear operator F around $x^\dagger$ in an acceptable manner. Nevertheless, under the source condition (3.29), the variational inequality of the form (3.37), with $0 \le \beta_1 < 1$ and $\beta_2 \ge 0$, can be established (cf. [104, 197, 210]), but only in combination with a *smallness condition*

$$\tilde{\eta}\|v\|_{Y^*} < 1 \tag{3.47}$$

in order to ensure the required condition $\beta_1 < 1$.

Proposition 3.35. *For $x^\dagger \in \mathcal{M}$ we assume that the conditions* (3.29) *with* (3.47), *concerning the solution smoothness and* (3.46), *concerning the structure of nonlinearity hold. Then a variational inequality of the form* (3.37) *with constants $\beta_1 = \tilde{\eta}\, \|v\|_{Y^*} < 1$ and $\beta_2 = \|v\|_{Y^*} \geq 0$ is valid for all $x \in \mathcal{M}$.*

Proof. Based on (3.29) and (3.46) we can estimate for all $x \in \mathcal{M}$

$$\begin{aligned}\langle \xi^\dagger, x^\dagger - x\rangle_{X^*\times X} &= \langle (F'(x^\dagger))^* v, x^\dagger - x\rangle_{X^*\times X} = \langle v, F'(x^\dagger)(x^\dagger - x)\rangle_{Y^*\times Y}\\ &\leq \langle v, (F(x) - F(x^\dagger)) - F'(x^\dagger)(x - x^\dagger))\rangle_{Y^*\times Y}\\ &\quad - \langle v, F(x) - F(x^\dagger)\rangle_{Y^*\times Y}\\ &\leq \|v\|_{Y^*}\, \|F(x) - F(x^\dagger) - F'(x^\dagger)(x - x^\dagger)\|\\ &\quad + \|v\|_{Y^*}\, \|F(x) - F(x^\dagger)\|\\ &\leq \tilde{\eta}\, \|v\|_{Y^*}\, D^\Omega_{\xi^\dagger}(x, x^\dagger) + \|v\|_{Y^*}\, \|F(x) - F(x^\dagger)\|\,.\end{aligned}$$

This proves the proposition. □

Remark 3.36. We note that, when the structure of nonlinearity is given only by (3.46), no results concerning either variational inequalities or convergence rates of Tikhonov-type regularization in Banach spaces (not even under corresponding conditions in Hilbert spaces) are known. Moreover the benchmark source condition (3.29) is violated.

In Section 1.4, for a Hilbert space X and $D^\Omega_{\xi^\dagger}(x, x^\dagger) = \|x - x^\dagger\|^2$, we made use of the fact that (3.37) can also be observed when the nonlinearity condition (3.46) is replaced with the weaker condition

$$\|F(x) - F(x^\dagger) - F'(x^\dagger)(x - x^\dagger)\|_{\tilde{Y}} \leq \tilde{\eta}\, D^\Omega_{\xi^\dagger}(x, x^\dagger) \qquad \text{for all} \quad x \in \mathcal{M},$$

where the Banach space $\tilde{Y}$ with $\tilde{Y} \supset Y$, possessing a weaker norm than Y, is chosen such that, with the dual inclusion $\tilde{Y}^* \subset Y^*$, the equation

$$\langle \tilde{v}, v\rangle_{Y^*\times Y} = \langle \tilde{v}, v\rangle_{\tilde{Y}^*\times\tilde{Y}} \qquad \text{for all} \quad \tilde{v} \in \tilde{Y}^*,\ v \in Y$$

is valid. For details we refer to [104, Remark 4.2].

Nearing the end of this chapter, we should mention that it makes sense to interpolate between the structural conditions (3.42) and (3.46) as follows

$$\|F(x) - F(x^\dagger) - F'(x^\dagger)(x - x^\dagger)\| \leq \hat{\eta}\, \|F(x) - F(x^\dagger)\|^{c_1} \left(D^\Omega_{\xi^\dagger}(x, x^\dagger)\right)^{c_2}. \tag{3.48}$$

Definition 3.37 (degree of nonlinearity). Let $0 \leq c_1, c_2 \leq 1$ with $0 < c_1 + c_2 \leq 1$. We define F to be *nonlinear of degree* (c_1, c_2) in $x^\dagger$ for the Bregman distance $D^\Omega_{\xi^\dagger}(\cdot, x^\dagger)$ of Ω with $\xi^\dagger \in \partial\Omega(x^\dagger)$, if there is a constant $\hat{\eta} > 0$ such that (3.48) holds for all $x \in \mathcal{M}$, where $\mathcal{M}$ is assumed to be starlike with respect to $x^\dagger$.

This definition, introduced for the first time in [97], extends the corresponding concept developed in [110] from Hilbert spaces to Banach spaces. Under condition (3.48), with $c_1 > 0$ replacing (3.43), the assertions of Propositions 3.32 and 3.33 change as follows:

Proposition 3.38. *Let F be nonlinear of degree (c_1, c_2) with $0 < c_1 \le 1 - c_2$ in $x^\dagger$ with corresponding subgradient $\xi^\dagger$, where the set $\mathcal{M}$, starlike with respect to $x^\dagger$, is such that $\|F(x) - F(x^\dagger)\| \le C$ for all $x \in \mathcal{M}$ and some constant $C > 0$. Then, the benchmark source condition* (3.29) *implies a variational inequality of the form* (3.32) *with $0 \le \beta_1 < 1$, $\beta_2 > 0$ and with the concave index function*

$$\varphi(t) = t^{\frac{c_1}{1-c_2}} . \tag{3.49}$$

If (3.29) *is violated, but the corresponding distance function satisfies the limit condition $d_{\xi^\dagger}(R) \to 0$ as $R \to \infty$, then, under condition* (3.44), *we alternatively arrive at* (3.32), *with an index function φ, defined using the auxiliary function*

$$\Psi(R) := \frac{\left(d_{\xi^\dagger}(R)\right)^{q^*}}{R^{\frac{1}{1-c_2}}}, \qquad R > 0, \qquad \frac{1}{q} + \frac{1}{q^*} = 1 ,$$

as

$$\varphi(0) = 0, \quad \varphi(t) = \left[d_{\xi^\dagger} \left(\Psi^{-1} \left(t^{\frac{c_1}{1-c_2}} \right) \right) \right]^{q^*}, \quad t > 0 . \tag{3.50}$$

Proof. Based on (3.29) and (3.48), we can estimate for all $x \in \mathcal{M}$ as

$$\begin{aligned}
\langle \xi^\dagger, x^\dagger - x \rangle_{X^* \times X} &= \langle (F'(x^\dagger))^* v, x^\dagger - x \rangle_{X^* \times X} \\
&= \langle v, F'(x^\dagger)(x^\dagger - x) \rangle_{Y^* \times Y} \\
&\le \langle v, (F(x) - F(x^\dagger)) - F'(x^\dagger)(x - x^\dagger)) \rangle_{Y^* \times Y} \\
&\quad - \langle v, F(x) - F(x^\dagger) \rangle_{Y^* \times Y} \\
&\le \|v\|_{Y^*} \|F(x) - F(x^\dagger) - F'(x^\dagger)(x - x^\dagger)\| \\
&\quad + \|v\|_{Y^*} \|F(x) - F(x^\dagger)\| \\
&\le \|v\|_{Y^*} \left(\hat{\eta} \, \|F(x) - F(x^\dagger)\|^{c_1} \left(D_{\xi^\dagger}^{\Omega}(x, x^\dagger) \right)^{c_2} \right. \\
&\qquad \left. + \|F(x) - F(x^\dagger)\| \right) .
\end{aligned}$$

This immediately yields φ from (3.49), with $\beta_1 = 0$ if $c_2 = 0$. For $0 < c_2 \le 1 - c_1$ with $c_1 > 0$ we can again exploit Young's inequality written as

$$ab \le \frac{a^\theta}{4} + \frac{b^{\theta^*}}{\theta^* \, (\theta/4)^{\theta^*/\theta}}, \qquad a, b \ge 0, \;\; 1/\theta + 1/\theta^* = 1,$$

by setting $\theta := 1/c_2 > 1$, $a := D^{\Omega}_{\xi^\dagger}(x, x^\dagger)$ and $b := \|v\|_{Y^*}\, \hat\eta\, \|F(x) - F(x^\dagger)\|^{c_1}$. This implies $\theta^* = 1/(1 - c_2)$ and hence a variational inequality, with φ from (3.49) and $\beta_1 = 1/4 < 1$, since the exponent $\kappa := c_1/(1 - c_2)$ is never larger than one.

If $\xi^\dagger$ fails to satisfy the benchmark source condition (3.29), we can argue as we did in the proof of Proposition 3.33, using the ansatz

$$\xi^\dagger = F'(x^\dagger)^* v_R + r_R \qquad \text{with} \qquad \|v_R\|_{Y^*} = R, \;\; \|r_R\|_{X^*} = d_{\xi^\dagger}(R)\,.$$

Under the nonlinearity condition (3.48) we obtain

$$\langle \xi^\dagger, x^\dagger - x\rangle_{X^*\times X} \le \langle r_R, x^\dagger - x\rangle_{X^*\times X}$$
$$+\|v_R\|_{Y^*}\left(\hat\eta\, \|F(x) - F(x^\dagger)\|^{c_1} \left(D^{\Omega}_{\xi^\dagger}(x, x^\dagger)\right)^{c_2} + \|F(x) - F(x^\dagger)\|\right).$$

Furthermore, under condition (3.44), by multiple use of Young's inequality and with $\kappa := c_1/(1 - c_2)$, we find constants $K, \tilde K, \hat K, \overline{K} > 0$ such that $\langle \xi^\dagger, x^\dagger - x\rangle_{X^*\times X}$ can be bounded from above by the following chain of expressions:

$$R\left(\hat\eta\, \|F(x) - F(x^\dagger)\|^{c_1} \left(D^{\Omega}_{\xi^\dagger}(x, x^\dagger)\right)^{c_2} + \|F(x) - F(x^\dagger)\|\right) + d_{\xi^\dagger}(R)\, \|x - x^\dagger\|$$
$$\le \frac{1}{4} D^{\Omega}_{\xi^\dagger}(x, x^\dagger) + \hat K\, R^{\frac{1}{1-c_2}}\, \|F(x) - F(x^\dagger)\|^{\kappa} + R\, \|F(x) - F(x^\dagger)\|$$
$$+\, \overline{K}\, d_{\xi^\dagger}(R) \left(D^{\Omega}_{\xi^\dagger}(x, x^\dagger)\right)^{1/q}.$$
$$\le \frac{1}{4} D^{\Omega}_{\xi^\dagger}(x, x^\dagger) + \tilde K\, R^{\frac{1}{1-c_2}}\, \|F(x) - F(x^\dagger)\|^{\kappa} + \overline{K}\, d_{\xi^\dagger}(R) \left(D^{\Omega}_{\xi^\dagger}(x, x^\dagger)\right)^{1/q}$$
$$\le \left(\frac{1}{4} + \frac{1}{q}\right) D^{\Omega}_{\xi^\dagger}(x, x^\dagger) + \tilde K\, R^{\frac{1}{1-c_2}}\, \|F(x) - F(x^\dagger)\|^{\kappa} + K\, \left(d_{\xi^\dagger}(R)\right)^{q^*}$$

with $\beta_1 := \frac{1}{4} + \frac{1}{q} < 1$ due to $q \ge 2$. The sum of the last two terms

$$\tilde K\, R^{\frac{1}{1-c_2}}\, \|F(x) - F(x^\dagger)\|^{\kappa} + K\, \left(d_{\xi^\dagger}(R)\right)^{q^*}$$

can be equilibrated analogously to the proof of Proposition 3.33 and bounded by the expression

$$\beta_2 \left[d_{\xi^\dagger}\left(\Psi^{-1}\left(\|F(x) - F(x^\dagger)\|^{\kappa}\right)\right)\right]^{q^*},$$

with some constant $\beta_2 > 0$, where we exploit the auxiliary function

$$\Psi(R) := \frac{\left(d_{\xi^\dagger}(R)\right)^{q^*}}{R^{\frac{1}{1-c_2}}}, \qquad R > 0,$$

for the equilibration process. Thus, φ in (3.32) attains the form (3.50) and the proof is complete. □

Remark 3.39. We emphasize that smallness conditions of the type (3.47), occurring in the case $c_1 = 0$, $c_2 = 1$, are not required whenever the degree of nonlinearity of F at $x^\dagger$ can be established to be $c_1 > 0$. Moreover we note that the condition

$$\|F(x) - F(x^\dagger)\| \le C < \infty \qquad \text{for all} \quad x \in \mathcal{M},$$

occurring in Proposition 3.38, is evident, if the operator F is norm-to-norm continuous and $\mathcal{M}$ can be embedded into a ball around $x^\dagger$, in case of norm-convergent regularized solutions. On the other hand, as we will see, the most serious requirement on $\mathcal{M}$ is that, for $0 < \delta \le \delta_{\max}$, all regularized solutions x_α^δ belong to such a set. However, $\|F(x_\alpha^\delta) - F(x^\dagger)\|$ is uniformly bounded for those regularized solutions, taking into account that they are minimizers of T_α.

The results of the Propositions 3.32, 3.33 and 3.38 will be used in Section 4.2.2 right away, in order to formulate convergence rates for the Tikhonov-type regularization of ill-posed operator equations.

Part III

Tikhonov-type regularization

In this part of the monograph we present some important components of the analysis of variational regularization, where the stable approximate solutions are found by minimizing the Tikhonov functional. Therefore, we call this approach Tikhonov-type regularization. The Tikhonov functional for nonlinear operator equations $F(x) = y$ with noisy data y^δ and a noise level $\delta > 0$ satisfying $\|y - y^\delta\| \le \delta$ is formed by the linear combination

$$T_\alpha(x) := \frac{1}{p}\|F(x) - y^\delta\|_Y^p + \alpha\,\Omega(x)$$

of a fidelity term $\frac{1}{p}\|F(x) - y^\delta\|_Y^p$, with $1 < p < \infty$ characterizing the misfit of the data y^δ and of a penalty term, expressed by a convex nonnegative functional $\Omega(x)$. In this situation, we assume a sufficiently smooth nonlinear forward operator

$$F : \mathcal{D}(F) \subseteq X \to Y,$$

mapping between the Banach spaces X and Y. The regularization parameter $\alpha > 0$ has to equilibrate both terms in an appropriate manner.

In Chapter 4 we outline the modern theory of Tikhonov-type regularization in Banach spaces with general convex penalties Ω. Based on a series of standing assumptions formulated in Chapter 3, we show results concerning existence and stability of regularized solutions in Section 4.1. Moreover, assertions about the convergence of Tikhonov-regularized to Ω-minimizing solutions $x^\dagger$ of the operator equation are formulated in this section. The most important assumption that we exploit, is that Ω is a *stabilizing functional* in the sense that its sublevel sets are weakly sequentially compact in the Banach space X. Convergence rates under assumptions on the solutions smoothness $x^\dagger$ and on the structure of the nonlinearity of the operator F at $x^\dagger$ are derived in Section 4.2. This section benefits from a long list of propositions formulated and partially proven in Chapter 3, concerning source conditions, approximate source conditions and variational inequalities. Based on those results, convergence rates for general error measures and in particular for the Bregman distance can be proven in the case of appropriate a priori parameter choices as well as in the case of the applicability of some discrepancy principle. Moreover, we combine conditional stability estimates and Tikhonov-type regularization in order to derive rate results. A brief visit of sparsity situations completes the chapter.

In Chapter 5 we will present essential ingredients for the analysis of the Tikhonov-type regularization for linear ill-posed operator equations $Ax = y$, with a bounded linear operator $A : X \to Y$ mapping between the Banach spaces X and Y. Here, the Tikhonov functional is of the form

$$T_\alpha(x) := \frac{1}{p}\|Ax - y^\delta\|_Y^p + \alpha\,\frac{1}{q}\|x\|_X^q,$$

where $p, q > 1$ and the associated penalty functional consists of a power of the norm in the Banach space X. The first section of this chapter is devoted to the character of source conditions for the specific case under consideration. In the subsequent Section 5.2, classes of rules regarding the choice of the regularization parameter $\alpha > 0$ are discussed in detail. The section presents assertions on a priori parameter choices, as well as on the discrepancy principle, frequently named after Morozov, as a particular a posteriori parameter choice and also on a modified version of this principle. This chapter is completed in Section 5.3 by answering the question of how to minimize the Tikhonov functional. It is an advantage of the specific version of Tikhonov-type regularization for linear problems with norm power penalties under consideration in this chapter, that for this case efficient numerical approaches to compute the Tikhonov-regularized solutions were developed recently. We will distinguish between a primal and a dual method for finding a minimizer of the Tikhonov functional T_α.

Chapter 4

Tikhonov regularization in Banach spaces with general convex penalties

4.1 Basic properties of regularized solutions

In this chapter we make use of the results of Chapter 3. More precisely, we will briefly summarize the properties of Tikhonov-regularized solutions x_α^δ defined, for all regularization parameters $\alpha > 0$, as minimizers of the extremal problem

$$T_\alpha(x) := \frac{1}{p}\|F(x) - y^\delta\|^p + \alpha\,\Omega(x) \to \min, \\ \text{subject to } x \in \mathcal{D}(F) \subseteq X, \tag{4.1}$$

based on the *Tikhonov functional* T_α for nonlinear operator equations (3.3). Throughout this chapter, we tacitly assume the Assumptions 3.11, 3.22 and 3.26, that make requirements on the Banach spaces X, Y, the forward operator F with its domain $\mathcal{D}(F)$ and its derivatives as well as on the stabilizing functional Ω to hold. Note that linear operator equations (3.1) can be considered as a special case of (3.3), where for $F := A$, with a bounded linear operator $A : X \to Y$ and domain $\mathcal{D}(F) = X$, the corresponding requirements in Assumption 3.11 are satisfied, taking into account that A is weak-to-weak continuous. On the other hand, for penalties (3.12) with $\mathcal{D}(\Omega) = X$, due to Corollary 3.13, the corresponding requirements in Assumption 3.22 are also satisfied.

In the following two subsections we discuss *existence*, *stability* and *convergence* of regularized solutions (cf. [104]).

4.1.1 Existence and stability of regularized solutions

Proposition 4.1 (existence of regularized solutions). *For all $\alpha > 0$ and $y^\delta \in Y$, there exists a regularized solution $x_\alpha^\delta \in \mathcal{D}(F)$, minimizing the Tikhonov functional $T_\alpha(x)$ in* (4.1) *over all $x \in \mathcal{D}(F)$.*

Proof. Since $\mathcal{D} \neq \emptyset$ there exists at least one $\hat{x} \in X$, such that $T_\alpha(\hat{x}) < \infty$. Moreover, there is a sequence $\{x_n\}$ in $\mathcal{D}$ such that

$$\lim_{n\to\infty} T_\alpha(x_n) = c := \inf\{T_\alpha(x) : x \in \mathcal{D}\} \geq 0\,.$$

Then, $\{T_\alpha(x_n)\}$ and, because of $\Omega(x_n) \leq T_\alpha(x_n)/\alpha$, also $\{\Omega(x_n)\}$ are bounded sequences in $\mathbb{R}$. From Assumption 3.22 (c), it follows that $\{x_k\}$ has a weakly convergent subsequence $\{x_{n_k}\}$, which has a weak limit element $\tilde{x}$, i.e., $x_{n_k} \rightharpoonup \tilde{x}$ in X. In particular, we have $\tilde{x} \in \mathcal{D}$, since $\mathcal{D}$ is convex and closed and therefore weakly closed. The weak-to-weak continuity of F (see Assumption 3.11 (c)) then implies weak convergence $F(x_{n_k}) - y^\delta \rightharpoonup F(\tilde{x}) - y^\delta$ in Y and, since the norm is weakly lower semi-continuous

$$\frac{1}{p}\|F(\tilde{x}) - y^\delta\|^p \le \liminf_{k\to\infty} \frac{1}{p}\|F(x_{n_k}) - y^\delta\|^p . \tag{4.2}$$

On the other hand, for the lower semi-continuous functional Ω, which is also weakly lower semi-continuous, we have

$$\Omega(\tilde{x}) \le \liminf_{k\to\infty} \Omega(x_{n_k}) . \tag{4.3}$$

Combination of (4.2) and (4.3) shows that $\tilde{x}$ minimizes T_α. □

Proposition 4.2 (stability of regularized solutions). *For all $\alpha > 0$ the minimizers of* (4.1) *are stable with respect to the data y^δ. More precisely, for a data sequence $\{y_n\}$ converging to y^δ with respect to the norm-topology of Y i.e., $\lim_{n\to\infty}\|y_n - y^\delta\| = 0$, every associated sequence $\{x_n\}$ of minimizers to the extremal problem*

$$\frac{1}{p}\|F(x) - y_n\|^p + \alpha\,\Omega(x) \to \min, \quad \textit{subject to } x \in \mathcal{D}(F) \subseteq X, \tag{4.4}$$

has a subsequence $\{x_{n_k}\}$, which converges in the weak topology of X, and the weak limit $\tilde{x}$ of each such subsequence is a minimizer x_α^δ of (4.1)*. Moreover, we have for every such weakly convergent subsequence $\lim_{k\to\infty}\Omega(x_{n_k}) = \Omega(x_\alpha^\delta)$.*

Proof. From the definition of x_n it follows that

$$\frac{1}{p}\|F(x_n) - y_n\|^p + \alpha\Omega(x_n) \le \frac{1}{p}\|F(x) - y_n\|^p + \alpha\Omega(x), \quad x \in \mathcal{D} .$$

With $\mathcal{D} \ne \emptyset$ we can select $\hat{x} \in \mathcal{D}$ such that

$$\begin{aligned}
\alpha\Omega(x_n) &\le \frac{1}{p}\|F(x_n) - y^\delta\|^p + \alpha\Omega(x_n) \\
&\le \frac{2^{p-1}}{p}\|F(x_n) - y_n\|^p + 2^{p-1}\alpha\Omega(x_n) + \frac{2^{p-1}}{p}\|y_n - y^\delta\|^p \\
&\le \frac{2^{p-1}}{p}\|F(\hat{x}) - y_n\|^p + 2^{p-1}\alpha\Omega(\hat{x}) + \frac{2^{p-1}}{p}\|y_n - y^\delta\|^p \\
&\le \frac{4^{p-1}}{p}\|F(\hat{x}) - y^\delta\|^p + 2^{p-1}\alpha\Omega(\hat{x}) + \frac{4^{p-1}}{p}\|y_n - y^\delta\|^p .
\end{aligned}$$

Since $\lim_{n\to\infty} \|y_n - y^\delta\| = 0$, it follows that, for every $\varepsilon > 0$, there exists $n_0 \in \mathbb{N}$ such that for $n \geq n_0$

$$\Omega(x_n) \leq \frac{4^{p-1}}{\alpha p} \|F(\hat{x}) - y\|^p + 2^{p-1}\Omega(\hat{x}) + \varepsilon = C < \infty .$$

Thus, the sequence $\{x_n\}$ is in $\mathcal{M}_\Omega(\max\{C, \Omega(x_1), ..., \Omega(x_{n_0-1})\}$ and, by Assumption 3.22 (c), has a weakly convergent subsequence $\{x_{n_k}\}$ with limit $\tilde{x} \in \mathcal{D}$. Since F is weak-to-weak continuous, it follows that $F(x_{n_k}) \rightharpoonup F(\tilde{x})$ leading us to obtain $F(x_{n_k}) - y_{n_k} \rightharpoonup F(\tilde{x}) - y^\delta$. Since the norm and Ω are both weakly lower semi-continuous functionals, it follows that

$$\begin{aligned} \frac{1}{p}\|F(\tilde{x}) - y^\delta\|^p &\leq \liminf_{k\to\infty} \frac{1}{p}\|F(x_{n_k}) - y_{n_k}\|^p , \\ \Omega(\tilde{x}) &\leq \liminf_{k\to\infty} \Omega(x_{n_k}) . \end{aligned} \tag{4.5}$$

From formula (4.5) we obtain

$$\begin{aligned} \frac{1}{p}\|F(\tilde{x}) - y^\delta\|^p + \alpha\Omega(\tilde{x}) &\leq \liminf_{k\to\infty} \frac{1}{p}\|F(x_{n_k}) - y_{n_k}\|^p + \alpha \liminf_{k\to\infty} \Omega(x_{n_k}) \\ &\leq \liminf_{k\to\infty} \left(\frac{1}{p}\|F(x_{n_k}) - y_{n_k}\|^p + \alpha\Omega(x_{n_k})\right) \\ &\leq \lim_{k\to\infty} \left(\frac{1}{p}\|F(x) - y_{n_k}\|^p + \alpha\Omega(x)\right) \\ &= \frac{1}{p}\|F(x) - y^\delta\|^p + \alpha\Omega(x), \quad x \in \mathcal{D} . \end{aligned}$$

This implies that $\tilde{x}$ is a minimizer of (4.1) and moreover, by taking $x = \tilde{x} \in \mathcal{D}$ on the right hand side, it follows that

$$\frac{1}{p}\|F(\tilde{x}) - y^\delta\|^p + \alpha\Omega(\tilde{x}) = \lim_{k\to\infty} \left(\frac{1}{p}\|F(x_{n_k}) - y_{n_k}\|^p + \alpha\Omega(x_{n_k})\right) . \tag{4.6}$$

Now assume that $\Omega(x_{n_k})$ does not converge to $\Omega(\tilde{x})$. Since Ω is weakly lower semi-continuous, it follows that

$$c := \limsup_{k\to\infty} \Omega(x_{n_k}) > \Omega(\tilde{x}) .$$

We take a subsequence $\{x_{n_{k_l}}\}$ such that $\Omega(x_{n_{k_l}}) \to c$ as $l \to \infty$. For this subsequence, as a consequence of (4.6), we find that

$$\begin{aligned} \lim_{l\to\infty} \frac{1}{p}\|F(x_{n_{k_l}}) - y_{n_{k_l}}\|^p &= \frac{1}{p}\|F(\tilde{x}) - y^\delta\|^p + \alpha\,(\Omega(\tilde{x}) - c) \\ &< \frac{1}{p}\|F(\tilde{x}) - y^\delta\|^p . \end{aligned}$$

This contradicts (4.5). Therefore we obtain $\Omega(x_{n_k}) \to \Omega(\tilde{x})$ as $k \to \infty$. □

4.1.2 Convergence of regularized solutions

In order to study the convergence of regularized solutions, we remember the concept of Ω-minimizing solutions (cf. Definition 3.25). Such solutions always exist under our standing Assumptions 3.11 and 3.22 and we can formulate the following convergence result followed by two corollaries dealing with a priori and a posteriori parameter choices.

Theorem 4.3 (convergence of regularized solutions). *Let $\{y_n := y^{\delta_n}\} \subset Y$ denote a sequence of perturbed data to the exact right-hand side $y \in F(\mathcal{D}(F))$ of the operator equation* (3.3) *and let $\|y - y_n\| \le \delta_n$ for a sequence $\{\delta_n > 0\}$ of noise levels converging monotonically to zero. Moreover, we consider a sequence $\{\alpha_n > 0\}$ of regularization parameters and an associated sequence $\{x_n := x_{\alpha_n}^{\delta_n}\}$ of regularized solutions that are minimizers of*

$$\frac{1}{p}\|F(x) - y_n\|^p + \alpha_n\, \Omega(x) \to \min, \quad \textit{subject to } x \in \mathcal{D}(F) \subseteq X. \tag{4.7}$$

Under the conditions

$$\limsup_{n\to\infty} \Omega(x_n) \le \Omega(x_0) \quad \textit{for all} \quad x_0 \in \{x \in \mathcal{D} : F(x) = y\} \tag{4.8}$$

and

$$\lim_{n\to\infty} \|F(x_n) - y_n\| = 0 \tag{4.9}$$

the sequence $\{x_n\}$ has a weakly convergent subsequence, where each weak limit of such a subsequence is an Ω-minimizing solution $x^\dagger \in \mathcal{D}$ of the operator equation (3.3)*. If, in addition, the Ω-minimizing solution $x^\dagger \in \mathcal{D}$ is unique, we have the weak convergence $x_n \rightharpoonup x^\dagger$ in X.*

Proof. Let $x^\dagger$ denote an Ω-minimizing solution to (3.3). From (4.8), it follows that

$$\limsup_{n\to\infty} \Omega(x_n) \le \Omega(x^\dagger). \tag{4.10}$$

On the other hand, from (4.9) and $\delta_n \to 0$ as $n \to \infty$, by the triangle inequality, we obtain that

$$\lim_{n\to\infty} \|F(x_n) - y\| = 0. \tag{4.11}$$

By (4.10), we have a constant $K > 0$ such that $x_n \in \mathcal{M}_\Omega(K)$ for all $n \in \mathbb{N}$. Consequently, by Assumption 3.22 (c), a subsequence $\{x_{n_k}\}$ converges weakly in X to some element $\tilde{x} \in \mathcal{D}$. Since F is weak-to-weak continuous, it follows from (4.11) that $F(\tilde{x}) = y$. From the weak lower semi-continuity of Ω we derive that

$$\Omega(\tilde{x}) \le \liminf_{k\to\infty} \Omega(x_{n_k}) \le \limsup_{k\to\infty} \Omega(x_{n_k}) \le \Omega(x^\dagger) \le \Omega(x_0) \tag{4.12}$$

for all $x_0 \in \mathcal{D}$ satisfying $F(x_0) = y$. Taking $x_0 := \tilde{x}$ shows that $\Omega(\tilde{x}) = \Omega(x^\dagger)$. Hence $\tilde{x}$ is an Ω-minimizing solution. Using this and (4.12) it follows, by a subsequence-subsequence argument, that

$$\lim_{n\to\infty} \Omega(x_n) = \Omega(x^\dagger). \tag{4.13}$$

If the Ω-minimizing solution $x^\dagger$ is uniquely determined, it follows that $\{x_n\}$ has a weakly convergent subsequence and the limit of any such subsequence equals $x^\dagger$. Therefore, again a subsequence-subsequence argument implies weak convergence of the whole sequence to $x^\dagger$. □

Remark 4.4. We strongly emphasize that the limit conditions (4.11) and (4.13), which are derived in the proof of Theorem 4.3, are important properties of the sequences of Tikhonov-regularized solutions on their own. Together with the weak convergence in X of regularized solutions x_n along subsequences to Ω-minimizing solutions $x^\dagger$, we even have norm convergence in Y of the corresponding image elements $F(x_n)$ to the exact right-hand side $y = F(x^\dagger)$ of the operator equation (3.3) and moreover we have convergence of the values $\Omega(x_n)$ of the stabilizing functional to $\Omega(x^\dagger)$. If the implication

$$x_n \rightharpoonup x^\dagger \text{ and } \Omega(x_n) \to \Omega(x^\dagger) \ \Rightarrow\ \|x_n - x^\dagger\| \to 0 \text{ as } n \to \infty \tag{4.14}$$

is true in X, we say that a *generalized Radon–Riesz property* holds. In this case, under the assumptions of Theorem 4.3, we even have norm convergence of regularized solutions,. The condition (4.14) is an extension of the Radon–Riesz property, which is also called *Kadec–Klee property*, where $\Omega(x)$ in (4.14) is simply taken as the norm $\|x\|$. A wide range of frequently used Banach spaces X satisfy this condition and are called Kadec–Klee spaces (cf. [186, Remark 3.1] and reference therein). In such spaces X and for functionals of the types (3.10) and (3.12), we obtain norm convergence under our assumptions, including the reflexivity of X and the condition $q > 1$. For functionals Ω violating (4.14), however, the convergence of regularized solutions derived from Theorem 4.3 is not necessarily a convergence with respect to the norm topology in X, but norm convergence can be shown under stronger assumptions on the type of convexity of Ω, and we refer to [210, pp.67–69] for sufficient conditions. In general, we note that norm convergence of regularized solutions holds true if *convergence rates* for the error norm can be proved, or if convergence rates for the Bregman distance can be proved and the Bregman distance can be bounded below by an index function of the error norm. It will be an essential goal of Section 4.2 to discuss convergence rates results.

Remark 4.5. If we relax the strong condition (4.8) in Theorem 4.3 to

$$\Omega(x_n) \le K \quad \text{for all } n \in \mathbb{N} \text{ and some constant } K > 0, \tag{4.15}$$

it also follows from (4.11) that $F(\tilde{x}) = y$, whenever $x_{n_k} \rightharpoonup \tilde{x}$, since F is weak-to-weak continuous. Each such weak limit $\tilde{x}$ is indeed a solution to (3.3), but not necessarily an Ω-minimizing solution.

Corollary 4.6 (convergence under a priori parameter choice). *For an a priori parameter choice $\alpha_n := \alpha(\delta_n)$, based on an index function (cf. Definition* 3.18*) $\alpha(\delta)$ that satisfies the limit conditions*

$$\alpha(\delta) \to 0 \quad \textit{and} \quad \frac{\delta^p}{\alpha(\delta)} \to 0 \quad \textit{as} \quad \delta \to 0 \tag{4.16}$$

and associated regularized solutions $x_n := x^\delta_{\alpha_n}$ Theorem 4.3 *applies and yields subsequences of $\{x_n\}$, which have Ω-minimizing solutions $x^\dagger$ of* (3.3) *as weak limits.*

Proof. From the definition of x_n as minimizers of (4.7), it follows that

$$\frac{1}{p}\|F(x_n) - y_n\|^p + \alpha_n \Omega(x_n) \le \frac{\delta_n^p}{p} + \alpha_n \Omega(x^\dagger), \tag{4.17}$$

where $x^\dagger$ denotes an Ω-minimizing solution to (3.3). The inequality (4.17) implies one the one hand

$$\Omega(x_n) \le \frac{\delta_n^p}{p\,\alpha_n} + \Omega(x^\dagger)$$

and, by (4.16) with $\delta_n^p/\alpha_n \to 0$ as $n \to \infty$, that (4.10) and consequently also (4.8) are valid. On the other hand, (4.17) with $\alpha_n \to 0$ also implies that

$$\|F(x_n) - y_n\| \le \left(\delta_n^p + \alpha_n\, p\, \Omega(x^\dagger)\right)^{1/p} \to 0 \text{ as } n \to \infty.$$

Hence, we have (4.9) and Theorem 4.3 applies. □

From the inequality (4.17) in the proof of Corollary 4.6, we immediately find the assertion of the following remark.

Remark 4.7. Given $\delta_{\max} > 0$, let $\alpha(\delta)$ denote an index function, satisfying (4.16). Moreover, set $C := \sup_{0<\delta\le\delta_{\max}} \frac{\delta^p}{\alpha(\delta)} < \infty$. Then, for all $0 < \delta \le \delta_{\max}$, the Tikhonov-regularized solutions $x^\delta_{\alpha(\delta)}$, as minimizers of (4.1), with regularization parameters $\alpha = \alpha(\delta)$, belong to a common sublevel set of Ω, which also contains the Ω-minimizing solutions $x^\dagger$ of (3.3). More precisely, we have

$$x^\delta_{\alpha(\delta)} \in \mathcal{M}_\Omega\left(C + \Omega(x^\dagger)\right) \cap \mathcal{D}(F), \quad 0 < \delta \le \delta_{\max}. \tag{4.18}$$

Moreover, for alternative a priori parameter choices, with index functions $\alpha = \alpha(\delta)$ that violate (4.16), but satisfy $C := \sup_{0<\delta\le\delta_{\max}} \frac{\delta^p}{\alpha(\delta)} < \infty$, the condition (4.15) is satisfied and Remark 4.5 can be exploited to show weak convergence of regularized solutions to solutions of (3.3), which are not necessarily Ω-minimizing solutions.

The most prominent a posteriori strategy for choosing the regularization parameter is the *discrepancy principle*, frequently named after Morozov (cf. [165]), where, for given $1 \le \tau_1 \le \tau_2 < \infty$, the parameter $\alpha_{\text{discr}} = \alpha_{\text{discr}}(y^\delta, \delta)$ satisfies the inequalities

$$\tau_1 \, \delta \le \| F(x^\delta_{\alpha_{\text{discr}}}) - y^\delta \| \le \tau_2 \, \delta \tag{4.19}$$

for regularized solutions x_α^δ, solving the extremal problem (4.1). Since, for fixed and sufficiently small $\delta > 0$, the discrepancy norm $\|F(x_\alpha^\delta) - y^\delta\|$ is monotonically increasing and continuous with respect to α, if T_α in (4.1) has a unique minimizer $x_\alpha^\delta \in \mathcal{D}$ for all $\alpha > 0$ and $y^\delta \in Y$ under consideration, then, under this assumption, the equation

$$\|F(x_\alpha^\delta) - y^\delta\| = \tau \, \delta, \quad \tau \ge 1, \tag{4.20}$$

has a uniquely determined solution $\alpha = \alpha(\delta)$. Then $x^\delta_{\alpha_{\text{discr}}} \in \mathcal{D}$ satisfying (4.19) exists for any given $1 \le \tau_1 \le \tau_2 < \infty$. In the case of injective and bounded linear operators $F = A : X \to Y$, this assumption is always satisfied.

However, for nonlinear forward operators $F : \mathcal{D}(F) \subseteq X \to Y$ and equations (3.3), parameters $\alpha > 0$ satisfying (4.20) and in the same manner (4.19) need not always exist, since the problem

$$\Omega(x) \to \min, \quad \text{subject to } \frac{1}{p} \|F(x) - y^\delta\|^p = \frac{(\tau\delta)^p}{p}, \quad x \in \mathcal{D}(F), \tag{4.21}$$

which is similar to the extremal problem (3.8) of the residual method, will in general be a nonconvex optimization problem, as a consequence of the fact that the set of admissible solutions in (4.21) can be nonconvex for nonlinear F. On the other hand, nonconvex optimization problems often possess *duality gaps* (cf., e.g., [23]), where the saddle point problem of the associated *Lagrangian functional* has no solution. The Tikhonov functional $T_\alpha(x)$ however, plays the role of the Lagrangian functional in problem (4.21) (cf. [100, Section 4.1.2]). For fixed $\alpha > 0$, the sets X_α^δ of regularized solutions x_α^δ need not be singletons. If we consider a trajectory of regularized solutions, i.e., we choose some element $x_\alpha^\delta \in X_\alpha^\delta$ for all $\alpha > 0$, then, for that trajectory, the function $T_\alpha(x_\alpha^\delta)$ is continuous and monotonically increasing with respect to $\alpha > 0$ and $\|F(x_\alpha^\delta) - y^\delta\|$ is also monotonically increasing. But $\|F(x_\alpha^\delta) - y^\delta\|$ can attain different values if x_α^δ varies through the set X_α^δ. As a consequence, the residual norm of regularized solutions may have jumps in the sense that there are $\alpha_0 > 0$ such that

$$\lim_{\alpha_n \nearrow \alpha_0 - 0} \; \sup_{x^\delta_{\alpha_n} \in X^\delta_{\alpha_n}} \|F(x^\delta_{\alpha_n}) - y^\delta\| < \lim_{\alpha_n \searrow \alpha_0 + 0} \; \inf_{x^\delta_{\alpha_n} \in X^\delta_{\alpha_n}} \|F(x^\delta_{\alpha_n}) - y^\delta\| .$$

Hence, the conditions (4.20) and (4.19) may be failed by all regularization parameters $\alpha > 0$. Mostly, sufficient conditions to avoid such gaps and to ensure the uniqueness of the minimizers x_α^δ seem to be based on some smallness of the penalty term $\Omega(x_\alpha^\delta)$

(see, e.g., [122, 123]). For an alternative kind of sufficient conditions ensuring the existence of α_{discr} we also refer to [8].

In the next corollary, we simply assume that α_{discr} exists for all $\delta > 0$ and all data elements y^δ under consideration.

Corollary 4.8 (convergence under a posteriori parameter choice). *For an a posteriori parameter choice $\alpha_n := \alpha_{discr}(y_n, \delta_n)$ and associated regularized solutions $x_n := x^\delta_{\alpha_n}$, satisfying the discrepancy principle*

$$\tau_1\, \delta_n \le \|F(x_n) - y_n\| \le \tau_2\, \delta_n \tag{4.22}$$

for some $1 \le \tau_1 \le \tau_2$ Theorem 4.3 applies and yields subsequences of $\{x_n\}$ having Ω-minimizing solutions $x^\dagger$ of (3.3) as weak limits.

Proof. From the definition of x_n as minimizers of (4.7) result (4.17) follows again, where $x^\dagger$ denotes an Ω-minimizing solution of (3.3). The inequality (4.17) here implies

$$\Omega(x_n) \le \frac{1}{\alpha_n\, p}\left(\delta_n^p - \|F(x_n) - y_n\|^p\right) + \Omega(x^\dagger) \le \Omega(x^\dagger)$$

and hence (4.10), as a consequence of the left-hand inequality in (4.22). On the other hand, (4.9) is an immediate consequence of the right-hand inequality in (4.22). Hence Theorem 4.3 applies. □

From the proof of Corollary 4.8 we see that $x^\delta_{\alpha_{\text{discr}}} \in \mathcal{M}_\Omega\left(\Omega(x^\dagger)\right) \cap \mathcal{D}(F)$ for all $\delta > 0$ and $y^\delta \in Y$, whenever the discrepancy principle can be realized, that is, if there exists some α_{discr} satisfying (4.19).

Remark 4.9. Relaxing our assumptions stated above, convergence theory for Tikhonov-regularized solutions can be extended twofold:

(i) On the one hand, some models of inverse problems in machine learning, mechanics and optimal control require Banach spaces X that are *not reflexive*, i.e. for which our standing Assumption 3.11 (a) is violated.

(ii) On the other hand, in imaging one often makes use of penalties Ω that are *not stabilizing* in the sense of Assumption 3.22 (c).

We note that there is a close relationship between both kinds of extensions. Taking into account Lemma 3.12 closed balls in a non-reflexive Banach space fail to be weakly sequentially compact. Then, the penalty functionals $\Omega(x) = \|x\|_X^q$, $q \ge 1$, as powers of the norm in a non-reflexive Banach space X, are not stabilizing functionals in the sense of Assumption 3.22 (c). As mentioned earlier, when L^q- or ℓ^q-spaces are used as Banach space X, there are good reasons to prefer the exponent q in Ω. Note that the exponent $q = 1$ expresses an exceptional case, because L^1 and ℓ^1 are non-reflexive Banach spaces, whereas L^q and ℓ^q are reflexive for all $1 < q < \infty$. If, however, Assumption 3.22 (c) fails, the weakly convergent subsequences of regularized solutions, required in the propositions and theorems of the Sections 4.1.1, 4.1.2 and 4.2 below, do not necessarily exist. This negative property can in general be seen when the penalty functional, with some convex index function χ and some reference element $\overline{x} \in X$, is of the form

$$\Omega(x) = \chi\left(\|x - \overline{x}\|_X\right) \tag{4.23}$$

for non-reflexive Banach spaces X like $L^1, L^\infty, \ell^1, \ell^\infty, \mathcal{C}^k$ and BV.

The standard approach for overcoming this shortcoming is based on exploiting the *sequential Banach–Alaoglu theorem* in the form of the following lemma.

Lemma 4.10. *The closed unit ball of a Banach space X is sequentially compact in the weak*-topology if there is a separable Banach space Z (predual space), with dual $Z^* = X$. Then, any bounded sequence $\{x_n\}$ in X has a subsequence $\{x_{n_k}\}$, such that $x_{n_k} \rightharpoonup^* x_0 \in X$ as $k \to \infty$.*

If we replace Assumption 3.22 (c) with the assumption

(c*) Ω is assumed to be a *weak*-stabilizing functional*, in the sense that the sublevel sets
$$\mathcal{M}_\Omega(c) := \{x \in X \;:\; \Omega(x) \le c\}$$
are weak* sequentially pre-compact in X, for all $c \ge 0$,

then due to Lemma 4.10 *weak*-convergent* subsequences of regularized solutions exist that are able to replace the weakly convergent ones (see, e.g., [234, Sect. 2.5] and [34]). Then, for non-reflexive Banach spaces X like $X = \ell^1$ and $X = BV$, which possess *separable* predual spaces (cf., e.g., [7, Remark B.7, p. 299]), the penalty functionals (3.10), (3.12) and more generally (4.23), can be included into the theory. In this context, however, the forward operator F is assumed to be *weak*-to-weak continuous* between X and Y, at least in the sense of subsequences (cf., e.g., [125]). For $X = BV$, it is common to use the semi-norm as penalty Ω, see (4.30) below and for more details [33]. We conclude this remark by noting that non-reflexive and non-separable Banach spaces like BV and L^∞ require some specific attention, when the *discretization* of Tikhonov-type approaches is under consideration, which is a necessary step for the numerical solution of the corresponding ill-posed equations, and we refer to the recent paper [186] for many details concerning this field.

4.2 Error estimates and convergence rates

In this chapter, we are going to collect assertions on error estimates and convergence rates of regularized solutions x_α^δ, that are minimizers of the extremal problem (4.1) for an a priori, as well as an a posteriori choice of the regularization parameter $\alpha > 0$. We aim at formulating the results as generally as possible for nonlinear operator equations (3.3) in Banach spaces, under the standing Assumptions 3.11, 3.22, and 3.26 on the Banach spaces X and Y, the forward operator F and its derivatives, its domain $\mathcal{D}(F)$, and on the stabilizing penalty functional Ω. If, as a special case, linear operator equations (3.1) are considered, we again use the symbol A to characterize the bounded linear forward operator. We denote by $x^\dagger$ the Ω-minimizing solutions to equation (3.3), in the sense of Definition 3.25 which always exist.

4.2.1 Error estimates under variational inequalities

Definition 4.11 (error measure). We call a non-negative functional, defined on a subset of the product space $X \times X$, an *error measure* for an approximation $x \in X$ of the element $\tilde{x} \in X$, if the value $E(x, \tilde{x})$ expresses the precision of this approximation with respect to $\tilde{x}$ in a well-defined sense. In general, the properties $E(\tilde{x}, \tilde{x}) = 0$ and the limit condition $\lim_{x \to \tilde{x}} E(x, \tilde{x}) = 0$ are required.

An error measure can be constructed, based on a *metric* in X, i.e., we have $E(\tilde{x}, \tilde{x}) = 0$ and $E(x, \tilde{x}) > 0$ if $x \neq \tilde{x}$, moreover $E(x, \tilde{x}) = E(\tilde{x}, x)$ and

$$E(x, \tilde{x}) \le E(x, \hat{x}) + E(\hat{x}, \tilde{x})$$

for all $x, \tilde{x}, \hat{x} \in X$. The simplest form is the error norm in the space X,

$$E(x, \tilde{x}) := \|x - \tilde{x}\|, \quad x, \tilde{x} \in X. \tag{4.24}$$

In Banach space theory, however, also non-symmetric measures E, like Kullback–Leibler and other divergences play an important role and we refer to [185] for a collection of such measures. In this work, however, our focus is on the non-symmetric *Bregman distance*

$$\begin{aligned} D_{\tilde{\xi}}^{\Omega}(x, \tilde{x}) &:= \Omega(x) - \Omega(\tilde{x}) - \langle \tilde{\xi}, x - \tilde{x} \rangle_{X^* \times X}, \\ & x \in \mathcal{D}, \quad \tilde{x} \in \mathcal{D}_B(\Omega), \quad \tilde{\xi} \in \partial\Omega(\tilde{x}) \end{aligned} \tag{4.25}$$

and the corresponding error measure

$$E(x, \tilde{x}) := D_{\tilde{\xi}}^{\Omega}(x, \tilde{x}). \tag{4.26}$$

Let us mention here that some authors, e.g. [34], prefer the corresponding symmetric analog

$$\begin{aligned} D_{\text{sym}}^{\Omega}(x, \tilde{x}) &:= \langle \xi - \tilde{\xi}, x - \tilde{x} \rangle_{X^* \times X} = D_{\xi}^{\Omega}(\tilde{x}, x) + D_{\tilde{\xi}}^{\Omega}(x, \tilde{x}), \\ & \xi \in \partial\Omega(x), \quad \tilde{\xi} \in \partial\Omega(\tilde{x}), \end{aligned}$$

to (4.25), which, however, acts on a more restricted domain.

We note that originally Bregman distances (4.25) were only formulated for *strictly convex* functionals Ω. In that case, the error measure (4.26) satisfies the advantageous property

$$E(x, \tilde{x}) = 0 \quad \text{if and only if} \quad x = \tilde{x}\,. \tag{4.27}$$

If a convex functional Ω, failing to be strictly convex, is chosen as penalty, approximate solutions x can be far from the exact one $\tilde{x}$ although $D^{\Omega}_{\tilde{\xi}}(x, \tilde{x}) = 0$. This is the case, for example, if Ω is taken from (3.10), with $q = 1$, and more generally when Ω is a *positively homogeneous functional.* In [147, §4], Bregman distances vanishing for $x \neq \tilde{x}$ are verified in a sparsity setting for the penalty functional

$$\Omega(x) := \sum_{n=1}^{\infty} w_n\,|x_n|, \qquad w_n \geq 0, \tag{4.28}$$

with $x = (x_1, x_2, \ldots)$ from the space $X = \ell^2$ of quadratically summable infinite sequences of real numbers. The same situation occurs for

$$\Omega(x) := \sum_{n=1}^{\infty} w_n\,|\langle \zeta_n, x\rangle|, \qquad w_n \geq 0, \tag{4.29}$$

and X being a separable Hilbert space with orthonormal basis $\{\zeta_n\}$ (see [79, 82] and Section 4.2.6), as well as for the Rudin–Osher–Fatemi-model (cf. [204]), using the semi-norm

$$\Omega(x) := |x|_{BV(\mathcal{G})} = \sup_{g \in C_0^{\infty}(\mathcal{G}),\, \|g\|_{L^{\infty}(\mathcal{G})} \leq 1} \int_{\mathcal{G}} x\, \nabla \cdot g\, dt \tag{4.30}$$

in the Banach space $X = BV(\mathcal{G})$ of functions of bounded variation embedded into the Hilbert space $L^2(\mathcal{G})$, with some bounded domain $\mathcal{G} \subset \mathbb{R}^k$, $k = 1, 2, \ldots$ (see, e.g., [11, 33, 34, 46, 179, 236]).

Definition 4.12 (error estimate and convergence rate). If, for given error measure E and given $\delta_{\max} > 0$, we have a non-negative functional Ξ such that an inequality of the form

$$E(x_\alpha^\delta, x^\dagger) \leq \Xi\left(\delta, \Omega(x_\alpha^\delta), \Omega(x^\dagger), \|F(x_\alpha^\delta) - F(x^\dagger)\|\right) \tag{4.31}$$

holds for $0 < \delta \leq \delta_{\max}$ and all $\alpha = \alpha(\delta)$, according to a given a priori parameter choice rule, or all $\alpha = \alpha(y^\delta, \delta)$, according to a given a posteriori parameter choice rule, then we call (4.31) an *error estimate* for regularized solutions, with respect to the Ω-minimizing solution $x^\dagger$ of (3.3). If, moreover, Ξ can be estimated from above by an index function χ, such that

$$E(x_\alpha^\delta, x^\dagger) \leq \chi(\delta) \tag{4.32}$$

for all $0 < \delta \le \delta_{\max}$ and all associated $\alpha > 0$ under consideration, we call an index function χ in (4.32) *convergence rate* and

$$E(x^\delta_{\alpha(y^\delta,\delta)}, x^\dagger) = O(\chi(\delta)) \quad \text{as} \quad \delta \to 0 \tag{4.33}$$

convergence rate result for the Tikhonov regularization.

In the classical regularization theory in Hilbert spaces, see [67], convergence rates results (4.33) for E from (4.24) and linear problems were obtained under source conditions expressing the solution smoothness of $x^\dagger$ with respect to A. See in particular the versions (3.18) and (3.20). These particular source conditions have (3.22) and (3.23) as counterparts in Banach spaces. For nonlinear problems A is replaced with $F'(x^\dagger)$ in the source conditions but, additionally, structural conditions on the nonlinearity of F (cf. Section 3.2.4) have to be handled, in order to get such rates.

In the past, research on convergence rates in Tikhonov-type regularization often used the idea of exploiting *variational inequalities*, which have the capability of expressing both solution smoothness and nonlinearity conditions. Following an approach developed simultaneously by Flemming (cf. [73, 74]) and Grasmair (cf. [78]), we first of all consider variational inequalities of the form

$$\beta\, E(x, x^\dagger) \le \Omega(x) - \Omega(x^\dagger) + \varphi(\|F(x) - F(x^\dagger)\|) \quad \text{for all} \;\; x \in \mathcal{M} \subseteq \mathcal{D} \tag{4.34}$$

with some index function φ and some $0 < \beta \le 1$. If, for $0 < \delta \le \delta_{\max}$, all associated regularized solutions x^δ_α belong to $\mathcal{M}$ (for an example see Remark 4.7, with sublevel sets $\mathcal{M} = \mathcal{M}_\Omega(C + \Omega(x^\dagger)) \cap \mathcal{D}(F)$), then (4.34), after multiplication by $1/\beta$, provides us with an error estimate in the sense of Definition 4.12.

Note that variational inequalities (4.34) for E from (4.24) were already mentioned in [27, Lemma 4.4]. The most prominent version in Banach space regularization, however, refers to the Bregman distance E from (4.26). For that case, by setting $\beta_1 := \frac{1-\beta}{\beta}$ and $\beta_2 = \frac{1}{\beta}$, (4.34) can be rewritten as

$$\langle \xi^\dagger, x^\dagger - x\rangle_{X^* \times X} \le \beta_1\, D^\Omega_{\xi^\dagger}(x, x^\dagger) + \beta_2\, \varphi(\|F(x) - F(x^\dagger)\|)$$
$$\text{for all} \;\; x \in \mathcal{M} \subseteq \mathcal{D}$$

with $\xi^\dagger \in \partial\Omega(x^\dagger)$, some index function φ and constants $0 \le \beta_1 < 1$, $\beta_2 \ge 0$, as was suggested originally for $\varphi(t) = t$ in [104] and extended to the case of general concave index functions φ in [20]. For more details see also the Sections 3.2.3, 3.2.4 and 4.2.2.

We now show that the right-hand side in inequality (4.34) has an appropriate structure to derive convergence rates for the Tikhonov-type regularization. More precisely, under the variational inequality (4.34), with some concave index function φ and some $0 < \beta \le 1$, we obtain an inequality (4.32), with $\chi(\delta) \le K\,\varphi(\delta)$, for some constant $K > 0$ independent of $\delta > 0$ and hence a convergence rate result

$E(x_\alpha^\delta, x^\dagger) = O(\varphi(\delta))$ for $\delta \to 0$, whenever α is chosen in an adapted manner. This parameter choice can be of a posteriori type $\alpha = \alpha_{\text{discr}}$, according to the discrepancy principle

$$\tau_1 \delta \le \|F(x^\delta_{\alpha_{\text{discr}}}) - y^\delta\| \le \tau_2 \delta \tag{4.35}$$

for prescribed $1 \le \tau_1 \le \tau_2$, or of a priori type, according to an appropriate choice $\alpha = \alpha(\delta)$, depending on $p > 1$ and φ .

Theorem 4.13. *Let $c \ge \Omega(x^\dagger)$ and let, for all $0 < \delta \le \delta_{max}$, regularization parameters $\alpha_{discr} = \alpha(y^\delta, \delta)$, satisfying the discrepancy principle* (4.35) *exist. Moreover, let the variational inequality* (4.34) *be valid for a concave index function φ and $\mathcal{M} := \mathcal{M}_\Omega(c) \cap \mathcal{D}(F)$. Then we obtain the convergence rate result*

$$E(x^\delta_{\alpha_{discr}}, x^\dagger) = O(\varphi(\delta)) \quad as \quad \delta \to 0. \tag{4.36}$$

Proof. From the proof of Corollary 4.8 we see that $\Omega(x^\delta_{\alpha_{\text{discr}}}) \le \Omega(x^\dagger) \le c$. Hence, as a consequence of (4.34), we have the error estimate

$$E(x^\delta_{\alpha_{\text{discr}}}, x^\dagger) \le \frac{\varphi(\|F(x^\delta_{\alpha_{\text{discr}}}) - F(x^\dagger)\|)}{\beta} \le \frac{\varphi(\|F(x^\delta_{\alpha_{\text{discr}}}) - y^\delta\| + \delta)}{\beta}$$

for $0 < \delta \le \delta_{\max}$. Then, the right-hand inequality of (4.35), together with the well-known property

$$\varphi(K\delta) \le \max(1, K)\, \varphi(\delta) \quad \text{for all} \quad \delta > 0 \tag{4.37}$$

of concave index functions φ, yield

$$E(x^\delta_{\alpha_{\text{discr}}}, x^\dagger) \le \frac{\varphi((1+\tau_2)\delta)}{\beta} \le \frac{1+\tau_2}{\beta}\,\varphi(\delta)$$

for $0 < \delta \le \delta_{\max}$, which gives (4.36) and proves the theorem. □

Note that Neubauer, in the recent paper [176], suggests an alternative approach for handling the discrepancy principle.

For a theorem analogous to Theorem 4.13, but involving an a priori parameter choice, we now translate Theorem 4.11 from [74] and its proof into our setting. We note that, for similar results in [78], the focus is on the *Fenchel conjugate function*

$$f^*(s) := \sup_{t>0} (st - f(t)), \quad s > 0,$$

of a convex function $f(t),\ t > 0$.

Theorem 4.14. *Let the variational inequality* (4.34) *be valid for a concave index function φ and $\mathcal{M} := \mathcal{M}_\Omega(c) \cap \mathcal{D}(F)$ with $c > \Omega(x^\dagger)$. Then, there is an a priori parameter choice $\alpha = \alpha(\delta)$, satisfying* (4.16) *that implies the convergence rate result*

$$E(x^\delta_{\alpha(\delta)}, x^\dagger) = O(\varphi(\delta)) \quad as \quad \delta \to 0. \tag{4.38}$$

Proof. Whenever $\alpha = \alpha(\delta)$ is chosen such that $\frac{\delta^p}{\alpha(\delta)}$ is small enough for implying $x_{\alpha(\delta)}^\delta \in \mathcal{M}$ for $0 < \delta \le \delta_{\max}$ (cf. Remark 4.7), by setting

$$\tilde{\varphi}(t) := \varphi\left(\left(p\, 2^{p-1}\right)^{1/p} t^{1/p}\right), \quad t > 0,$$

and due to

$$(a+b)^p \le 2^{p-1}\left(a^p + b^p\right), \quad a, b \ge 0,$$

we have the following chain of inequalities, where we note that $\tilde{\varphi}$ is a strictly concave index function, satisfying the limit condition

$$\lim_{t \to +0} \frac{\tilde{\varphi}(t)}{t} = +\infty, \tag{4.39}$$

since φ is concave:

$$\begin{aligned}
\beta\, E(x_\alpha^\delta, x^\dagger) &\le \Omega(x_\alpha^\delta) - \Omega(x^\dagger) + \varphi\left(\|F(x_\alpha^\delta) - F(x^\dagger)\|\right) \\
&= \frac{1}{\alpha}\left(\mathrm{T}_\alpha(x_\alpha^\delta) - \alpha\Omega(x^\dagger) - \frac{1}{p}\|F(x_\alpha^\delta) - y^\delta\|^p\right) \\
&\quad + \varphi\left(\|F(x_\alpha^\delta) - F(x^\dagger)\|\right) \\
&\le \frac{1}{\alpha}\left(\frac{1}{p}\|F(x^\dagger) - y^\delta\|^p - \frac{1}{p}\|F(x_\alpha^\delta) - y^\delta\|^p\right) \\
&\quad + \varphi\left(\|F(x_\alpha^\delta) - y^\delta\| + \|F(x^\dagger) - y^\delta\|\right) \\
&\le \frac{1}{\alpha}\left(-\frac{1}{p}\|F(x_\alpha^\delta) - y^\delta\|^p + \frac{1}{p}\delta^p\right) + \varphi\left(\|F(x_\alpha^\delta) - y^\delta\| + \delta\right) \\
&\le \frac{2\delta^p}{\alpha p} + \varphi\left(\|F(x_\alpha^\delta) - y^\delta\| + \delta\right) - \frac{1}{\alpha p}\left(\|F(x_\alpha^\delta) - y^\delta\|^p + \delta^p\right) \\
&\le \frac{2\delta^p}{\alpha p} + \varphi\left(\|F(x_\alpha^\delta) - y^\delta\| + \delta\right) - \frac{1}{\alpha p\, 2^{p-1}}\left(\|F(x_\alpha^\delta) - y^\delta\| + \delta\right)^p .
\end{aligned}$$

Setting again

$$\psi(\delta) := \frac{\delta^p}{p}, \quad \delta > 0,$$

this gives

$$\beta\, E(x_\alpha^\delta, x^\dagger) \le \frac{2\psi(\delta)}{\alpha} + \sup_{t>0}\left(\tilde{\varphi}(t) - \frac{1}{\alpha}t\right) = \frac{2\psi(\delta)}{\alpha} + (-\tilde{\varphi})^*\left(-\frac{1}{\alpha}\right). \tag{4.40}$$

Now the strict concavity of the index function $\tilde{\varphi}$ ensures that

$$0 < \sup_{s \in (\psi(\delta), \infty)} \frac{\tilde{\varphi}(s) - \tilde{\varphi}(\psi(\delta))}{s - \psi(\delta)} \le \inf_{s \in [0, \psi(\delta))} \frac{\tilde{\varphi}(\psi(\delta)) - \tilde{\varphi}(s)}{\psi(\delta) - s} < \infty .$$

Consequently, there exists $\alpha = \alpha(\delta) > 0$ such that for $0 < \delta \le \delta_{\max}$

$$\sup_{s \in (\psi(\delta),\infty)} \frac{\tilde{\varphi}(s) - \tilde{\varphi}(\psi(\delta))}{s - \psi(\delta)} \le \frac{1}{\alpha(\delta)} \le \inf_{s \in [0,\psi(\delta))} \frac{\tilde{\varphi}(\psi(\delta)) - \tilde{\varphi}(s)}{\psi(\delta) - s}. \tag{4.41}$$

Note that, for differentiable φ, this a priori parameter choice (4.41) simplifies to

$$\alpha(\delta) = \frac{1}{\tilde{\varphi}'(\psi(\delta))}. \tag{4.42}$$

From (4.39), we obtain that $\sup_{s \in (t,\infty)} \frac{\tilde{\varphi}(s) - \tilde{\varphi}(t)}{s-t} \to \infty$ as $t \to +0$ and hence $\alpha(\delta) \to 0$ as $\delta \to 0$. On the other hand, from both inequalities in (4.41) we simply derive that

$$\tilde{\varphi}(s) - \frac{s}{\alpha(\delta)} \le \tilde{\varphi}(\psi(\delta)) - \frac{\psi(\delta)}{\alpha(\delta)} \quad \text{for all} \quad s > 0.$$

Together with (4.40) we have

$$\beta\, E(x^\delta_{\alpha(\delta)}, x^\dagger) \le \frac{\psi(\delta)}{\alpha(\delta)} + \tilde{\varphi}(\psi(\delta)) \le \psi(\delta) \inf_{s \in [0,\psi(\delta))} \frac{\tilde{\varphi}(\psi(\delta)) - \tilde{\varphi}(s)}{\psi(\delta) - s} + \tilde{\varphi}(\psi(\delta))$$

and taking $s := 0$ also (4.16) in the form $\frac{\psi(\delta)}{\alpha(\delta)} \le \tilde{\varphi}(\psi(\delta)) \to 0$ as $\delta \to 0$ since $\tilde{\varphi}$ and ψ are both index functions. Consequently we arrive at the error estimate

$$E(x^\delta_{\alpha(\delta)}, x^\dagger) \le \frac{2}{\beta}\, \tilde{\varphi}(\psi(\delta)).$$

Hence, with the a priori parameter choice (4.41) and due to the property (4.37) of concave index functions, we obtain

$$E(x^\delta_{\alpha(\delta)}, x^\dagger) = O\left(\tilde{\varphi}(\delta^p)\right) = O\left(\varphi(\delta)\right) \quad \text{as} \quad \delta \to 0. \quad \square$$

4.2.2 Convergence rates for the Bregman distance

In this section we consider rates for the error measure $E(x^\delta_\alpha, x^\dagger) = D^\Omega_\xi(x^\delta_\alpha, x^\dagger)$, preferably under variational inequalities of the form

$$\begin{aligned} &\langle \xi^\dagger, x^\dagger - x\rangle_{X^* \times X} \le \beta_1\, D^\Omega_{\xi^\dagger}(x, x^\dagger) + \beta_2\, \varphi(\|F(x) - F(x^\dagger)\|) \\ &\text{for all} \quad x \in \mathcal{M} \subseteq \mathcal{D} \end{aligned} \tag{4.43}$$

with $\xi^\dagger \in \partial\Omega(x^\dagger)$, some index function φ and constants $0 \le \beta_1 < 1$, $\beta_2 \ge 0$.

When we recognize the powerful role of variational inequalities (4.43) for obtaining convergence rates in Tikhonov regularization, it is of some interest to verify and compare various sufficient conditions that imply such inequalities with index functions φ. Variational inequalities represent a sophisticated tool for expressing solution smoothness and to combine it with the nonlinearity structure of the forward operator in case of nonlinear problems. A main advantage is the fact that the function φ immediately expresses the occurring convergence rate. A series of such sufficient conditions and cross connections between source conditions and variational inequalities were already formulated in Section 3.2 (see Propositions 3.30, 3.31, 3.32, 3.33) and can also be found in [9, 20, 104, 210].

For *concave* index functions φ, we can immediately apply Theorem 4.13, yielding the convergence rate result

$$D_{\xi^\dagger}^{\Omega}(x_\alpha^\delta, x^\dagger) = O(\varphi(\delta)) \quad \text{as} \quad \delta \to 0 \tag{4.44}$$

if (4.43) is valid for $\mathcal{M} := \mathcal{M}_\Omega(\Omega(x^\dagger)) \cap \mathcal{D}(F)$ and if the discrepancy principle (4.35) with $\alpha = \alpha_{\text{discr}}$ can be realized for sufficiently small $\delta > 0$.

The same rate can be obtained by applying Theorem 4.14 directly, when for concave index functions φ the a priori parameter choice $\alpha = \alpha(\delta)$ is made in accordance with (4.41) and if (4.43) is valid for $\mathcal{M} := \mathcal{M}_\Omega(\Omega(x^\dagger) + C) \cap \mathcal{D}(F)$, with some $C > 0$. An alternative proof of this convergence rate result for differentiable φ and $\alpha(\delta)$ in accordance with (4.42), but also extended to more general convex index functions g in the misfit term $S(F(x), y^\delta) = g(\|F(x) - y^\delta\|)$, was given in [20, Theorem 4.3], based on the *generalized Young inequality*

$$ab \le \int_0^a f(t)dt + \int_0^b f^{-1}(t)dt, \quad a, b \ge 0,$$

which is true for any index function f.

Now, we can formulate corollaries for the cases that the benchmark source condition is satisfied or violated. In the latter case, we additionally require q-coercivity of the Bregman distance.

Corollary 4.15. *From Proposition* 3.32, *and due to the Theorems* 4.13 *and* 4.14 *we have the convergence rate result* (4.44), *with the concave index function φ, for appropriate choices of the regularization parameter $\alpha > 0$, whenever the source condition* (3.29) *is satisfied for some $\xi^\dagger \in \partial\Omega(x^\dagger)$ and the nonlinearity condition* (3.43) *holds, with $\varphi = \sigma$ for all $x \in \mathcal{M} := \mathcal{M}_\Omega(\Omega(x^\dagger) + C) \cap \mathcal{D}(F)$ and some constant $C > 0$.*

Corollary 4.16. *From Proposition* 3.33 *and due to the Theorems* 4.13 *and* 4.14, *we have the convergence rate result* (4.44), *with an index function φ, defined as*

$$\varphi(0) = 0, \quad \varphi(t) = \left[d_{\xi^\dagger}\left(\Psi^{-1}(\sigma(t))\right)\right]^{q^*}, \quad t > 0,$$

using the auxiliary function

$$\Psi(R) := \frac{(d_{\xi^\dagger}(R))^{q^*}}{R}, \quad R > 0, \quad \frac{1}{q} + \frac{1}{q^*} = 1,$$

for appropriate choices of the regularization parameter $\alpha > 0$*, whenever for* $\xi^\dagger \in \partial\Omega(x^\dagger)$ *the source condition* (3.29) *is violated, but the distance function* $d_{\xi^\dagger}(R)$ *tends to zero as* $R \to \infty$ *and, moreover, the conditions* (3.43) *and* (3.44) *hold, with some concave index function* σ *and with* $q \geq 2$ *for all* $x \in \mathcal{M} := \mathcal{M}_\Omega(\Omega(x^\dagger) + C) \cap \mathcal{D}(F)$ *and some constant* $C > 0$.

Note that, in Corollary 4.16, φ can be replaced by a concave majorant index function $\tilde{\varphi}$, in case that (3.45) itself fails to be concave. Such concave majorants always exist.

Example 4.17. In this example, concerning *logarithmic convergence rates*

$$D_{\xi^\dagger}^{\Omega}(x_\alpha^\delta, x^\dagger) = O\left(\frac{1}{\left[\log\left(\frac{1}{\delta}\right)\right]^\mu}\right) \quad \text{as} \quad \delta \to 0 \tag{4.45}$$

for some $\mu > 0$, we show that such rate result, as a special case of (4.44), can be derived along the lines of the Corollaries 4.15 and 4.16, for two completely different extreme situations, with respect to the assumptions on solution smoothness of $x^\dagger$ and nonlinearity of F.

Supposed that the source condition (3.29) is satisfied by Corollary 4.15, we immediately derive, for all $\mu > 0$, the logarithmic convergence rate (4.45), if we have

$$\sigma(t) = \frac{1}{\left[\log\left(\frac{1}{\delta}\right)\right]^\mu}, \qquad t > 0,$$

in the nonlinearity condition (3.43). This now means, that the nonlinearity condition expressing the smoothness of F at $x^\dagger$ is extremely poor, but the solution smoothness is good enough. For any $\mu > 0$, the rate in (4.45) is slower than a Hölder rate (4.48), for arbitrarily small $\nu > 0$.

The converse situation occurs if $\sigma(t) = t$, i.e. the η-condition (3.42) represents an advantageous structure of nonlinearity, but the solution smoothness is poor in the sense that a logarithmic decay rate

$$d_{\xi^\dagger}(R) \leq \overline{C}\,(\log R)^{-\lambda},$$

for sufficiently large $R > 0$ and with some constant $\overline{C} > 0$, expresses that $\xi^\dagger$ strongly violates the source condition (3.29), independent of the choice of $\lambda > 0$. However, since we have, for all such R and for $\varepsilon > 0$, a constant $\overline{K} > 0$ with

$$\Psi(R) = \frac{1}{R(\log R)^{\lambda q^*}} \geq \frac{\overline{K}}{R^{1+\varepsilon}},$$

which implies $\Psi^{-1}(t) \geq \hat{K} t^{-1/(1+\varepsilon)}$, for some constant $\hat{K} > 0$ and sufficiently small $t > 0$. Hence, by Corollary 4.16 we have a rate result (4.45) with

$$\mu = \lambda q^*.$$

The following proposition, which was proved as Proposition 12.10 in [74], will show that strictly concave index functions φ do not make sense in (4.43). Furthermore, higher rates than

$$D_{\xi^\dagger}^{\Omega}(x_\alpha^\delta, x^\dagger) = O(\delta) \quad \text{as} \quad \delta \to 0 \tag{4.46}$$

cannot be expected based on such a variational inequality alone. We refer to [80] for further ideas.

Proposition 4.18. *Under the standing assumptions, in particular Assumption* 3.26, *let the variational inequality* (4.43) *hold for an index function φ satisfying the limit condition $\lim_{t\to+0}\frac{\varphi(t)}{t} = 0$ and for some set $\mathcal{M}$ starlike with respect to $x^\dagger$. Then we have $\Omega(x^\dagger) \le \Omega(x)$ for all $x \in \mathcal{M}$.*

The message of this proposition is, that the occurrence of index functions φ with $\lim_{t\to+0}\frac{\varphi(t)}{t} = 0$ in (4.43) represents a singular case, which is really not of interest in the solution theory of operator equations (3.3). Note that a pre-stage of Proposition 4.18 for the Hölder (power) case $\varphi(t) = t^\kappa,\ \kappa > 1$ was already presented in [113, Proposition 4.3]. The proof of such propositions is always based on the existence of a bounded linear operator $F'(x^\dagger)$ with Gâteaux-derivative properties (3.9).

We will now highlight the case of Hölder rates a bit more. From Theorem 4.14, we easily derive the following proposition, taking into account that $p > 1$ is assumed. Sufficient conditions for obtaining (4.43) can be found in Propositions 3.32 and 3.33, and we refer to the Corollaries 4.15 and 4.16 for cross connections.

Proposition 4.19. *From a variational inequality* (4.43), *with $\varphi(t) = t^\nu,\ 0 < \nu \le 1$, and $\mathcal{M} := \mathcal{M}_\Omega(\Omega(x^\dagger) + C) \cap \mathcal{D}(F),\ C > 0$, for the a priori parameter choice*

$$\underline{c}\,\delta^{p-\nu} \le \alpha(\delta) \le \overline{c}\,\delta^{p-\nu}, \quad \delta > 0, \tag{4.47}$$

and constants $0 < \underline{c} \le \overline{c} < \infty$, we obtain the Hölder convergence rate result

$$D_{\xi^\dagger}^{\Omega}(x_{\alpha(\delta)}^\delta, x^\dagger) = O\left(\delta^\nu\right) \quad \textit{as} \quad \delta \to 0. \tag{4.48}$$

In Tikhonov regularization for Hilbert spaces X and Y, at least for linear ill-posed problems, Hölder rates

$$\|x_{\alpha(\delta)}^\delta - x^\dagger\| = O\left(\delta^{\frac{\nu}{2}}\right) \quad \text{as} \quad \delta \to 0$$

and Hölder source conditions (3.17), with $\kappa(t) = t^{\frac{\nu}{2}},\ 0 < \nu \le 1$, are closely connected, in the sense of nearly being a one-to-one correspondence. See the so-called *converse results*, presented in [67, 173]. In the Banach space situation, such a scale of intermediate (Hölder) source conditions is missing and we only have the benchmark source condition, corresponding to the exponent $\nu = 1$, which for the case of nonlinear problems can be written as (3.29). However, approximate source conditions, based on distance functions $d_{\xi^\dagger}(R)$, with respect to the benchmark (3.29), can again play the role of a replacement tool concerning Hölder convergence rates, as was done by Corollary 4.16 for general concave rate functions.

All the convergence rates results of this subsection, where the error is measured by the Bregman distance, were obtained with the help of variational inequalities (4.43). Whenever the benchmark source condition (3.29) is violated, we had to assume q-coercive Bregman distances in order to get a rate at all. The subsequent result avoids using (3.44) and hence cannot take a route via variational inequalities (4.43). Nevertheless, distance functions $d_{\xi^\dagger}(R)$ play a prominent role in getting the result, which was presented, together with a stringent proof, using Young's inequality and some extensions and corollaries, in the paper [97].

Theorem 4.20. *Let F be nonlinear of degree (c_1, c_2) with $0 < c_1 \le 1 - c_2$ at $x^\dagger$, with corresponding subgradient $\xi^\dagger$, where the set $\mathcal{M}$, starlike with respect to $x^\dagger$, is such that $\|F(x) - F(x^\dagger)\| \le C$ for all $x \in \mathcal{M}$ and some constant $C > 0$. We now set $\mu := \frac{c_1}{1-c_2} > 0$ and assume that the benchmark source condition (3.29) is violated, but the corresponding distance function satisfies the limit condition $d_{\xi^\dagger}(R) \to 0$ as $R \to \infty$. Then we have the convergence rate result*

$$D_{\xi^\dagger}^{\Omega}(x_{\alpha(\delta)}^{\delta}, x^\dagger) = O\left(d_{\xi^\dagger}(\Psi^{-1}(\delta))\right) \quad as \quad \delta \to 0 \tag{4.49}$$

with the auxiliary function

$$\Psi(R) := \frac{(d_{\xi^\dagger}(R))^{\frac{1}{\mu}}}{R^{\frac{1}{c_1}}}, \qquad R > 0,$$

when, for sufficiently small $\delta > 0$, the regularization parameter $\alpha = \alpha(\delta)$ satisfies the equation

$$\left(\alpha\, d_{\xi^\dagger}\left(\Phi^{-1}(\alpha)\right)\right)^{1/p} = \delta,$$

with

$$\Phi(R) := \frac{(d_{\xi^\dagger}(R))^{\frac{p-\mu}{\mu}}}{R^{\frac{p}{c_1}}}, \qquad R > 0.$$

Remark 4.21. By thorough inspection, one can verify that the rate in (4.49) is lower than the comparable rate φ in formula (3.50) of Proposition 3.38, which was essentially raised by the additional q-coercivity requirement (3.44). Following the ideas of Theorem 4.14 we easily see that (3.50) indeed represents a convergence rate for the Tikhonov-type regularization.

4.2.3 Tikhonov regularization under convex constraints

The case where $\mathcal{D}(F) := \mathcal{C}$ is a *convex* and closed subset of the Banach space X refers to a situation of great practical interest. Here, Tikhonov-type regularization searches for minimizers x_α^δ to the extremal problem

$$T_\alpha(x) := \frac{1}{p}\|F(x) - y^\delta\|^p + \alpha\,\Omega(x) \to \min, \quad \text{subject to } x \in \mathcal{C}, \tag{4.50}$$

with *convex constraints*. In the following, we briefly mention results on projected source conditions in Banach spaces, published in the joint paper [76] with J. Flemming that complemented the earlier results of [172] and [45] in Hilbert spaces.

In Banach spaces, the *normal cone* of $\mathcal{C}$ at $x^\dagger$ is defined as

$$N_{\mathcal{C}}(x^\dagger) := \{\xi \in X^* : \langle \xi, x^\dagger - x\rangle_{X^*\times X} \ge 0 \quad \text{for all } \; x \in \mathcal{C}\}. \tag{4.51}$$

We say that $x^\dagger$ satisfies a *projected source condition* with respect to the operator $F'(x^\dagger)$ if there are $\xi^\dagger \in \partial\Omega(x^\dagger)$ and $v \in Y^*$ such that

$$(F'(x^\dagger))^* v - \xi^\dagger \in N_{\mathcal{C}}(x^\dagger). \tag{4.52}$$

If $x^\dagger$ is an interior point of $\mathcal{C}$, we have $N_{\mathcal{C}}(x^\dagger) = \{0\}$ and (4.52) reduces to the ordinary source condition (3.29). Under the assumption that X is not only reflexive, but also strictly convex, the metric projection

$$P_{\mathcal{C}} : X \to \mathcal{C} \subseteq X$$

onto the convex set $\mathcal{C}$ is well-defined whenever $\mathcal{C}$ is additionally closed. Then, for given $x \in \mathcal{C}$, there is a uniquely determined element $x_{\mathcal{C}} = P_{\mathcal{C}}x \in \mathcal{C}$ such that

$$\|x_{\mathcal{C}} - x\| = \min_{\hat{x}\in\mathcal{C}} \|\hat{x} - x\|.$$

From [132, Proposition 2.2], with the bijective duality mapping $J_2^X : X \to X^*$, we further have the equivalence

$$x^\dagger = P_{\mathcal{C}}(x) \quad \Longleftrightarrow \quad J_2^X(x - x^\dagger) \in N_{\mathcal{C}}(x^\dagger),$$

for each $x \in X$. Taking into account the equation

$$(F(x^\dagger))^* v - \xi^\dagger = J_2^X\left(x^\dagger + J_2^{X^*}((F'(x^\dagger))^* v - \xi^\dagger) - x^\dagger\right),$$

where $J_2^{X^*} = (J_2^X)^{-1} : X^* \to X$ is the inverse duality mapping, this provides us with the fact that the projected source condition (4.52) can be rewritten as

$$x^\dagger = P_{\mathcal{C}}\left(x^\dagger + J_2^{X^*}((F'(x^\dagger))^* v - \xi^\dagger)\right),$$

which motivates the term 'projected' in condition (4.52).

We now prove a corollary of Theorems 4.13 and 4.14, with respect to the situation of projected source conditions.

Corollary 4.22. *Let* $\mathcal{C} = \mathcal{D}(F)$ *denote a convex and closed subset of* X *and let the* Ω*-minimizing solution* $x^\dagger$ *of* (3.3), *with* $\xi^\dagger \in \partial\Omega(x^\dagger)$, *satisfy a projected source condition* (4.52) *with respect to the operator* $F'(x^\dagger)$. *Furthermore, let* F *satisfy the nonlinearity condition*

$$\|F'(x^\dagger)(x - x^\dagger)\| \le K\,\sigma(\|F(x) - F(x^\dagger)\|) \quad \text{for all} \quad x \in \mathcal{C} \tag{4.53}$$

and some concave index function σ. *Then Theorems* 4.13 *and* 4.14 *both apply and supply the convergence rates*

$$D_{\xi^\dagger}^{\Omega}(x_\alpha^\delta, x^\dagger) = O(\sigma(\delta)) \quad \text{as} \quad \delta \to 0 \tag{4.54}$$

for the corresponding choices of the regularization parameter α.

Proof. By the definition of $N_{\mathcal{C}}(x^\dagger)$ and owing to (4.53), we have, for all $x \in \mathcal{C}$,

$$\begin{aligned}\langle \xi^\dagger, x^\dagger - x\rangle_{X^*\times X} &\le \langle (F'(x^\dagger))^* v, x^\dagger - x\rangle_{X^*\times X} \le \langle v, F'(x^\dagger)(x^\dagger - x)\rangle_{Y^*\times Y} \\ &\le \|v\|_{Y^*}\,\|F^*(x^\dagger)(x - x^\dagger)\| \le K\,\|v\|_{Y^*}\,\sigma(\|F(x) - F(x^\dagger)\|)\,.\end{aligned}$$

Hence, we find a variational inequality (4.34) with $E(x, x^\dagger) = D_{\xi^\dagger}^{\Omega}(x, x^\dagger)$, $\mathcal{M} = \mathcal{C}$, $\beta = 1$ and the concave index function $\varphi(t) = K\,\|v\|_{Y^*}\sigma(t)$. Then both theorems apply and yield (4.54). □

Remark 4.23. We can handle ordinary source conditions (3.29) along the lines of the proof of Corollary 4.22, by replacing inequalities with equations at some points, which is done in the proof of Proposition 3.32 above. Doing this, we see the same rate (4.54) appear. For linear forward operators A, Proposition 3.30 mentioned such a result, implying the best possible rate (4.46), since nonlinearity requirements are not necessary in this case. Note that, for nonlinear forward operators F, the condition (4.53) need not hold on the whole domain $\mathcal{D}(F)$. In general, it is sufficient to have such a nonlinearity condition on sublevel sets $\mathcal{M}$, covering the regularized solutions under consideration.

4.2.4 Higher rates briefly visited

As we have seen in Section 4.2.2, convergence rates for the Bregman distance, measuring the error of regularized solutions with respect to an Ω-minimizing solution $x^\dagger \in \mathcal{D}_B(\Omega)$ to (3.3), can be obtained up to the limiting rate (4.46), based on source conditions, approximate source conditions or variational inequalities and based on the ansatz

$$\mathrm{T}_\alpha(x_\alpha^\delta) \le \mathrm{T}_\alpha(x^\dagger), \quad \alpha > 0,$$

with T_α from (4.1). This is the so-called *low-rate region*. To obtain higher rates, belonging to the so-called *enhanced-rate region*, higher smoothness of an element

$\xi^\dagger \in \partial\Omega(x^\dagger)$ from the subdifferential with respect to the operator $F'(x^\dagger)$, expressed by stronger source conditions, is required. Moreover, for the proof of convergence rates results, a new ansatz

$$\mathrm{T}_\alpha(x_\alpha^\delta) \le \mathrm{T}_\alpha(x^\dagger - z) \quad \text{with appropriately chosen} \quad z \in X$$

is necessary. This idea has already been suggested by Tautenhahn in [136] and has been used in different works, see for example [67, 231] for the Hilbert space and [93, 175, 196] for the Banach space setting. In the following theorem, we sketch convergence rates results for exponents $p > 1$ in the misfit term of (4.1) that have been recently published in the joint paper [177] with A. Neubauer et al., where also the marginal case $p = 1$ was studied in detail. For the proof of Theorem 4.25, we refer to the proof of Theorem 3.3 in [177].

Before we start, we have to pose a set of additional assumptions:

Assumption 4.24.

(a) The Banach space Y is s-smooth, with $s > 1$.

(b) For the element $x^\dagger \in \mathcal{D}_B(\Omega)$, the set $\partial\Omega(x^\dagger)$ contains exactly one element $\xi^\dagger \in X^*$.

(c) There is an exponent $r > 1$ and there are constants $c_r > 0$ and $\varrho_r > 1$, such that

$$D_\xi^\Omega(x, x^\dagger) \le c_r \, \|x - x^\dagger\|^r \quad \text{for all } x \in \mathcal{D}(\Omega) \text{ with } \|x - x^\dagger\| \le \varrho_r .$$

(d) There are constants $c_F > 0$ and $\varrho_F > 0$ such that the nonlinearity condition

$$\|F(x) - F(x^\dagger) - F'(x^\dagger)(x - x^\dagger)\| \le c_F \, D_{\xi^\dagger}^\Omega(x, x^\dagger)$$

holds for all $x \in \mathcal{D}$ with $\Omega(x) \le \Omega(x^\dagger) + \varrho_F$ and $\|F(x) - F(x^\dagger)\| \le \varrho_F$.

Then we are ready to formulate the main theorem of this section.

Theorem 4.25. *Under the Assumptions* 3.11, 3.22, 3.26 *and* 4.24, *let* $\xi^\dagger \neq 0$ *denote the uniquely determined subgradient of* Ω *at the solution* $x^\dagger \in \mathcal{D}_B(\Omega)$ *of equation* (3.3), *satisfying a source condition*

$$\xi^\dagger = (F'(x^\dagger))^* J_2^Y(F'(x^\dagger)\, w), \quad w \in X, \tag{4.55}$$

accompanied by a corresponding smallness condition

$$c_F \, \|J_2^Y(F'(x^\dagger)\, w)\| < 1 , \tag{4.56}$$

such that $z_\beta := x^\dagger - \beta\, w \in \mathcal{D}$ *for sufficiently small* $\beta > 0$. *Then for the a priori parameter choice*

$$\underline{c}\, \delta^{\frac{(p-1)s}{r+s-1}} \le \alpha(\delta) \le \overline{c}\, \delta^{\frac{(p-1)s}{r+s-1}}, \quad \delta > 0 , \tag{4.57}$$

and constants $0 < \underline{c} \le \overline{c} < \infty$, *we obtain the convergence rate result*

$$D^{\Omega}_{\xi^\dagger}(x^{\delta}_{\alpha(\delta)}, x^\dagger) = O\left(\delta^{\frac{rs}{r+s-1}}\right) \quad as \quad \delta \to 0. \tag{4.58}$$

From Theorem 4.25 we find, for all $p > 1$ that, with appropriate parameter choices and in case of $r = 2$ in Assumption 4.24 (c), the Hölder rate result

$$D^{\Omega}_{\xi^\dagger}(x^{\delta}_{\alpha(\delta)}, x^\dagger) = O\left(\delta^{\frac{2s}{s+1}}\right) \quad \text{as} \quad \delta \to 0$$

can be established. According to Assumption 4.24 (a), the associated rate exponent grows, with the smoothness $s \in (1, 2]$ of the space Y, from 1 to 4/3. The maximal rate exponent 4/3 characterizes the limiting case $s = 2$ of a Hilbert space situation. As an aside, we note that the a priori choice (4.57) of the regularization parameter in the enhanced-rate region satisfies the condition (4.16), but the decay rate of that function $\alpha(\delta) \to 0$ as $\delta \to 0$ is slower than for the choice $\alpha \sim \delta^{p-1}$, yielding the best rate $O(\delta)$ in the low-rate region (cf. Proposition 4.19).

4.2.5 Rate results under conditional stability estimates

In this section, following and extending the ideas of [47] and [113, §6.2], we will combine Tikhonov-type regularization and conditional well-posedness (cf. Section 3.1.3), based on *conditional stability estimates*, which are of significant interest for parameter identification problems in partial differential equations and here attain the form

$$\|x_1 - x_2\| \le C(R)\,\varphi(\|F(x_1) - F(x_2)\|) \quad \text{for all} \quad x_1, x_2 \in \mathcal{M}_R. \tag{4.59}$$

The radius-dependent families of sets

$$\mathcal{M}_R := \{x \in \mathcal{D}(F) \cap Z : \|x\|_Z \le R\}, \tag{4.60}$$

on which the conditional stability (4.59) has been established, use a Banach space Z with norm $\|\cdot\|_Z$. This subspace Z of X is densely defined and continuously embedded in X. Moreover, let there exist constants $C(R) > 0$, for all radii $R > 0$ and an index function φ such that (4.59) is valid. We use the Tikhonov-type regularization to find stable approximate solutions to the nonlinear ill-posed operator equation (3.3).

Theorem 4.26. *Under the conditional stability estimate* (4.59)–(4.60), *with a concave index function* φ *and constants* $C(R) > 0$ *for all radii* $R > 0$, *consider Tikhonov-regularized solutions* x^δ_α *as minimizers of the extremal problem*

$$\begin{aligned} \mathrm{T}_\alpha(x) &:= \frac{1}{p}\|F(x) - y^\delta\|^p + \alpha\,\frac{1}{q}\|x\|_Z^q \to \min, \\ &\textit{subject to } x \in \mathcal{D}(F) \cap Z \subseteq X, \end{aligned} \tag{4.61}$$

with exponents $1 < p, q < \infty$, *where the regularization parameter is chosen a priori as*

$$\underline{c}\,\delta^p \le \alpha(\delta) \le \overline{c}\,\delta^p, \quad \delta > 0, \tag{4.62}$$

for constants $0 < \underline{c} \le \overline{c} < \infty$. *If there is a solution* $x^\dagger \in \mathcal{D}(F) \cap Z$ *of equation* (3.3), *then we have a convergence rate result of the form*

$$\|x^\delta_{\alpha(\delta)} - x^\dagger\| = O(\varphi(\delta)) \quad as \quad \delta \to 0. \tag{4.63}$$

Proof. We have

$$\|F(x^\delta_\alpha) - y^\delta\| \le \left(p\,\mathrm{T}_\alpha(x^\delta_\alpha)\right)^{1/p} \le \left(p\,\mathrm{T}_\alpha(x^\dagger)\right)^{1/p} \le \left(p\,\delta^p + \alpha\,\frac{p}{q}\|x^\dagger\|^q_Z\right)^{1/p}.$$

For the parameter choice (4.62), with focus on the upper bound and with the constant

$$K := \left(p + \overline{c}\,\frac{p}{q}\|x^\dagger\|^q_Z\right)^{1/p},$$

this yields the discrepancy norm estimate

$$\|F(x^\delta_{\alpha(\delta)}) - y^\delta\| \le K\,\delta.$$

Also from the extremal properties of the regularized solutions, one derives

$$\alpha\|x^\delta_\alpha\|^q_Z \le q\,\mathrm{T}_\alpha(x^\dagger) \le q\,\delta^p + \alpha\,\|x^\dagger\|^q_Z$$

and further, with the lower bound in (4.62),

$$\|x^\delta_{\alpha(\delta)}\|_Z \le \left(\frac{q\,\delta^p}{\underline{c}\,\delta^p} + \|x^\dagger\|^q_Z\right)^{1/q} = \left(\frac{q}{\underline{c}} + \|x^\dagger\|^q_Z\right)^{1/q}.$$

By setting $R := (\frac{q}{\underline{c}} + \|x^\dagger\|^q_Z)^{1/q}$, we then have $\|x^\delta_{\alpha(\delta)}\|_Z \le R$, as well as $\|x^\dagger\|_Z \le R$ and hence, with $x^\dagger, x^\delta_{\alpha(\delta)} \in \mathcal{M}_R$, the conditional stability estimate (4.59), together with (4.37), as a consequence of the concavity of φ, provide us with the chain of inequalities

$$\|x^\delta_{\alpha(\delta)} - x^\dagger\| \le C(R)\,\varphi(\|F(x^\delta_{\alpha(\delta)}) - y^\delta\|) \le C(R)\,\varphi(K\delta) \le K\,C(R)\,\varphi(\delta).$$

This implies (4.63) and completes the proof. □

Remark 4.27. As we see the convergence rate (4.63) is imposed on the problem by the character of the conditional stability estimate. In this context, Tikhonov-type regularization only acts as an auxiliary tool to ensure that all regularized solutions lie within a common stability domain $\mathcal{M}_R$. So, for the case $\varphi(t) = O(t)$ as $t \to 0$, it is also possible to obtain a convergence rate

$$\|x^\delta_{\alpha(\delta)} - x^\dagger\| = O(\delta) \quad \text{as} \quad \delta \to 0\,. \tag{4.64}$$

Note that such convergence rates (4.64), typical for well-posed problems but atypical for ill-posed problems, in exceptional cases can also be proved when regularization is applied to ill-posed operator equations under sparsity constraints, as briefly mentioned in the next subsection. From this point of view, sparsity in the sense that the solutions $x^\dagger$ belong to a finite dimensional subspace of X, is a specific form of conditional well-posedness.

4.2.6 A glimpse of rate results under sparsity constraints

Motivated by numerous applications in imaging and other fields, the amount of literature on convergence and convergence rates for Tikhonov-regularized solutions of linear and nonlinear ill-posed operator equations with forward operators A or F, mapping between Banach spaces X and Y, under *sparsity constraints*, assumed vast proportions in the past few years. Sparsity constraints hold whenever the solution $x^\dagger$ of the operator equation under consideration can be written as a linear combination of a *finite* number of functions from a basis (frame) in X, where it is unfortunately unknown which frame functions are active for the present solution.

In this subsection, we consider the Tikhonov-type regularization (4.1) of the operator equation (3.3) under sparsity constraints with in general nonlinear operators F, mapping from an infinite dimensional and separable Hilbert space X, with the orthonormal basis $\{\zeta_n\}_{n=1}^\infty$, to the reflexive Banach space Y, by using the family of convex penalty functionals

$$\Omega(x) := \sum_{n=1}^{\infty} w_n \,|\langle \zeta_n, x\rangle|^q \quad \text{with} \quad 1 \le q \le 2 \quad \text{and} \quad w_n \ge w_{\min} > 0\,. \tag{4.65}$$

Again, the symbol x^δ_α denotes a regularized solution, associated with the regularization parameter $\alpha > 0$. Existence, stability and convergence results along the lines of Section 4.1 can be found for Ω from (4.65) since, with

$$\|x\| = \left(\sum_{n=1}^{\infty} |\langle \zeta_n, x\rangle|^2\right)^{1/2} \le \left(\sum_{n=1}^{\infty} |\langle \zeta_n, x\rangle|^q\right)^{1/q} \le w_{\min}^{-1/q}\,\Omega(x),$$

for all $x \in X$, this penalty functional is stabilizing in the sense of Assumption 3.22 (c). For details we refer to [210, Section 3.3], where also the explicit form of the subdifferential $\partial\Omega(x^\dagger)$ (cf. formula (3.52) on p. 81 of [210]) and of the Bregman distance $D^\Omega_{\xi^\dagger}(x, x^\dagger)$ are given. In the case where $x^\dagger$ is sparse and $1 < q \le 2$ one can even show the 2-coercivity of the Bregman distance for Ω from (4.65).

With respect to convergence rates, for a birds-eye view of this topic, we select a rate result by Grasmair, Haltmeier and Scherzer from [81], which came as a surprise immediately after its publication. In this paper, it was shown that, for the special case $q = 1$ in the exponent of the penalty functional Ω, the super-rate (4.64) occurs, which can generally only be seen for well-posed problems. We present this result, in a slightly modified version, as Theorem 4.30, where we sketch the proof only with respect to the modification, presupposing the assertions and the corresponding proofs of Theorems 14 and 15 in [81]. For further studies on regularization under sparsity, we refer to the papers [55,82,189,191,192] with numerous references therein, as well as to the handbook [209].

For the sparsity situation, we further complement our standing assumptions on F, $\mathcal{D}(F)$ and the condition $p > 1$ for the exponent in the fidelity term by Assumption 4.24, in order to obtain the rate results of Theorem 4.30.

Assumption 4.28.

(a) We have an Ω-minimizing solution $x^\dagger$ of (3.3), which is *sparse* in the sense that there is a $k \in \mathbb{N}$ and a finite subset $(n_1, n_2, \dots, n_k)$ of indices $n_i \in \mathbb{N}$, such that $x^\dagger \in \operatorname{span}(\zeta_{n_1}, \zeta_{n_2}, \dots, \zeta_{n_k})$, i.e., $x^\dagger$ belongs to a k-dimensional subspace of X.

(b) The operator F is Gâteaux-differentiable at $x^\dagger$, and for every finite set $J \subset \mathbb{N}$ the restriction of its derivative $F'(x^\dagger)$ to the finite dimensional subspace $\{\zeta_j,\ j \in J\}$ is injective.

(c) There exists some $\xi^\dagger \in \partial\Omega(x^\dagger)$, such that the benchmark source condition (3.29) is satisfied.

(d) There exist constants $\gamma_1, \gamma_2, \geq 0$ such that

$$\begin{aligned}\|F(x)-F(x^\dagger) - F'(x^\dagger)(x - x^\dagger)\| \\ \leq \gamma_1\, D^{\Omega}_{\xi^\dagger}(x, x^\dagger) + \gamma_2\, \|F(x) - F(x^\dagger)\|\end{aligned} \tag{4.66}$$

holds for $\xi^\dagger$ from (c) and all $x \in \mathcal{M}_{\mathrm{T}_\alpha}(\varrho)$ (cf. (3.33)) for sufficiently large $\varrho > 0$.

Remark 4.29. Taking into account the Gâteaux-differentiability of F at $x^\dagger$ due to Propositions 3.30 and 3.31, the requirement of Assumption 4.28 (c), concerning the solution smoothness, is equivalent to a variational inequality (3.37), which needs to hold for all $x \in \mathcal{M}_{\mathrm{T}_\alpha}(\varrho)$ and sufficiently large $\varrho > 0$. On the other hand, the nonlinearity condition, expressed in other terms by Assumption 4.28 (d), indicates a degree of nonlinearity (c_1, c_2) of F at $x^\dagger$, with $\max(c_1, c_2) = 1$ for the Bregman distance $D_{\xi^\dagger}(\cdot, x^\dagger)$ of Ω, with $\xi^\dagger \in \partial\Omega(x^\dagger)$ (cf. Definition 3.37). This is not a too strong condition and includes the common cases (3.42) with $c_1 = 1,\ c_2 = 0$ and (3.46) with $c_1 = 0,\ c_2 = 1$.

Now, we are ready to formulate rate results with respect to the error norm in X. In our formulation, we suppress the fact that parts of Assumption 4.28 are only required for $q = 1$. For details, we refer to [81] and [210, Section 3.3].

Theorem 4.30. *If* $1 \le q \le 2$ *in the penalty functional* (4.65) *then, under Assumption* 4.28 *and for an a priori parameter choice* $\alpha(\delta) \sim \delta^{p-1}$ *with* p *from* (4.61), *we obtain the convergence rate*

$$\|x^\delta_{\alpha(\delta)} - x^\dagger\| = O\left(\delta^{1/q}\right) \quad \text{as} \quad \delta \to 0 .$$

In particular, for $q = 1$ *in* (4.65), *this even ensures the super-rate*

$$\|x^\delta_{\alpha(\delta)} - x^\dagger\| = O\,(\delta) \quad \text{as} \quad \delta \to 0 .$$

Proof. The rate results immediately follow from the proofs of Theorems 14 and 15 in [81], if we can show that the items (c) and (d) of Assumption 4.28 imply the existence of constants $\gamma_3, \gamma_4, \gamma_5, \gamma_6 > 0$ such that the inequalities

$$\begin{aligned} &\|F(x) - F(x^\dagger) - F'(x^\dagger)(x - x^\dagger)\| \\ &\qquad \le \gamma_3\,(\Omega(x) - \Omega(x^\dagger)) \,+\, \gamma_4\,\|F(x) - F(x^\dagger)\| \end{aligned} \tag{4.67}$$

and

$$\gamma_5\,\langle \xi^\dagger, x^\dagger - x\rangle_{X^* \times X} \,\le\, \Omega(x) - \Omega(x^\dagger) + \gamma_6\,\|F(x) - F(x^\dagger)\| \tag{4.68}$$

are valid for all $x \in \mathcal{M}_{\mathrm{T}_\alpha}(\varrho)$. Evidently, (4.66) implies (4.67), because

$$D^\Omega_{\xi^\dagger}(x, x^\dagger) \le \beta_3\,(\Omega(x) - \Omega(x^\dagger)) \,+\, \beta_4\,\|F(x) - F(x^\dagger)\|,$$

with some constants $\beta_3 > 0$ and $\beta_4 \ge 0$, as a reformulation of (3.37) (cf. Section 4.2.1), is a consequence of the benchmark source condition (3.29). It follows directly from the variational inequality (3.37), valid for $0 \le \beta_1 < 1$ and for all $x \in \mathcal{M}_{\mathrm{T}_\alpha}(\varrho)$, that (4.68) holds, with $\gamma_5 = 1 - \beta_1 > 0$. This completes the proof. □

Taking into account the results of the previous subsection on conditional stability, we see that the super-rate (4.64) in the case $q = 1$ is not such a sensation after all: the chance to obtain such a rate is given for problems which are originally ill-posed but possess a hidden stability potential or, in other words, they are well-posed shrouded in an ill-posed environment. As in the case of conditional stability in Section 4.2.5, the Tikhonov-type regularization is only an auxiliary tool, thriving on the hidden well-posedness of the problem, also under sparsity constraints.

Chapter 5

Tikhonov regularization of linear operators with power-type penalties

Now that we have presented the general results concerning the regularization properties for a wide range of nonlinear operators and penalty terms, we turn our focus, in this chapter, to the important special case of a bounded linear forward operator. Moreover, we restrict our considerations to convex penalty functionals

$$\Omega(x) = \frac{1}{q}\|x\|_X^q,$$

which are powers of the norm in X, with exponents $q > 1$. The aim of this chapter is to collect all information about this case in one place. We will, therefore, once more discuss source conditions and the resulting estimates for the error measure, being a Bregman distance. Moreover, we will revisit the parameter choice rules, discussed in Chapter 4, in the sense of a priori parameter choices and of the discrepancy principle. We will also introduce a modified version of the discrepancy principle and, finally, present two minimization schemes for the Tikhonov functional, together with a convergence analysis for the corresponding minimization schemes. Some results that were presented in Chapter 4 will be restated in a form adapted to the setting considered in this chapter.

Throughout this chapter the Tikhonov functional is given by

$$T_\alpha(x) := \frac{1}{p}\|Ax - y^\delta\|_Y^p + \alpha\frac{1}{q}\|x\|_X^q,$$

where $p, q > 1$ and A is a bounded (continuous) linear operator.

First, we recall that, due to the results proved in Chapter 4, the minimizers x_α^δ of the Tikhonov functional are well-defined and provide a regularization method, i.e. $x_{\alpha(\delta)}^\delta$ converges to $x^\dagger$ as $\delta \to 0$ if the regularization parameter $\alpha = \alpha(\delta)$ is chosen appropriately. However, the convergence of $x_{\alpha(\delta)}^\delta$ to $x^\dagger$ can be arbitrarily slow. Therefore, additional conditions, namely source conditions, have to be assumed, in order to obtain convergence rates with respect to some error measure $E(x_\alpha^\delta, x^\dagger)$. In the previous chapter, Bregman distances proved to be the appropriate tool for measuring the error. For the reader's convenience, we omit the indices of the dual pairing $\langle\cdot,\cdot\rangle$, since it is clear from the context whether we mean $\langle\cdot,\cdot\rangle_{X^*\times X}$ or $\langle\cdot,\cdot\rangle_{Y^*\times Y}$. Furthermore we set $j_q = j_q^X$, $J_q = J_q^X$, $j_p = j_p^Y$ and $J_p = J_p^Y$.

5.1 Source conditions

We begin our analysis with the presentation of two source conditions and the resulting error estimates, with respect to the Bregman distance. The corresponding *low-order source condition* is given here by

$$j_q(x^\dagger) = A^* w \quad \text{for some } j_q \in J_q \text{ and } w \in Y^*, \tag{5.1}$$

whereas the *high order source condition* is here defined by

$$j_q(x^\dagger) = A^* j_p(Aw) \quad \text{for some } j_q \in J_q,\, j_p \in J_p \text{ and } w \in X. \tag{5.2}$$

We mention that such source conditions have already been introduced by formulas (3.22) and (3.23).

Using these source conditions, we will provide estimates for the error measure, given by the Bregman distance $D_{j_q}(x_\alpha^\delta, x^\dagger)$. Moreover, by exploiting these estimates, we will be able to show convergence rates of the form

$$D_{j_q}(x_\alpha^\delta, x^\dagger) \le C \cdot \delta^\kappa,$$

i.e., convergence rates with respect to the Bregman distance. In particular, we will give an alternative proof that Morozov's discrepancy principle, together with the low order source condition, yields the convergence rate $D_{j_q}(x_\alpha^\delta, x^\dagger) \le C \cdot \delta$. We note that this result was already proved in Theorem 4.13 in a more general setting, since the low order source condition implies the variational inequality (4.34), with $\varphi(t) = t$, cf. [81, Proposition 11].

For the low order source condition (5.1) we obtain the following estimate:

Theorem 5.1. *Let the low order source condition* (5.1) *hold. Then we have*

$$D_{j_q}(x_\alpha^\delta, x^\dagger) \le \alpha^{-1} \cdot \tfrac{1}{p}(\delta^p - \|Ax_\alpha^\delta - y^\delta\|^p) + \|w\|(\|Ax_\alpha^\delta - y^\delta\| + \delta)$$

and

$$D_{j_q}(x_\alpha^\delta, x^\dagger) \le \tfrac{1}{p^*}\|w\|^{p^*} \cdot \alpha^{p^*-1} + \tfrac{1}{p}\alpha^{-1} \cdot \delta^p + \|w\|\delta.$$

Proof. Since x_α^δ minimizes the Tikhonov functional, we have

$$\begin{aligned}
&\tfrac{1}{p}\|Ax_\alpha^\delta - y^\delta\|^p + \alpha D_{j_q}(x_\alpha^\delta, x^\dagger) \\
&\quad = \tfrac{1}{p}\|Ax_\alpha^\delta - y^\delta\|^p + \alpha\tfrac{1}{q}\|x_\alpha^\delta\|^q - \alpha\tfrac{1}{q}\|x^\dagger\|^q - \alpha\langle j_q(x^\dagger), x_\alpha^\delta - x^\dagger\rangle \\
&\quad \le \tfrac{1}{p}\|Ax^\dagger - y^\delta\|^p + \alpha\tfrac{1}{q}\|x^\dagger\|^q - \alpha\tfrac{1}{q}\|x^\dagger\|^q - \alpha\langle j_q(x^\dagger), x_\alpha^\delta - x^\dagger\rangle \\
&\quad \le \tfrac{1}{p}\delta^p - \alpha\langle j_q(x^\dagger), x_\alpha^\delta - x^\dagger\rangle.
\end{aligned}$$

Due to the source condition, we get

$$\begin{aligned}
-\alpha\langle j_q(x^\dagger), x_\alpha^\delta - x\rangle &= \alpha\langle -w, Ax_\alpha^\delta - Ax^\dagger\rangle \le \alpha\|w\|\|Ax_\alpha^\delta - y\| \\
&\le \alpha\|w\|\|Ax_\alpha^\delta - y^\delta\| + \alpha\|w\|\delta,
\end{aligned}$$

which proves the first claim. By Young's inequality we have

$$\alpha\|w\|\|Ax_\alpha^\delta - y^\delta\| \le \tfrac{1}{p}\|Ax_\alpha^\delta - y^\delta\|^p + \tfrac{1}{p^*}\|w\|^{p^*}\alpha^{p^*}$$

and therefore

$$\begin{aligned}&\tfrac{1}{p}\|Ax_\alpha^\delta - y^\delta\|^p + \alpha D_{j_q}(x_\alpha^\delta, x^\dagger)\\ &\quad\le \tfrac{1}{p}\|Ax_\alpha^\delta - y^\delta\|^p + \tfrac{1}{p^*}\|w\|^{p^*}\cdot\alpha^{p^*} + \tfrac{1}{p}\delta^p + \alpha\|w\|\delta,\end{aligned}$$

which is equivalent to the second claim. □

The astonishing fact concerning the above theorem is that no properties of the space X are used in the proof. However, to translate the convergence with respect to the Bregman distance into norm convergence, one has to assume a convexity property on X, like uniform convexity or convexity of power type.

Next, we consider the high order source condition (5.2). For this source condition, Hein (cf. [94, Lemma 3.1]) proved an estimate of the Bregman distance.

Theorem 5.2. *Let the high order source condition* (5.2) *be valid. Then, we have with* $\gamma = \alpha^{1/(p-1)}$

$$D_{j_q}(x_\alpha^\delta, x^\dagger) \le D_{j_q}(x^\dagger - \gamma w, x^\dagger) + \alpha^{-1}\cdot D_{j_p}(A(x^\dagger - \gamma w) - y^\delta, -A(\gamma w)).$$

Proof. Since

$$\tfrac{1}{q}\|y\|^q - \tfrac{1}{q}\|x\|^q = D_{j_q}(y, x) + \langle j_q(x), y - x\rangle,$$

we have

$$\begin{aligned}\tfrac{1}{q}\|x_\alpha^\delta\|^q - \tfrac{1}{q}\|x^\dagger - \gamma w\|^q &= \tfrac{1}{q}\|x_\alpha^\delta\|^q - \tfrac{1}{q}\|x^\dagger\|^q - (\tfrac{1}{q}\|x^\dagger - \gamma w\|^q - \tfrac{1}{q}\|x^\dagger\|^q)\\ &= D_{j_q}(x_\alpha^\delta, x^\dagger) + \langle j_q(x^\dagger), x_\alpha^\delta - x^\dagger\rangle\\ &\quad - D_{j_q}(x^\dagger - \gamma w, x^\dagger) - \langle j_q(x^\dagger), (x^\dagger - \gamma w) - x^\dagger\rangle\\ &= D_{j_q}(x_\alpha^\delta, x^\dagger) - D_{j_q}(x^\dagger - \gamma w, x^\dagger)\\ &\quad + \langle j_q(x^\dagger), x_\alpha^\delta + \gamma w - x^\dagger\rangle.\end{aligned}$$

Since x_α^δ is the minimizer of the Tikhonov functional, with the above equality, we get the estimate

$$\begin{aligned}&\tfrac{1}{p}\|Ax_\alpha^\delta - y^\delta\|^p + \alpha D_{j_q}(x_\alpha^\delta, x^\dagger)\\ &\qquad = \tfrac{1}{p}\|Ax_\alpha^\delta - y^\delta\|^p + \alpha\tfrac{1}{q}\|x_\alpha^\delta\|^q - \alpha\tfrac{1}{q}\|x^\dagger - \gamma w\|^q\\ &\qquad\quad + \alpha D_{j_q}(x^\dagger - \gamma w, x^\dagger) - \alpha\langle j_q(x^\dagger), x_\alpha^\delta + \gamma w - x^\dagger\rangle\\ &\qquad \le \tfrac{1}{p}\|A(x^\dagger - \gamma w) - y^\delta\|^p\\ &\qquad\quad + \alpha D_{j_q}(x^\dagger - \gamma w, x^\dagger) - \alpha\langle j_q(x^\dagger), x_\alpha^\delta + \gamma w - x^\dagger\rangle.\end{aligned}$$

Moreover, we have

$$\begin{aligned}\tfrac{1}{p}\|A(x^\dagger - \gamma w) - y^\delta\|^p &= D_{j_p}(A(x^\dagger - \gamma w) - y^\delta, -A(\gamma w)) + \tfrac{1}{p}\| - A(\gamma w)\|^p \\ &\quad + \langle j_p(-A(\gamma w)), (A(x^\dagger - \gamma w) - y^\delta) - (-A(\gamma w))\rangle \\ &= D_{j_p}(A(x^\dagger - \gamma w) - y^\delta, -A(\gamma w)) + \tfrac{1}{p}\|A(\gamma w)\|^p \\ &\quad - \langle j_p(A(\gamma w)), Ax^\dagger - y^\delta\rangle\end{aligned}$$

and

$$\begin{aligned}&-\langle j_p(A(\gamma w)), Ax^\dagger - y^\delta\rangle + \tfrac{1}{p}\|A(\gamma w)\|^p \\ &\quad = -\langle j_p(A(\gamma w)), Ax^\dagger - Ax_\alpha^\delta + Ax_\alpha^\delta - y^\delta\rangle + \tfrac{1}{p}\|A(\gamma w)\|^p \\ &\quad = -\langle A^* j_p(A(\gamma w)), x^\dagger - x_\alpha^\delta\rangle - \langle j_p(A(\gamma w)), Ax_\alpha^\delta - y^\delta\rangle + \tfrac{1}{p}\|A(\gamma w)\|^p .\end{aligned}$$

Due to Cauchy's inequality and Young's inequality we derive

$$\begin{aligned}&-\langle -j_p(A(\gamma w)), Ax_\alpha^\delta - y^\delta\rangle + \tfrac{1}{p}\|A(\gamma w)\|^p \\ &\quad \le \tfrac{1}{p^*}\|j_p(A(\gamma w))\|^{p^*} + \tfrac{1}{p}\|A(\gamma w)\|^p + \tfrac{1}{p}\|Ax_\alpha^\delta - y^\delta\|^p .\end{aligned}$$

By the properties of the duality mapping it follows that

$$\begin{aligned}\tfrac{1}{p^*}\|j_p(A(\gamma w))\|^{p^*} + \tfrac{1}{p}\|A(\gamma w)\|^p &= \tfrac{1}{p^*}\|A(\gamma w)\|^{p^*(p-1)} + \tfrac{1}{p}\|A(\gamma w)\|^p \\ &= \|A(\gamma w)\|^p \\ &= \langle j_p(A(\gamma w), A(\gamma w)\rangle \\ &= -\langle A^* j_p(A\gamma w), -\gamma w\rangle .\end{aligned}$$

Hence, we obtain

$$\begin{aligned}&- \langle j_p(A(\gamma w)), Ax^\dagger - y^\delta\rangle + \tfrac{1}{p}\|A(\gamma w)\|^p \\ &\le -\langle A^* j_p(A(\gamma w)), -x_\alpha^\delta - \gamma w + x^\dagger\rangle + \tfrac{1}{p}\|Ax_\alpha^\delta - y^\delta\|^p .\end{aligned}$$

Altogether we then have

$$\begin{aligned}\tfrac{1}{p}\|A(x^\dagger - \gamma w) - y^\delta\|^p \le\; & D_{j_p}(-A(\gamma w), A(x^\dagger - \gamma w) - y^\delta) \\ & - \langle A^* j_p(A(\gamma w)), -x_\alpha^\delta - \gamma w + x^\dagger\rangle \\ & + \tfrac{1}{p}\|Ax_\alpha^\delta - y^\delta\|^p.\end{aligned}$$

Due to $\gamma = \alpha^{1/(p-1)}$ and with the source condition, we get

$$A^* j_p(A(\gamma w)) = \alpha A^* j_p(Aw) = \alpha j_q(x^\dagger)$$

yielding

$$\begin{aligned}
&\tfrac{1}{p}\|Ax_\alpha^\delta - y^\delta\|^p + \alpha D_{j_q}(x^\dagger, x_\alpha^\delta) \\
&\le \tfrac{1}{p}\|A(x^\dagger - \gamma w) - y^\delta\|^p + \alpha D_{j_q}(x^\dagger - \gamma w, x^\dagger) - \alpha\langle j_q(x^\dagger), x_\alpha^\delta + \gamma w - x^\dagger\rangle \\
&\le \alpha D_{j_q}(x^\dagger - \gamma w, x^\dagger) - \alpha\langle j_q(x^\dagger), x_\alpha^\delta + \gamma w - x^\dagger\rangle \\
&\quad + D_{j_p}(A(x^\dagger - \gamma w) - y^\delta, -A(\gamma w)) \\
&\quad - \alpha\langle j_q(x^\dagger), -x_\alpha^\delta - \gamma w + x^\dagger\rangle + \tfrac{1}{p}\|Ax_\alpha^\delta - y^\delta\|^p \\
&= \alpha D_{j_q}(x^\dagger - \gamma w, x^\dagger) + D_{j_p}(A(x^\dagger - \gamma w) - y^\delta, -A(\gamma w)) + \tfrac{1}{p}\|Ax_\alpha^\delta - y^\delta\|^p.
\end{aligned}$$

This proves the theorem. □

Remark 5.3. If Y is a Hilbert space then, with $p = 2$, we can estimate

$$D_{j_p}(A(x^\dagger - \gamma w) - y^\delta, -A(\gamma w)) = \tfrac{1}{2}\|y - y^\delta\|^2 \le \tfrac{1}{2}\delta^2.$$

Hence, Theorem 5.2 is a generalization of the results, presented in [196, p.1308] as a remark and a corollary.

Corollary 5.4. *Let $x^\dagger$ be a minimum norm solution of $Ax = y$ and let there exist $w \in X$, $j_q \in J_q$ and $j_p \in J_p$ satisfying*

$$j_q(x^\dagger) = A^* j_p(Aw).$$

Furthermore, let x_α^δ be a minimizer of the Tikhonov functional

$$T_\alpha(x) := \tfrac{1}{p}\|Ax - y^\delta\|^p + \alpha \cdot \tfrac{1}{q}\|x\|^q$$

with $\|y - y^\delta\| \le \delta$ and let Y be p-smooth, as well as X be q-smooth . Then, there exists a constant $C > 0$, independent of $x^\dagger$ and x_α^δ, such that

$$D_{j_q}(x_\alpha^\delta, x^\dagger) \le C(\alpha^{q/(p-1)}\|w\|^q + \alpha^{-1}\delta^p).$$

Proof. Due to Theorem 5.2 and the smoothness of power type of X and Y, we have with $\gamma = \alpha^{1/(p-1)}$

$$\begin{aligned}
D_{j_q}(x_\alpha^\delta, x^\dagger) &\le D_{j_q}(x^\dagger - \gamma w, x^\dagger) + \alpha^{-1} \cdot D_{j_p}(A(x^\dagger - \gamma w) - y^\delta, -A(\gamma w)) \\
&\le C(\|\gamma w\|^q + \alpha^{-1}\|Ax^\dagger - y^\delta\|^p).
\end{aligned}$$

□

5.2 Choice of the regularization parameter

The aim of this section is to show convergence rates of the form $D_{j_q}(x^\delta_{\alpha(\delta)}, x^\dagger) \leq C \cdot \delta^\kappa$ of the minimizers $x^\delta_{\alpha(\delta)}$ to the minimum norm solution $x^\dagger$. To achieve this, we will consider the source conditions introduced in the previous section, together with an appropriate parameter choice rule for α.

First, we consider a priori choice rules $\alpha(\delta) \sim \delta^\nu$, where $\nu > 0$. Then, we will discuss the discrepancy principle, which results in the convergence rate $D_{j_q}(x^\delta_\alpha, x^\dagger) \leq C \cdot \delta$, cf. Theorem 4.13.

To obtain better convergence rates, H. W. Engl, in [66], proposed an alternative discrepancy principle. In the last part of this section, we will show that the discrepancy principle of Engl can be generalized to the setting of Banach spaces, in the best case yielding better convergence rates than for the discrepancy principle of Morozov.

5.2.1 A priori parameter choice

For a priori parameter choice rules of the form $\alpha(\delta) \sim \delta^\nu$, with $\nu > 0$, convergence rates results with respect to the Bregman distance are straightforward consequences of Proposition 4.19, for the low order source condition and of Theorem 4.25 for the high order source condition, respectively.

In this section, we will draw the conclusions of these theorems by means of the Bregman distance estimates, proved in the previous section.

Theorem 5.5. *Let $x^\dagger$ be the minimum norm solution of $Ax = y$ and x^δ_α a minimizer of the Tikhonov functional $T_\alpha(x) = \frac{1}{p}\|Ax - y^\delta\|^p + \alpha \cdot \frac{1}{q}\|x\|^q$ with $\|y - y^\delta\| \leq \delta$ and $p > 1$.*

(a) *If $j_q(x^\dagger) = A^* w$ for some $j_q \in J_q$ and α is given by*

$$\alpha \sim \delta^{p-1},$$

then

$$D_{j_q}(x^\delta_{\alpha(\delta)}, x^\dagger) \leq C \cdot \delta\,.$$

(b) *If $j_q(x^\dagger) = A^* j_p(Aw)$ for some $j_q \in J_q$, $j_p \in J_p$ and if the space X is smooth of power type q, as well as Y smooth of power type p, then for α, given by*

$$\alpha \sim \delta^{p(p-1)/(p+q-1)}$$

we have

$$D_{j_q}(x^\delta_{\alpha(\delta)}, x^\dagger) \leq C \cdot \delta^{pq/(p+q-1)}.$$

In both cases, the constant C neither depends on x^δ_α nor on α.

Proof. First, consider the low order source condition $j_q(x^\dagger) = A^*w$. By Theorem 5.1, we have

$$D_{j_q}(x_\alpha^\delta, x^\dagger) \le \tfrac{1}{p^*}\|w\|^{p^*} \cdot \alpha^{p^*-1} + \tfrac{1}{p}\alpha^{-1} \cdot \delta^p + \|w\|\delta.$$

Then, the optimal choice of $\alpha(\delta)$ is given by $\alpha \sim \delta^{p-1}$, resulting in the convergence rate $D_{j_q}(x_{\alpha(\delta)}^\delta, x^\dagger) \le C \cdot \delta$. Alternatively, the claim can be regarded as a direct consequence of Proposition 4.19 for $\nu = 1$.

Next, we consider the high order source condition $j_q(x^\dagger) = A^* j_p(Aw)$. Since the space X is smooth of power type q and Y is smooth of power type p, by Corollary 5.4, we get that

$$D_{j_q}(x_\alpha^\delta, x^\dagger) \le C(\alpha^{q/(p-1)}\|w\|^q + \alpha^{-1}\delta^p).$$

Therefore, the optimal rate $\delta^{pq/(p+q-1)}$ is achieved for $\alpha \sim \delta^{p(p-1)/(p+q-1)}$. □

Alternatively, assertion (b) of Theorem 5.5 can be seen as a consequence of Theorem 4.25.

Remark 5.6. If, in addition to the assumptions of the latter theorem, the space X is convex of power type, say κ-convex, then both convergence rates can be expressed in terms of the norm of X.

First, we see that

$$\alpha\tfrac{1}{q}\|x_\alpha^\delta\|^q \le T_\alpha(x_\alpha^\delta) \le T_\alpha(x^\dagger) \le \tfrac{1}{p}\delta^p + \alpha \cdot \tfrac{1}{p}\|x^\dagger\|^p.$$

Hence, for all sufficiently small α and δ, the minimizers x_α^δ are uniformly bounded.

(a) First, assume that the low order source condition $j_q(x^\dagger) = A^*w$ holds. If $\kappa \le q$, then, by Corollary 2.61,

$$C\|x^\dagger - x_{\alpha(\delta)}^\delta\|^q \le D_{j_q}(x_{\alpha(\delta)}^\delta, x^\dagger) \le C \cdot \delta$$

and consequently

$$\|x^\dagger - x_{\alpha(\delta)}^\delta\| \le C \cdot \delta^{1/q}\ .$$

Notice that this is the rate mentioned in Theorem 4.30. If, on the other hand, $q \le \kappa$ then, again by Corollary 2.61,

$$C(\|x^\dagger\| + \|x_{\alpha(\delta)}^\delta\|)^{q-\kappa}\|x^\dagger - x_{\alpha(\delta)}^\delta\|^\kappa \le D_{j_q}(x_{\alpha(\delta)}^\delta, x^\dagger) \le C \cdot \delta$$

and, by the boundedness of $x_{\alpha(\delta)}^\delta$, we have

$$\|x^\dagger - x_{\alpha(\delta)}^\delta\| \le C \cdot \delta^{1/\kappa}$$

for all sufficiently small δ.

(b) Assume, on the other hand, that the high order source condition $j_q(x^\dagger) = A^* j_p(Aw)$ holds, where X is q-smooth. Then, by Theorems 2.50 and 2.51, we know that $q \le \kappa$. Hence, once again by Corollary 2.61, we get

$$\|x^\dagger - x^\delta_{\alpha(\delta)}\| \le C \cdot \delta^{pq/[(p+q-1)\kappa]}$$

for sufficiently small δ.

5.2.2 Morozov's discrepancy principle

As discussed in Chapter 4, the discrepancy principle from formula (4.19) is a powerful way of choosing the regularization parameter. In this section, we want to show that the rate obtained in Theorem 4.13 in a more general setting, can also be found by exploiting the distance estimates of Section 5.1.

Without loss of generality, we will consider the discrepancy principle with $\tau_1 = \tau_2 =: \tau$. Hence, the parameter $\alpha = \alpha(\delta, y^\delta)$ is chosen via

$$\|Ax^\delta_{\alpha(\delta,y^\delta)} - y^\delta\| = \tau\delta,$$

where the parameter τ has to be chosen so as to ensure that such $\alpha(\delta, y^\delta)$ exists. For the remaining part of the section, we will assume that this is the case. However, we note that, for linear A, the proof of existence can be carried out along the lines of the proofs in [121] or [21]. We also refer to the first part of the proof of Theorem 7.13 in Chapter 7 below and of Theorem 3 in [129].

Under these assumptions, the following convergence rate can be shown:

Theorem 5.7. *Let $x^\dagger$ be a minimum norm solution of $Ax = y$, such that $j_q(x^\dagger) = A^* w$ for some $j_q \in J_q$ and $w \in Y^*$. Let x^δ_α be a minimizer of the Tikhonov functional $T_\alpha(x) := \frac{1}{p}\|Ax - y^\delta\|^p + \alpha \cdot \frac{1}{q}\|x\|^q$, where $p > 1$. Moreover, let $\tau > 1$ be chosen such that, for y^δ, with $\|y - y^\delta\| \le \delta$, there exists an $\alpha(\delta, y^\delta)$ such that*

$$\|Ax^\delta_{\alpha(\delta,y^\delta)} - y^\delta\| = \tau\delta.$$

Then,

$$D_{j_q}(x^\delta_{\alpha(\delta,y^\delta)}, x^\dagger) \le (1+\tau)\|w\|\,\delta\,.$$

Proof. By Theorem 5.1 we have

$$D_{j_q}(x^\delta_\alpha, x^\dagger) \le \alpha^{-1} \cdot \frac{1}{p}(\delta^p - \|Ax^\delta_\alpha - y^\delta\|^p) + \|w\|(\|Ax^\delta_\alpha - y^\delta\| + \delta)\,.$$

Since $\|Ax^\delta_{\alpha(\delta,y^\delta)} - y^\delta\| = \tau\delta$, with $\tau > 1$, we have

$$\delta^p - \|Ax^\delta_{\alpha(\alpha,y^\delta)} - y^\delta\|^p = \delta^p - (\tau\delta)^p \le 0.$$

This proves the claim, since $\|Ax^\delta_{\alpha(\delta,y^\delta)} - y^\delta\| + \delta = \tau\delta + \delta$. □

5.2.3 Modified discrepancy principle

It is well known that, at least if X and Y are Hilbert spaces and for $q = 2$, the convergence rate obtained in the previous section is the best one possible for Morozov's discrepancy principle (cf. [67, Section 4.3]). This fact is known as *early saturation* of Morozov's discrepancy principle and is its fundamental drawback. Therefore, we will consider a modified version of the discrepancy principle, which does not exhibit this phenomenon.

In [66], it was shown that the choice of $\alpha(\delta, y^\delta)$ from

$$\|A^* j_p^Y(Ax^\delta_{\alpha(\delta,y^\delta)} - y^\delta)\|_{X^*}^{q^*} = \delta^r \alpha^{-s}, \quad r, s > 0, \tag{5.3}$$

leads to an optimal convergence rate in the Hilbert space setting, if the parameters r and s are chosen appropriately. Hence, this modified version of the discrepancy principle does not suffer from early saturation.

Since [66] was published, the modified discrepancy principle was extended to nonlinear operator equations in Hilbert spaces in [124]. In this section, we will show that this modified approach can also be extended to linear operators mapping between Banach spaces.

We will present the claims as a series of theorems. First, in Theorem 5.9, we indicate that, in Banach spaces, the modified discrepancy principle is well-defined. In Theorem 5.11, we will emphasize that it is a regularization and, finally, in Theorem 5.12, we will show that the same convergence rates are achieved as for a priori parameter choice, if one of the source conditions

$$j_q(x^\dagger) = A^* w \quad w \in Y^*$$

or

$$j_q(x^\dagger) = A^* j_p(Aw) \quad w \in X$$

holds.

Remark 5.8. We remark that all results of this section are also true if $\alpha(\delta, y^\delta)$ is chosen in a more general way. Namely, if $\alpha(\delta, y^\delta)$ is selected such that

$$\|A^* j_p^Y(Ax^\delta_{\alpha(\delta,y^\delta)} - y^\delta)\|_{X^*}^{q^*} \sim \delta^r \alpha^{-s},$$

i.e.

$$c \cdot \delta^r \alpha^{-s} \le \|A^* j_p^Y(Ax^\delta_{\alpha(\delta,y^\delta)} - y^\delta)\|_{X^*}^{q^*} \le C \cdot \delta^r \alpha^{-s},$$

where the constants $0 < c \le C$ are independent of δ and y^δ.

However, for the sake of clarity, we will prove the claims for c and C being both set to one.

We shall show that the for $\alpha > 0$, (5.3) is well-defined.

Theorem 5.9. *Let X be convex of power type and Y be smooth of power type. Then, for every $r, s > 0$, $p, q > 1$, $\delta > 0$, $y^\delta \in Y$, with $A^* j_p(y^\delta) \neq 0$, there exists a regularization parameter $\alpha > 0$ such that*

$$\|A^* j_p^Y(Ax_\alpha^\delta - y^\delta)\|_{X^*}^{q^*} = \delta^r \alpha^{-s}$$

is satisfied and hence the parameter choice (5.3) *is well-defined.*

Proof. We will prove the claim in three steps:

(a) First, we show that $\alpha^s \|A^* j_p(Ax_\alpha^\delta - y^\delta)\|^{q^*} \to 0$ as $\alpha \to 0$.

(b) Then, we will show that $\alpha^s \|A^* j_p(Ax_\alpha^\delta - y^\delta)\|^{q^*} \to \infty$ as $\alpha \to \infty$.

(c) Finally, we will show that the function $\alpha \mapsto \alpha^s \|A^* j_p(Ax_\alpha^\delta - y^\delta)\|^{q^*}$ is continuous.

First, we notice that

$$A^* J_p(Ax_\alpha^\delta - y^\delta) = \partial(\tfrac{1}{p}\|A \cdot - y^\delta\|^p)(x_\alpha^\delta)$$

and, due to smoothness of power type of Y, the duality mapping J_p is single valued.

The optimality condition for T_α is given by $0 \in A^* J_p(Ax_\alpha^\delta - y^\delta) + \alpha\, J_q(x_\alpha^\delta)$. Hence, there exists a $j_q \in J_q$, such that

$$A^* j_p(Ax_\alpha^\delta - y^\delta) + \alpha \cdot j_q(x_\alpha^\delta) = 0$$

and we have

$$\|A^* j_p(Ax_\alpha^\delta - y^\delta)\|^{q^*} = \alpha^{q^*} \|x_\alpha^\delta\|^q \leq \alpha^{q^*-1} q T_\alpha(x_\alpha^\delta) \leq \alpha^{q^*-1} q T_\alpha(x^\dagger).$$

For bounded α, the term $\|A^* j_p(Ax_\alpha^\delta - y^\delta)\|^{q^*}$ is bounded and the limit condition $\alpha^s \|A^* j_p(Ax_\alpha^\delta - y^\delta)\|^{q^*} \longrightarrow 0$ as $\alpha \longrightarrow 0$ holds.

Moreover, we have

$$\begin{aligned} \|x_\alpha^\delta\|^{q-1} &\leq \alpha^{-1} \|A^*\| \|Ax_\alpha^\delta - y^\delta\|^{p-1} \\ &\leq C \cdot \alpha^{-1} T_\alpha(x_\alpha^\delta)^{\frac{p-1}{p}} \\ &\leq C \cdot \alpha^{-1} T_\alpha(x^\dagger)^{\frac{p-1}{p}} \\ &\leq C \cdot \alpha^{-1} (C + \alpha \cdot C)^{\frac{p-1}{p}} \\ &\leq C \cdot (\alpha^{-1} + \alpha^{-1/p}) \end{aligned}$$

This implies that $x_\alpha^\delta \longrightarrow 0$ as $\alpha \longrightarrow \infty$.

The space Y is assumed to be smooth of power type, say ϱ-smooth. Then, by the Xu–Roach inequalities (cf. Theorem 2.42), we obtain

$$\begin{aligned} \|A^* j_p(Ax_\alpha^\delta - y^\delta) - A^* j_p(y^\delta)\| &\leq C \cdot \|j_p(Ax_\alpha^\delta - y^\delta) - j_p(y^\delta)\| \\ &\leq C \cdot (\max\{\|Ax_\alpha^\delta - y^\delta\|, \|y^\delta\|\})^{p-\varrho} \|x_\alpha^\delta\|^{\varrho-1}. \end{aligned}$$

If $p \geq \varrho$ then, by $x_\alpha^\delta \longrightarrow 0$ as $\alpha \longrightarrow \infty$, there exists a constant $C > 0$ such that, for all sufficiently large $\alpha > 0$

$$(\max\{\|Ax_\alpha^\delta - y^\delta\|, \|y^\delta\|\})^{p-\varrho} \leq (\|A\|\|x_\alpha^\delta\| + \|y^\delta\|)^{p-\varrho} \leq C.$$

If, on the other hand, $p < \varrho$, then we know, as a consequence of $A^* j_p(y^\delta) \neq 0$, that $y^\delta \neq 0$. Then there exists a constant $C > 0$ such that

$$(\max\{\|Ax_\alpha^\delta - y^\delta\|, \|y^\delta\|\})^{p-\varrho} \leq 1/\|y^\delta\|^{\varrho-p} \leq C.$$

Hence, for $\alpha \to \infty$, the convergence

$$\|A^* j_p(Ax_\alpha^\delta - y^\delta)\|^{q^*} \longrightarrow \|A^* j_p(y^\delta)\|^{q^*} > 0$$

and consequently $\alpha^s\|A^* j_p(Ax_\alpha^\delta - y^\delta)\|^{q^*} \longrightarrow \infty$ as $\alpha \longrightarrow \infty$ hold.

Finally, we show that the mapping $\alpha \mapsto \alpha^s\|A^* j_p(Ax_\alpha^\delta - y^\delta)\|^{q^*}$ is continuous. Because of $\|A^* j_p(Ax_\alpha^\delta - y^\delta)\| = \alpha\|x_\alpha^\delta\|^{q-1}$, this is true if $\alpha \mapsto \|x_\alpha^\delta\|$ is continuous.

Now let $\alpha > 0$ be fixed. Since X is convex of power type and since J_p is the subgradient of $\frac{1}{p}\|\cdot\|^p$, we have, for every $\beta > 0$,

$$\frac{1}{p}\|Ax_\alpha^\delta - y^\delta\|^p \geq \frac{1}{p}\|Ax_\beta^\delta - y^\delta\|^p + \langle j_p(Ax_\beta^\delta - y^\delta), Ax_\alpha^\delta - Ax_\beta^\delta\rangle$$
$$\alpha\frac{1}{q}\|x_\alpha^\delta\|^q \geq \alpha\frac{1}{q}\|x_\beta^\delta\|^q + \alpha\langle j_q(x_\beta^\delta), x_\alpha^\delta - x_\beta^\delta\rangle + \alpha\sigma_q(x_\beta^\delta, x_\beta^\delta - x_\alpha^\delta),$$

where the function $\sigma_q > 0$ is defined as in Theorem 2.40. Summing up the inequalities and taking into account the optimality condition, we get

$$T_\alpha(x_\alpha^\delta) \geq T_\alpha(x_\beta^\delta) + (\alpha - \beta)\langle j_q(x_\beta^\delta), x_\alpha^\delta - x_\beta^\delta\rangle + \alpha\sigma_q(x_\beta^\delta, x_\beta^\delta - x_\alpha^\delta).$$

Since x_α^δ minimizes T_α, we have $0 \geq T_\alpha(x_\alpha^\delta) - T_\alpha(x_\beta^\delta)$. Hence,

$$0 \leq \alpha\sigma_p(x_\beta^\delta, x_\beta^\delta - x_\alpha^\delta) \leq (\alpha - \beta)\langle j_q(x_\beta^\delta), x_\alpha^\delta - x_\beta^\delta\rangle \leq (\alpha - \beta)\|x_\beta^\delta\|^{q-1}\|x_\alpha^\delta - x_\beta^\delta\|.$$

Moreover, for all sufficiently small $\delta > 0$, and all β sufficiently close to α, we have$\|x_\beta^\delta\| \leq C$ (cf. Remark 5.6). Therefore,

$$\alpha\sigma_q(x_\beta^\delta, x_\beta^\delta - x_\alpha^\delta) \leq |\beta - \alpha| \cdot \|x_\beta^\delta\|^{q-1}\|x_\alpha^\delta - x_\beta^\delta\| \leq C|\beta - \alpha| \cdot \|x_\alpha^\delta - x_\beta^\delta\|.$$

If $q - \kappa \leq 0$, we have, by Corollary 2.61 of the Xu–Roach inequalities for convexity of power type and by the uniform boundedness of the minimizers of the Tikhonov function, for all sufficiently small $\alpha, \beta, \delta > 0$,

$$\sigma_q(x_\beta^\delta, x_\beta^\delta - x_\alpha^\delta) \geq C(\|x_\alpha^\delta\| + \|x_\beta^\delta\|)^{p-\kappa}\|x_\alpha^\delta - x_\beta^\delta\|^\kappa \geq C\|x_\alpha^\delta - x_\beta^\delta\|^\kappa$$

and we get

$$\alpha \cdot C\|x_\alpha^\delta - x_\beta^\delta\|^{\kappa-1} \leq |\beta - \alpha|.$$

Thus, $\|x_\beta^\delta\| \longrightarrow \|x_\alpha^\delta\|$ as $\beta \longrightarrow \alpha > 0$.

If, on the other hand, $q - \kappa > 0$, then, again by Corollary 2.61,

$$\sigma_q(x^\delta_\beta, x^\delta_\beta - x^\delta_\alpha) \geq C\|x^\delta_\alpha - x^\delta_\beta\|^q$$

and thus

$$\alpha \cdot C\|x^\delta_\alpha - x^\delta_\beta\|^{q-1} \leq |\beta - \alpha|.$$

Therefore, again $\|x^\delta_\beta\| \longrightarrow \|x^\delta_\alpha\|$ as $\beta \longrightarrow \alpha > 0$ is valid. This completes the proof. □

Remark 5.10. If Y is smooth of power type and $A^* J_p(y) \neq 0$, then for all sufficiently small $\delta > 0$, we also have $A^* J_p(y^\delta) \neq 0$, since J_p is Hölder continuous on bounded subsets of spaces smooth of power type (cf. Corollary 2.44). Hence, if $A^* J_p(y) \neq 0$, the assumption of the latter theorem is satisfied for sufficiently small $\delta > 0$. In the next theorem, we will see that this condition is also sufficient to show the regularization property.

Theorem 5.11. *Let X be convex of power type, Y be smooth of power type, $x^\dagger$ be the minimum norm solution of $Ax = y$ with $y \in \mathcal{R}(A)$, where $A^* j_p(y) \neq 0$. Moreover, let $r, s > 0$, $p, q > 1$ be such that*

$$(s + q^* - 1)p - r \geq 0.$$

Then

$$x^\delta_{\alpha(\delta, y^\delta)} \longrightarrow x^\dagger \quad as \quad \delta \longrightarrow 0.$$

i.e., the parameter choice under consideration leads to a regularization method.

Proof. Since Y is smooth of power type, the duality mapping J_p is single valued and continuous on bounded sets (cf. Corollary 2.44). Therefore, by Theorem 5.9, the parameter choice (5.3) is well-defined for all sufficiently small $\delta > 0$.

We will prove the claim in two steps:

(a) First, we show that $\alpha(\delta, y^\delta) \to 0$ as $\delta \to 0$.

(b) Next, we show that $\delta^p / \alpha(\delta, y^\delta) \to 0$ as $\delta \to 0$.

Hence, the claim is a consequence of the regularization property of the minimizers of the Tikhonov functional.

Due to the optimality condition, we have

$$\alpha(\delta, y^\delta)^{q^*} \|x^\delta_{\alpha(\delta, y^\delta)}\|^q = \|A^* j_p(Ax^\delta_{\alpha(\delta, y^\delta)} - y^\delta)\|^{q^*} = \delta^r \alpha(\delta, y^\delta)^{-s}. \tag{5.4}$$

Assume there is a sequence $\delta_n \to 0$ as $n \to 0$ such that, for $\alpha_n := \alpha(\delta_n, y^{\delta_n})$, we have $\limsup_{n\to\infty} \alpha_n \geq c > 0$. Without loss of generality, we can assume that $\lim_{n\to\infty} \alpha_n = c$ and that $\alpha_n > c/2$ for all $n \in \mathbb{N}$. Due to (5.4) we have

$$\|x^{\delta_n}_{\alpha_n}\|^q = \delta_n^r \alpha_n^{-(s+q^*)} \leq \delta_n^r (c/2)^{-(s+q^*)} \longrightarrow 0.$$

Since j_p is continuous on bounded sets, we get

$$\begin{aligned} 0 = \lim_{n\to\infty} \delta_n^r (c/2)^{-s} &\geq \lim_{n\to\infty} \delta_n^r \alpha_n^{-s} \\ = \lim_{n\to\infty} \|A^* j_p(Ax_{\alpha_n}^{\delta_n} - y^{\delta_n})\|^{q^*} &= \|A^* j_p(y)\|^{q^*} > 0. \end{aligned}$$

Hence, we get $\alpha(\delta, y^\delta) \longrightarrow 0$ as $\delta \longrightarrow 0$.

Furthermore, we have

$$\begin{aligned} \delta^r &= \alpha(\delta, y^\delta)^{s+q^*} \|x_{\alpha(\delta,y^\delta)}^\delta\|^q \leq \alpha(\delta, y^\delta)^{s+q^*-1} q T_{\alpha(\delta,y^\delta)}(x_{\alpha(\delta,y^\delta)}^\delta) \\ &\leq \alpha(\delta, y^\delta)^{s+q^*-1} q T_{\alpha(\delta,y^\delta)}(x^\dagger) \end{aligned}$$

and

$$T_{\alpha(\delta,y^\delta)}(x^\dagger) \leq \tfrac{1}{p}\delta^p + \alpha(\delta, y^\delta) \cdot \tfrac{1}{q}\|x^\dagger\|^q \longrightarrow 0 \quad \text{as} \quad \delta \longrightarrow 0.$$

Consequently,

$$(\delta^p/\alpha(\delta, y^\delta))^r \leq (q T_{\alpha(\delta,y^\delta)}(x^\dagger))^p \cdot \alpha(\delta, y^\delta)^{(s+q^*-1)p-r} \longrightarrow 0 \quad \text{as} \quad \delta \to 0,$$

since we assume $(s + q^* - 1)p - r \geq 0$. This implies $\delta^p/\alpha(\delta, y^\delta) \longrightarrow 0$ for $\delta \longrightarrow 0$. □

Theorem 5.12. *Let the parameter choice (5.3) be well-defined and regularizing, i.e., $\alpha(\delta, y^\delta)$ exists and we have $x_{\alpha(\delta,y^\delta)}^\delta \to x^\dagger$ for $\delta \to 0$. Then, the following assertions are true:*

(a) *If $j_q(x^\dagger) = A^* w$ for some $j_q \in J_q$, $w \in Y^*$, and $r, s > 0$ are chosen such that*

$$\frac{r}{s + q^*} = p - 1\,,$$

we have, for sufficiently small $\delta > 0$,

$$D_{j_q}(x_{\alpha(\delta,y^\delta)}^\delta, x^\dagger) \leq C \cdot \delta.$$

(b) *If Y is p-smooth, X is q-smooth, where p, q are the powers in*

$$T_\alpha(x) = \tfrac{1}{p}\|Ax - y^\delta\|^p + \alpha\tfrac{1}{q}\|x\|^q,$$

and, moreover, the source condition $j_q(x^\dagger) = A^* j_p(Aw)$ holds for some $j_q \in J_q$, $j_p \in J_p$, $w \in X$, and $r, s > 0$ are chosen such that

$$\frac{r}{s + q^*} = \frac{p}{p + q - 1} \cdot (p - 1),$$

then we have for sufficiently small $\delta > 0$

$$D_{j_q}(x_{\alpha(\delta,y^\delta)}^\delta, x^\dagger) \leq C \cdot \delta^{\frac{pq}{p+q-1}}.$$

Proof. First, we consider the case $x^\dagger \neq 0$. We note that the optimality condition of T_α leads to

$$\begin{aligned}\delta^r \alpha(\delta, y^\delta)^{-(s+q^*)} &= \alpha(\delta, y^\delta)^{-q^*} \|A^* j_p(Ax^\delta_{\alpha(\delta,y^\delta)} - y^\delta)\|^{q^*} \\ &= \alpha(\delta, y^\delta)^{-q^*} \alpha(\delta, y^\delta)^{q^*} \|x^\delta_{\alpha(\delta,y^\delta)}\|^q = \|x^\delta_{\alpha(\delta,y^\delta)}\|^q .\end{aligned}$$

Therefore, due to the regularization property, we obtain

$$\lim_{\delta\to 0} \delta^r \cdot \alpha(\delta, y^\delta)^{-(s+q^*)} = \lim_{\delta\to 0} \|x^\delta_{\alpha(\delta,y^\delta)}\|^q = \|x^\dagger\|^q > 0 .$$

Hence, for all sufficiently small $\delta > 0$, we have

$$0 < c \leq \delta^r \cdot \alpha(\delta, y^\delta)^{-(s+q^*)} \leq C < \infty. \tag{5.5}$$

We now show the first claim. By Theorem 5.1 and (5.5), we deduce

$$\begin{aligned}D_{j_q}(x^\delta_{\alpha(\delta,y^\delta)}, x^\dagger) &\leq C\left(\delta^p \alpha(\delta, y^\delta)^{-1} + \alpha(\delta, y^\delta)^{\frac{1}{p-1}} + \delta\right) \\ &\leq C\left(\delta^{p-\frac{r}{s+q^*}} + \delta^{\frac{r}{s+q^*}\cdot\frac{1}{p-1}} + \delta\right).\end{aligned}$$

Since $\frac{r}{s+q^*} = p - 1$ this shows (a).

Next, we show (b). Using Corollary 5.4 of Theorem 5.2 and (5.5) yields

$$\begin{aligned}D_{j_q}(x^\delta_{\alpha(\delta,y^\delta)}, x^\dagger) &\leq C\left(\alpha(\delta, y^\delta)^{\frac{q}{p-1}} + \alpha(\delta, y^\delta)^{-1} \cdot \delta^p\right) \\ &\leq C\left(\delta^{\frac{r}{s+q^*}\cdot\frac{q}{p-1}} + \delta^{p-\frac{r}{s+q^*}}\right).\end{aligned}$$

The claim follows, since here $\frac{r}{s+q^*} = \frac{p}{p+q-1} \cdot (p-1)$.

Finally, we consider the case $x^\dagger = 0$. We have then

$$\lim_{\delta\to 0} \delta^r \cdot \alpha(\delta, y^\delta)^{-(s+q^*)} = \lim_{\delta\to 0} \|x^\delta_{\alpha(\delta,y^\delta)}\|^{q^*} = \|x^\dagger\|^{q^*} = 0$$

and hence, for all sufficiently small $\delta > 0$,

$$0 \leq \delta^r \cdot \alpha(\delta, y^\delta)^{-(s+q^*)} \leq C < \infty.$$

As in the proof of Theorem 5.1, we may conclude that

$$\alpha D_{j_q}(x^\delta_\alpha, x^\dagger) \leq \tfrac{1}{p}\delta^p - \langle j_q(x^\dagger), x^\delta_\alpha - x^\dagger\rangle .$$

Since $x^\dagger = 0$ we get $j_q(x^\dagger) = 0$ and, therefore,

$$D_{j_q}(x^\delta_{\alpha(\delta,y^\delta)}, x^\dagger) \leq \tfrac{1}{p}\delta^p \cdot \alpha(\delta, y^\delta)^{-1} \leq C\delta^{p-r/(s+q^*)}.$$

Hence, the (a) and (b) hold in this case too. □

5.3 Minimization of the Tikhonov functionals

In this section we present two methods for iteratively minimizing Tikhonov functionals. More precisely, we solve

$$T_\alpha(x) \to \min, \quad \text{subject to } x \in X, \tag{5.6}$$

where the Tikhonov functional $T_\alpha : X \to \mathbb{R}$ is given by

$$T_\alpha(x) = \frac{1}{p}\|Ax - y\|_Y^p + \alpha\frac{1}{q}\|x\|_X^q,$$

with a bounded (continuous) linear operator $A : X \to Y$, mapping between Banach spaces X and Y. The functional T_α is strictly convex on X for all regularization parameters $\alpha > 0$ and hence we have a unique minimizer $x_\alpha^\delta \in X$, such that

$$T_\alpha(x_\alpha^\delta) = \min_{x \in X} T_\alpha(x).$$

Whenever X and Y are Hilbert spaces, there exist comprehensive results concerning solution methods for (5.6), as well as results on convergence and stability. On the other hand, a wide range of parameter choice rules for α can be found in the literature. In the case where only Y is a Hilbert space, the minimization of the Tikhonov functional has been thoroughly studied and many solvers have been established, see [55, 145]. One way to get an approximate solution for (5.6) in Banach spaces is to use steepest descent methods. The two methods that we describe and analyze in this section are essentially of this type. For smooth T_α, we consider the iteration process

$$x_{n+1} = x_n - \mu_n J_{q^*}^{X^*}\left(\nabla T_\alpha(x_n)\right), \quad n = 0, 1, \ldots, \tag{5.7}$$

which we call the *primal method*, since the new iterate is computed in X directly. In contrast to (5.7) we also investigate the iteration scheme

$$\begin{aligned} x_{n+1}^* &= x_n^* - \mu_n \psi_n \quad \text{with} \quad \psi_n \in \partial T_\alpha(x_n), \\ x_{n+1} &= J_{q^*}^{X^*}(x_{n+1}^*), \end{aligned} \tag{5.8}$$

which first computes an iterate in the dual space X^*. This new iterate is then pulled back to X by the dual mapping $J_{q^*}^{X^*}$. This is why we call (5.8) the *dual method*. Note that we do not postulate T_α to be smooth in order to perform this method. We will prove strong convergence to the unique minimizer x_α^δ for both methods.

Since we consider smooth Banach spaces X, we write D_q for the Bregman distance $D_{J_q}^X$ on X throughout this chapter. We write J_p, J_q for J_p^Y, J_q^X and $J_{p^*}^*$, $J_{q^*}^*$ for $J_{p^*}^{Y^*}$, $J_{q^*}^{X^*}$, respectively. We omit the subscripts of the norms $\|\cdot\|$ when it is clear from the context. Furthermore, by $C > 0$ we always denote a generic positive constant.

In [3], Alber, Iusem and Solodov presented an algorithm for the minimization of convex and not necessarily smooth functionals on uniformly smooth and uniformly convex Banach spaces, which looks very similar to the dual method, where the authors impose summation conditions on the step sizes μ_n. However, only weak convergence of the proposed scheme is shown there. Another interesting approach for obtaining convergence results of descent methods in general Banach spaces can be found in [193, 194].

5.3.1 Primal method

Let X be uniformly convex and uniformly smooth and Y be uniformly smooth. Then, the Tikhonov functional

$$T_\alpha(x) := \frac{1}{p}\|Ax - y\|_Y^p + \frac{\alpha}{q}\|x\|_X^q \tag{5.9}$$

is strictly convex, weakly lower semi-continuous, coercive and Gâteaux differentiable, with derivative

$$\nabla T_\alpha(x) = A^* J_p(Ax - y) + \alpha J_q(x)\,. \tag{5.10}$$

Hence, the unique minimizer x_α^δ of T_α is characterized by

$$T_\alpha(x_\alpha^\delta) = \min_{x \in X} T_\alpha(x) \quad \Longleftrightarrow \quad \nabla T_\alpha(x_\alpha^\delta) = 0\,. \tag{5.11}$$

In [36, 182], it has already been proved that, for a general continuously differentiable functional T_α, every cluster point of a steepest descent method like (5.7), is a stationary point. Recently, Canuto and Urban [39] have shown strong convergence under the additional assumption of ellipticity, which our T_α in (5.9) would satisfy, if we required X to be q-convex. Here, we prove strong convergence for uniformly convex and uniformly smooth X and uniformly smooth Y.

Algorithm 5.13 (Primal method).

(1) Choose an arbitrary starting point $x_0 \in X$ and set $n = 0$.

(2) If $\nabla T_\alpha(x_n) = 0$ then STOP else do a line search to find $\mu_n > 0$ such that

$$T_\alpha\Big(x_n - \mu_n J_{p^*}^*\big(\nabla T_\alpha(x_n)\big)\Big) = \min_{\mu \in \mathbb{R}} T_\alpha\Big(x_n - \mu J_{p^*}^*\big(\nabla T_\alpha(x_n)\big)\Big)\,.$$

(3) Set

$$x_{n+1} := x_n - \mu_n J_{p^*}^*\big(\nabla T_\alpha(x_n)\big)\,,$$

$n \leftarrow (n+1)$ and go to step (1).

Remark 5.14.

(a) If the stopping criterion $\nabla T_\alpha(x_n) = 0$ is satisfied for some $n \in \mathbb{N}$, then by (5.11) we already have $x_n = x_\alpha^\delta$ and we can stop the iteration process.

(b) Due to the properties of T_α, the function $f_n : \mathbb{R} \longrightarrow [0, \infty)$, defined by

$$f_n(\mu) := T_\alpha\Big(x_n - \mu J_{q^*}^*(\nabla T_\alpha(x_n))\Big)$$

appearing in the line search of step (1), is strictly convex and differentiable, with the continuous derivative

$$f_n'(\mu) = -\langle \nabla T_\alpha\Big(x_n - \mu J_{q^*}^*(\nabla T_\alpha(x_n))\Big), J_{q^*}^*(\nabla T_\alpha(x_n))\rangle\,.$$

Since, by the monotonicity of the duality mappings, $f_n'(0) = -\|\nabla T_\alpha(x_n)\|^{q^*} < 0$ and f_n' is increasing, we know that μ_n must in fact be positive.

Theorem 5.15. *If X is uniformly convex and uniformly smooth, Y uniformly smooth and $p, q \geq 2$, then the sequence $\{x_n\}$, generated by the primal method of Algorithm* 5.13, *converges strongly to the unique minimizer x_α^δ of T_α.*

Before we are able to prove Theorem 5.15, we need an additional result, based on the article of Xu and Roach [239]. Let $\overline{\rho}_X : (0, \infty) \longrightarrow (0, 1]$ be the function

$$\overline{\rho}_X(\tau) := \frac{\rho_X(\tau)}{\tau}\,,$$

where ρ_X is the modulus of smoothness of a Banach space X. The function $\overline{\rho}_X$ is known to be continuous and nondecreasing, see [71, 143].

Lemma 5.16. *Let X be a uniformly smooth Banach space with duality mapping J_q and weight $q \geq 2$. Then, for all $x, y \in X$, we have*

$$\|J_q(x) - J_q(y)\| \leq C \max\left\{1, (\|x\| \vee \|y\|)^{q-1}\right\} \overline{\rho}_X(\|x - y\|) \tag{5.12}$$

and

$$\|x - y\|^q \leq \|x\|^q - q\,\langle J_q(x), y\rangle + C\Big(1 \vee (\|x\| + \|y\|)^{q-1}\Big)\rho_X(\|y\|)\,. \tag{5.13}$$

The estimate (5.12) *implies that J_q is uniformly continuous on bounded sets of X.*

Proof. By formula (3.1)' in [239] we have

$$\|J_q(x) - J_q(y)\| \leq C(\|x\| \vee \|y\|)^{q-1}\overline{\rho}_X\left(\frac{\|x - y\|}{\|x\| \vee \|y\|}\right).$$

If $\frac{1}{\|x\| \vee \|y\|} \leq 1$, we get, by the monotonicity of $\overline{\rho}_X$,

$$\overline{\rho}_X\left(\frac{\|x-y\|}{\|x\| \vee \|y\|}\right) \leq \overline{\rho}_X\big(\|x-y\|\big),$$

which gives (5.12) for this case.

In the case $\frac{1}{\|x\| \vee \|y\|} \geq 1$, we use the fact that ρ_X is equivalent to a decreasing function, i.e.,

$$\frac{\rho_X(\eta)}{\eta^2} \leq L\frac{\rho_X(\tau)}{\tau^2}$$

for $\eta \geq \tau > 0$ [143] and we get

$$\rho_X\left(\frac{\|x-y\|}{\|x\| \vee \|y\|}\right) \leq \frac{L}{\big(\|x\| \vee \|y\|\big)^2}\rho_X\big(\|x-y\|\big),$$

which implies

$$\overline{\rho}_X\left(\frac{\|x-y\|}{\|x\| \vee \|y\|}\right) \leq \frac{L}{\|x\| \vee \|y\|}\,\overline{\rho}_X\big(\|x-y\|\big)\,.$$

For $q \geq 2$, we thus arrive at

$$\begin{aligned}\|J_q(x) - J_q(y)\| &\leq C\,L\big(\|x\| \vee \|y\|\big)^{q-2}\overline{\rho}_X\big(\|x-y\|\big)\\ &\leq C\,L\,\overline{\rho}_X\big(\|x-y\|\big),\end{aligned}$$

and (5.12) is also valid if $\frac{1}{\|x\| \vee \|y\|} \geq 1$.

As in [239], we consider the continuously differentiable function $f : [0,1] \longrightarrow \mathbb{R}$, with

$$\begin{aligned}f(t) := \|x - t\,y\|^q, \quad & f'(t) = -q\,\langle J_q(x - t\,y), y\rangle\,,\\ f(0) = \|x\|^q, \quad f(1) = \|x-y\|^q, \quad & f'(0) = -q\,\langle J_q(x), y\rangle\end{aligned}$$

and obtain

$$\begin{aligned}\|x-y\|^q - \|x\|^q + q\,\langle J_q(x), y\rangle &= f(1) - f(0) - f'(0)\\ &= \int_0^1 f'(t) - f'(0)\,dt\\ &= q\int_0^1 \langle J_q(x) - J_q(x - t\,y), y\rangle\,dt\\ &\leq q\int_0^1 \|J_q(x) - J_q(x - t\,y)\|\,\|y\|\,dt\,.\end{aligned}$$

For $t \in [0,1]$ we set $\tilde{y} := x - ty$ and get $x - \tilde{y} = ty$, $\|\tilde{y}\| \le \|x\| + \|y\|$, and thus $\|x\| \vee \|\tilde{y}\| \le \|x\| + \|y\|$. The monotonicity of $\overline{\rho}_X$ yields

$$\overline{\rho}_X(t\|y\|)\,\|y\| \le \overline{\rho}_X(\|y\|)\,\|y\| = \rho_X(\|y\|),$$

and, by (5.12), we obtain

$$\begin{aligned}
&\|x-y\|^q - \|x\|^q + q\,\langle J_q(x), y\rangle \\
&\quad\le q \int_0^1 C \max\left\{1, (\|x\| + \|y\|)^{q-1}\right\} \overline{\rho}_X(t\|y\|)\,\|y\|\,dt \\
&\quad\le C \max\left\{1, (\|x\| + \|y\|)^{q-1}\right\} \rho_X(\|y\|).
\end{aligned}$$

This completes the proof. □

Proof of Theorem 5.15. We fix $\gamma \in (0,1)$, $\overline{\mu} > 0$ and for $n \in \mathbb{N}$ we choose $\tilde{\mu}_n \in (0, \overline{\mu}]$ such that

$$\phi_n(\tilde{\mu}_n) = \phi_n(\overline{\mu}) \wedge \gamma\,. \tag{5.14}$$

Here, the function $\phi_n : (0,\infty) \longrightarrow (0,\infty)$ is defined by

$$\begin{aligned}
\phi_n(\mu) :=& \frac{C_Y}{p}\left(1 \vee \left(\|Ax_n - y\| + \overline{\mu}\left\|AJ^*_{q^*}(\nabla T_\alpha(x_n))\right\|\right)^{p-1}\right) \\
&\times \frac{\left\|AJ^*_{q^*}(\nabla T_\alpha(x_n))\right\|}{\|\nabla T_\alpha(x_n)\|^{q^*}} \overline{\rho}_Y\left(\mu\left\|AJ^*_{q^*}(\nabla T_\alpha(x_n))\right\|\right) \\
&+ \alpha \frac{C_X}{q}\left(1 \vee \left(\|x_n\| + \overline{\mu}\,\|\nabla T_\alpha(x_n)\|^{q^*-1}\right)^{q-1}\right) \\
&\times \frac{\overline{\rho}_X\left(\mu\,\|\nabla T_\alpha(x_n)\|^{q^*-1}\right)}{\|\nabla T_\alpha(x_n)\|}
\end{aligned} \tag{5.15}$$

with the constants C_X, C_Y being the ones appearing in the respective characteristic inequalities (5.13). This choice of $\tilde{\mu}_n$ is possible since, by the properties of $\overline{\rho}_Y$ and $\overline{\rho}_X$, the function ϕ_n is continuous, increasing and satisfies the limit condition $\lim_{\mu\to 0}\phi_n(\mu) = 0$. At first, we prove an estimate of the form

$$T_\alpha(x_{n+1}) \le T_\alpha(x_n) - \tilde{\mu}_n \|\nabla T_\alpha(x_n)\|^{q^*}(1-\gamma)\,,$$

which will finally assure convergence. We use the characteristic inequalities (5.13) to estimate

$$\begin{aligned}T_\alpha(x_{n+1}) \le T_\alpha(x_n) &- \tilde{\mu}_n \|\nabla T_\alpha(x_n)\|^{q^*} \\ &+ \frac{C_Y}{p}\left(1 \vee \left(\|Ax_n - y\| + \left\|\tilde{\mu}_n A J_{q^*}^*(\nabla T_\alpha(x_n))\right\|\right)^{p-1}\right) \\ &\times \rho_Y\left(\left\|\tilde{\mu}_n A J_{q^*}^*(\nabla T_\alpha(x_n))\right\|\right) \\ &+ \alpha\frac{C_X}{q}\left(1 \vee \left(\|x_n\| + \left\|\tilde{\mu}_n J_{q^*}^*(\nabla T_\alpha(x_n))\right\|\right)^{q-1}\right) \\ &\times \rho_X\left(\left\|\tilde{\mu}_n J_{q^*}^*(\nabla T_\alpha(x_n))\right\|\right).\end{aligned}$$

By $\tilde{\mu}_n \le \overline{\mu}$, taking into account the definition of ϕ_n (5.15), we derive

$$\begin{aligned}T_\alpha(x_{n+1}) \le T_\alpha(x_n) &- \tilde{\mu}_n \|\nabla T_\alpha(x_n)\|^{q^*} \\ &+ \frac{C_Y}{p}\left(1 \vee \left(\|Ax_n - y\| + \overline{\mu}\left\|A J_{q^*}^*(\nabla T_\alpha(x_n))\right\|\right)^{p-1}\right) \\ &\times \rho_Y\left(\tilde{\mu}_n\left\|A J_{q^*}^*(\nabla T_\alpha(x_n))\right\|\right) \\ &+ \alpha\frac{C_X}{q}\left(1 \vee \left(\|x_n\| + \overline{\mu}\left\|J_{q^*}^*(\nabla T_\alpha(x_n))\right\|\right)^{q-1}\right) \\ &\times \rho_X\left(\tilde{\mu}_n\left\|J_{q^*}^*(\nabla T_\alpha(x_n))\right\|\right) \\ = T_\alpha(x_n) &- \tilde{\mu}_n \|\nabla T_\alpha(x_n)\|^{q^*}\left(1 - \phi_n(\tilde{\mu}_n)\right).\end{aligned}$$

The choice of $\tilde{\mu}_n$ (5.14) eventually yields

$$T_\alpha(x_{n+1}) \le T_\alpha(x_n) - \tilde{\mu}_n \|\nabla T_\alpha(x_n)\|^{q^*}(1-\gamma). \tag{5.16}$$

The next step of the proof consists in proving the convergence

$$\lim_{n\to\infty} \|\nabla T_\alpha(x_n)\| = 0.$$

From (5.16), we infer that

$$\lim_{n\to\infty} \tilde{\mu}_n \|\nabla T_\alpha(x_n)\|^{q^*} = 0 \tag{5.17}$$

and that the sequences $\{x_n\}$ and $\{\nabla T_\alpha(x_n)\}$ are bounded.

Suppose $\limsup_{n\to\infty} \|\nabla T_\alpha(x_n)\| = \varepsilon > 0$ and let $\|\nabla T_\alpha(x_{n_k})\| \to \varepsilon$, for $k \to \infty$. By (5.17), then we have $\lim_{k\to\infty} \tilde{\mu}_{n_k} = 0$. We show that this leads to a contradiction. On the one hand, by (5.15), we get

$$\phi_{n_k}(\tilde{\mu}_{n_k}) \le \frac{L_1}{\|\nabla T_\alpha(x_{n_k})\|^{q^*}}\overline{\rho}_Y\left(\tilde{\mu}_{n_k} L_2\right) + \frac{C_1}{\|\nabla T_\alpha(x_{n_k})\|}\overline{\rho}_X\left(\tilde{\mu}_{n_k} C_2\right).$$

Since the right-hand side converges to zero for $k \to \infty$, so does $\phi_{n_k}(\tilde{\mu}_{n_k})$. On the other hand, by (5.14), we have

$$\phi_{n_k}(\tilde{\mu}_{n_k}) = \phi_{n_k}(\overline{\mu}) \wedge \gamma$$

and

$$\phi_{n_k}(\overline{\mu}) \geq 0 + C\overline{\rho}_X\left(\overline{\mu}\left\|\nabla T_\alpha(x_{n_k})\right\|^{q^*-1}\right).$$

Hence, $\phi_{n_k}(\tilde{\mu}_{n_k}) \geq L > 0$, for all sufficiently large k, which contradicts

$$\lim_{k\to\infty} \phi_{n_k}(\tilde{\mu}_{n_k}) = 0\,.$$

So, we have $\limsup_{n\to\infty} \|\nabla T_\alpha(x_n)\| = 0$ and thus $\lim_{n\to\infty} \|\nabla T_\alpha(x_n)\| = 0$.

We finally show that $\{x_n\}$ converges strongly to x_α^δ. By (5.11) and the monotonicity of the duality mapping J_r, we obtain

$$\begin{aligned}
\|\nabla T_\alpha(x_n)\|\,\|x_n - x_\alpha^\delta\| &\geq \langle \nabla T_\alpha(x_n), x_n - x_\alpha^\delta\rangle \\
&= \langle \nabla T_\alpha(x_n) - \nabla T_\alpha(x_\alpha^\delta), x_n - x_\alpha^\delta\rangle \\
&= \langle J_p(Ax_n - y) - J_p(Ax_\alpha^\delta - y), (Ax_n - y) - (Ax_\alpha^\delta - y)\rangle \\
&\quad + \alpha\langle J_q(x_n) - J_q(x_\alpha^\delta), x_n - x_\alpha^\delta\rangle \\
&\geq \alpha\langle J_q(x_n) - J_q(x_\alpha^\delta), x_n - x_\alpha^\delta\rangle\,.
\end{aligned}$$

Since $\{x_n\}$ is bounded and $\lim_{n\to\infty} \|\nabla T_\alpha(x_n)\| = 0$, this yields

$$\lim_{n\to\infty} \langle J_q(x_n) - J_q(x_\alpha^\delta), x_n - x_\alpha^\delta\rangle = 0\,,$$

from which we infer that $\{x_n\}$ converges strongly to x_α^δ whenever X is uniformly convex (cf. [48, Th. II.2.17]). □

Next, we will show what kind of convergence rates are obtained by this choice of the step size, if we impose additional assumptions on the spaces X and Y.

Remark 5.17. We will see that the convergence results, achieved in this section, also hold for the step size μ_n, being a minimizer of

$$-\mu\|\psi_n\|^{q^*} + |\mu|^p \frac{G_Y}{p}\|AJ_{q^*}^{X^*}(\psi_n)\|^p + \alpha|\mu|^q \frac{G_X}{q}\|\psi_n\|^{q^*} \to \min,$$
$$\text{subject to } \mu > 0,$$

where G_X and G_Y are the constants in the definition of smoothness of power type for X and Y. We notice that, to compute the above modified step size, we only have to solve a one-dimensional minimization problem. However, we also must know the values of the constants G_X and G_Y, which are not needed for the original step size choice.

Theorem 5.18. *Let X be a q-smooth and c-convex Banach space, as well as Y be a p-smooth Banach space satisfying the condition*

$$s := \min\{p, q\} < c.$$

Then, we have

$$\|x_n - x_\alpha^\delta\| \le C \cdot n^{-(s-1)/(c-s)}.$$

Proof. Since X is q-smooth and Y is p-smooth, there exist constants $G_X, G_Y > 0$, such that

$$\frac{1}{p}\|x-y\|_Y^p \le \frac{1}{p}\|x\|_Y^p - \langle J_p^Y(x), y\rangle + \frac{G_Y}{p}\|y\|_Y^p \quad \forall x, y \in Y \quad \text{and}$$
$$\frac{1}{q}\|x-y\|_X^q \le \frac{1}{q}\|x\|_X^q - \langle J_q^X(x), y\rangle + \frac{G_X}{q}\|y\|_X^q \quad \forall x, y \in X.$$

Hence, for every μ, we get that

$$\begin{aligned}
&\frac{1}{p}\|A(x_n - \mu J_{q^*}^{X^*}(\psi_n)) - y^\delta\|^p \\
&\le \frac{1}{p}\|Ax_n - y^\delta\|^p - \mu\langle J_p^Y(Ax_n - y^\delta), AJ_{q^*}^{X^*}(\psi_n)\rangle \\
&\quad + |\mu|^p \frac{G_Y}{p}\|AJ_{q^*}^{X^*}(\psi_n)\|^p \\
&= \frac{1}{p}\|Ax_n - y^\delta\|^p - \mu\langle A^* J_p^Y(Ax_n - y^\delta), J_{q^*}^{X^*}(\psi_n)\rangle \\
&\quad + |\mu|^p \frac{G_Y}{p}\|AJ_{q^*}^{X^*}(\psi_n)\|^p
\end{aligned}$$

and

$$\begin{aligned}
&\frac{1}{q}\|x_n - \mu J_{q^*}^{X^*}(\psi_n)\|^q \\
&\le \frac{1}{q}\|x_n\|^q - \mu\langle J_q^X(x_n), J_{q^*}^{X^*}(\psi_n)\rangle + |\mu|^q \frac{G_X}{q}\|J_{q^*}^{X^*}(\psi_n)\|_X^q \\
&= \frac{1}{q}\|x_n\|^q - \mu\langle J_q^X(x_n), J_{q^*}^{X^*}(\psi_n)\rangle + |\mu|^q \frac{G_X}{q}\|\psi_n\|_X^{q^*}.
\end{aligned}$$

With the equalities above, we find that

$$\begin{aligned}
T_\alpha(x_n - \mu\psi_n) \le{}& T_\alpha(x_n) - \mu\|\psi_n\|^{q^*} \\
&+ |\mu|^p \frac{G_Y}{p}\|AJ_{q^*}^{X^*}(\psi_n)\|^p + \alpha|\mu|^q \frac{G_X}{q}\|\psi_n\|^{q^*}
\end{aligned} \tag{5.18}$$

based on $\langle A^* J_p^Y(Ax_n - y^\delta) + \alpha J_q^X(x_n), J_{q^*}^{X^*}(\psi_n)\rangle = \langle \psi_n, J_{q^*}^{X^*}(\psi_n)\rangle = \|\psi_n\|^{q^*}$. The iterates x_n are uniformly bounded because of

$$\alpha\frac{1}{q}\|x_n\|^q \le T_\alpha(x_n) \le T_\alpha(x_{n-1}) \le \cdots \le T_\alpha(x_0).$$

Since X and Y are smooth of power type q and p respectively, the mappings $x \mapsto J_q(x)$ and $x \mapsto A^* J_p(Ax - y^\delta)$ are $(q-1)$ and $(p-1)$ Hölder continuous, respectively (cf. Theorem 2.42). Therefore, we have

$$\|\psi_n\| \le C \cdot (\|x_n - x_\alpha^\delta\|^{p-1} + \|x_n - x_\alpha^\delta\|^{q-1}).$$

Consequently, the gradients ψ_n are uniformly bounded. With $s = \min\{p, q\}$, we get

$$\min(p(q^* - 1), q^*) = \min(p(q^* - 1), q(q^* - 1)) = s(q^* - 1).$$

Since $\|\psi_n\|$ are uniformly bounded, we have the estimate

$$\begin{aligned}\max\{\frac{G_Y}{p}\|A J_{q^*}^{X^*}(\psi_n)\|^p, \alpha\frac{G_X}{q}\|\psi_n\|^{q^*}\} &\le C \cdot \max\{\|A\|\|\psi_n\|^{(q^*-1)p}, \|\psi_n\|^{q^*}\}\\ &\le C \cdot \max(\{\|\psi_n\|^{(q^*-1)p}, \|\psi_n\|^{q^*}\}\\ &\le C \cdot \|\psi_n\|^{s(q^*-1)}.\end{aligned}$$

Moreover, we get

$$\begin{aligned}T_\alpha(x_{n+1}) &= \min_\mu T_\alpha(x_n - \mu\psi_n)\\ &\le \min_\mu\{T_\alpha(x_n) - \mu\|\psi_n\|^{q^*} + (|\mu|^p + |\mu|^q) \cdot \|\psi_n\|^{s(q^*-1)} \cdot C\}\\ &\le \min_{0<\mu<1}\{T_\alpha(x_n) - \mu\|\psi_n\|^{q^*} + (\mu^p + \mu^q) \cdot \|\psi_n\|^{s(q^*-1)} \cdot C\}.\end{aligned}$$

For $0 < \mu < 1$, we have $\mu^p \le \mu^s$ and $\mu^q \le \mu^s$, since $s \le p$ and $s \le q$. Summarizing this, the estimate

$$T_\alpha(x_{n+1}) \le \min_{0<\mu<1}\{T_\alpha(x_n) - \mu\|\psi_n\|^{q^*} + \mu^s \cdot \|\psi_n\|^{s(q^*-1)} \cdot C_0\} \tag{5.19}$$

is true for some $C_0 > 0$. The minimization problem on the right-hand side of the last inequality is solved by

$$m_n := \min\left\{1, 1/(sC_0)^{1/(s-1)} \cdot \|\psi_n\|^{s^*-q^*}\right\}.$$

If $m_n = 1/(sC_0)^{1/(s-1)} \cdot \|\psi_n\|^{s^*-q^*}$, then

$$-m_n \cdot \|\psi_n\|^{q^*} + m_n^s \cdot \|\psi_n\|^{s(q^*-1)} \cdot C_0 \le -C\|\psi_n\|^{s^*}.$$

If $m_n = 1$, then $s C_0 \|\psi_n\|^{s(q^*-1)-q^*} \le 1$, and thus

$$-m_n \cdot \|\psi_n\|^{q^*} + m_n^s \cdot \|\psi_n\|^{s(q^*-1)} \cdot C_0 \le -C\|\psi_n\|^{q^*}.$$

Then, we have

$$T_\alpha(x_{n+1}) \le T_\alpha(x_n) - C \cdot \min\{\|\psi_n\|^{s^*}, \|\psi_n\|^{q^*}\}.$$

Since T_α is bounded from below, we can conclude that

$$\|\psi_n\| \longrightarrow 0 \quad \text{as} \quad n \to \infty.$$

Thus, for $n \in \mathbb{N}$ sufficiently large, we have

$$\|\psi_n\| \le 1$$

enabling the estimate

$$T_\alpha(x_{n+1}) \le T_\alpha(x_n) - C \cdot \|\psi_n\|^{s^*}, \tag{5.20}$$

due to $s^* \ge q^*$. Now, let us introduce the numbers

$$r_n := T_\alpha(x_n) - T_\alpha(x_\alpha^\delta),$$

where x_α^δ is the minimizer of the Tikhonov functional. Due to the assumptions about X and Y, we know that x_α^δ is unique, and therefore the numbers r_n are well-defined. We also note that the numbers r_n may be regarded as a Bregman distance with respect to the functional T_α, i.e.

$$r_n = D_{\psi_\alpha^\delta}^{T_\alpha}(x_n, x_\alpha^\delta) := T_\alpha(x_n) - T_\alpha(x_\alpha^\delta) - \langle \psi_\alpha^\delta, x_n - x_\alpha^\delta \rangle.$$

Evidently $\psi_\alpha^\delta = \nabla T_\alpha(x_\alpha^\delta) = 0$, i.e., the gradient of T_α at the minimizer x_α^δ of the Tikhonov functional vanishes.

The next step of the proof is to connect r_n to $\|\psi_n\|$. Since, by Theorem 2.40, the space X is c-convex, we have

$$\langle J_q^X(x_n) - J_q^X(x_\alpha^\delta), x_n - x_\alpha^\delta \rangle \ge C(\max\{\|x_n\|, \|x_\alpha^\delta\|\})^{q-c}\|x_n - x_\alpha^\delta\|^c.$$

Theorem 2.50, together with Theorem 2.51, show that $q \le c$. Therefore, since the sequence $\{x_n\}$ is uniformly bounded, we get

$$\langle J_q^X(x_n) - J_q^X(x_\alpha^\delta), x_n - x_\alpha^\delta \rangle \ge C\|x_n - x_\alpha^\delta\|^c.$$

Since $\psi_\alpha^\delta = \nabla T_\alpha(x_\alpha^\delta) = 0$ and the subgradient of a convex functional is monotone (cf. Theorem 2.26), we obtain

$$\begin{aligned}\|\psi_n\|\|x_n - x_\alpha^\delta\| &\ge \langle \psi_n - \psi_\alpha^\delta, x_n - x_\alpha^\delta \rangle \\ &\ge \alpha\langle J_q^X(x_n) - J_q^X(x_\alpha^\delta), x_n - x_\alpha^\delta \rangle \ge C\|x_n - x_\alpha^\delta\|^c.\end{aligned}$$

Hence,

$$\|\psi_n\| \geq C\|x_n - x_\alpha^\delta\|^{c-1}.$$

The above inequality indicates that

$$\|x_n - x_\alpha^\delta\| \longrightarrow 0 \quad \text{as} \quad n \to \infty,$$

since we have already proved that $\|\psi_n\| \to 0$ as $n \to \infty$ and $c - 1 > 0$. Thus

$$\|x_n - x_\alpha^\delta\| \leq 1$$

for $n \in \mathbb{N}$ sufficiently large. We already know that

$$\|x_n - x_\alpha^\delta\| \leq C\|\psi_n\|^{c^*-1}.$$

A view on the definition of the subgradient gives

$$r_n \leq \langle \psi_n, x_n - x_\alpha^\delta \rangle \leq \|\psi_n\|\|x_n - x_\alpha^\delta\|.$$

Hence, we get

$$r_n \leq C\|\psi_n\| \cdot \|\psi_n\|^{c^*-1} = C \cdot \|\psi_n\|^{c^*}.$$

By (5.20), we find that, for all sufficiently large n,

$$\begin{aligned} r_{n+1} &\leq r_n - C \cdot \|\psi_n\|^{s^*} \\ &\leq r_n - C \cdot r_n^{s^*/c^*}. \end{aligned}$$

Hence,

$$r_n \longrightarrow 0 \quad \text{as} \quad n \to \infty.$$

Next, we use a trick developed by Dunn [59, 60], which is common in the literature with respect to convergence rates in Banach spaces (cf., e.g., [26, 131]). We set

$$a := (s^*/c^*) - 1.$$

Exploiting the mean value theorem, we have

$$\frac{1}{r_{n+1}^a} - \frac{1}{r_n^a} = -a\frac{1}{\rho^{a+1}}(r_{n+1} - r_n)$$

with $\rho \in (r_{n+1}, r_n)$. Therefore, for all sufficiently large $n \in \mathbb{N}$, we also have

$$\frac{1}{r_{n+1}^a} - \frac{1}{r_n^a} \geq C > 0$$

and consequently, for all sufficiently large n and N,

$$\frac{1}{r_n^a} \geq \frac{1}{r_n^a} - \frac{1}{r_N^a} = \sum_{k=N}^{n-1} \frac{1}{r_{k+1}^a} - \frac{1}{r_k^a} \geq C \cdot (n - N).$$

Hence, for all $n \geq 0$, we get

$$r_n \leq C \cdot n^{-1/a}.$$

Since the functional $\frac{1}{p}\|Ax - y^\delta\|^p$ is convex and $\frac{1}{q}\|x\|^q$ is c-convex, this yields for the Tikhonov functional T_α that

$$r_n \geq C\,\|x_n - x_\alpha^\delta\|^c. \tag{5.21}$$

For a proof of (5.21) we refer to Lemma 5.19. Finally, the estimate

$$\|x_n - x_\alpha^\delta\| \leq C \cdot n^{-1/(a\cdot c)}$$

completes the proof. □

Lemma 5.19. *Under the assumptions of Theorem* 5.18, *the Tikhonov functional* T_α *is* c*-convex in the sense that*

$$T_\alpha(x_n) - T_\alpha(x_\alpha^\delta) = r_n \geq C\,\|x_n - x_\alpha^\delta\|^c. \tag{5.22}$$

Proof. The c-convexity of T_α seems to be intuitively reasonable, since it is common knowledge that the sum of two convex functionals always inherits the best of both convexity properties. The proof is done as follows: Since X is c-convex, we get, from Corollary 2.61,

$$\begin{aligned} D_{j_q}(x_\alpha^\delta, x_n) &= \frac{1}{q}\|x_n\|^q - \frac{1}{q}\|x_\alpha^\delta\|^q - \langle j_q(x_\alpha^\delta), x_n - x_\alpha^\delta\rangle \\ &\geq C \cdot (\|x_\alpha^\delta\| + \|x_n\|)^{q-c}\|x_\alpha^\delta - x_n\|^c \end{aligned}$$

and

$$\begin{aligned} \frac{1}{q}\|x_n\|^q \geq &\frac{1}{q}\|x_\alpha^\delta\|^q + \langle j_q(x_\alpha^\delta), x_n - x_\alpha^\delta\rangle \\ &+ C \cdot (\|x_\alpha^\delta\| + \|x_n\|)^{q-c}\|x_\alpha^\delta - x_n\|^c. \end{aligned} \tag{5.23}$$

Furthermore, we have

$$\frac{1}{p}\|Ax_n - y^\delta\|^p \geq \frac{1}{p}\|Ax_\alpha^\delta - y^\delta\|^p + \langle A^* j_p(Ax_\alpha^\delta - y^\delta), x_n - x_\alpha^\delta\rangle. \tag{5.24}$$

Multiplying (5.23) by α and summing, with (5.24), we get

$$T_\alpha(x_n) \geq T_\alpha(x_\alpha^\delta) + \langle \psi_\alpha^\delta, x_n - x_\alpha^\delta\rangle + C \cdot (\|x_\alpha^\delta\| + \|x_n\|)^{q-c}\|x_\alpha^\delta - x_n\|^c,$$

where $\psi_\alpha^\delta \in \partial T_\alpha(x_\alpha^\delta)$. Since in Theorem 5.18 we also assumed that both X and Y are smooth spaces, we also find that the subgradient is in fact a gradient. Since x_α^δ is the minimizer of T_α, we have $\psi_\alpha^\delta = 0$. This finally leads to

$$r_n \geq C \cdot (\|x_\alpha^\delta\| + \|x_n\|)^{q-c}\|x_\alpha^\delta - x_n\|^c.$$

For a q-smooth and c-convex space X, we have $q \leq 2$ and $c \geq 2$. Hence,

$$q - c \leq 0.$$

Finally, the boundedness of $\{x_n\}$ implies (5.22). □

Theorem 5.18 is not applicable if

$$s := \min\{p, q\} = c. \tag{5.25}$$

Since $p \leq 2, q \leq 2$ and $c \geq 2$ (cf. Theorems 2.51 and 2.50), the condition (5.25) holds only if

$$p = q = c = 2.$$

In such a case, stronger convergence results can be proved.

Theorem 5.20. *Let X be a 2-smooth and 2-convex Banach space as well as Y a 2-smooth Banach space, i.e., $p = q = 2$. Then,*

$$\|x_n - x_\alpha^\delta\| \leq C \cdot \exp(-n/C). \tag{5.26}$$

Proof. As in the proof of Theorem 5.18, we arrive at the inequality

$$\begin{aligned} T_\alpha(x_n - \mu\psi_n) &\leq T_\alpha(x_n) - \left(\mu - \frac{1}{2}|\mu|^2\left[G_Y\|A\|^2 + \alpha G_X\right]\right) \cdot \|\psi_n\|^2 \\ &\leq T_\alpha(x_n) - (\mu - \frac{1}{2}|\mu|^2 \cdot C_0) \cdot \|\psi_n\|^2, \end{aligned}$$

which is an analogue of equation (5.18). The right-hand side of the last inequality is minimal, if μ is chosen as

$$m_n = 1/C_0.$$

Hence,

$$T_\alpha(x_{n+1}) \leq T_\alpha(x_n - m_n\psi_n) \leq T_\alpha(x_n) - C \cdot \|\psi_n\|^2.$$

Since X is 2-smooth and 2-convex, we get that

$$\|\psi_n\|\|x_n - x_\alpha^\delta\| \geq \alpha\langle J_2(x_n) - J_2(x_\alpha^\delta), x_n - x_\alpha^\delta\rangle \geq C\|x_n - x_\alpha^\delta\|^2$$

resulting in

$$\|x_n - x_\alpha^\delta\| \leq C\|\psi_n\|.$$

With $r_n := T_\alpha(x_n) - T_\alpha(x_\alpha^\delta)$, this implies that

$$r_n \leq \langle\psi_n, x_n - x_\alpha^\delta\rangle \leq \|\psi_n\|\|x_n - x_\alpha^\delta\| \leq C\|\psi_n\|^2$$

and hence

$$r_{n+1} \leq r_n - C \cdot \|\psi_n\|^2 \leq (1 - C)r_n.$$

Since the functional $\frac{1}{p}\|Ax - y^\delta\|^p$ is convex and $\frac{1}{q}\|x\|^q$ is 2-convex, we get for T_α that

$$r_n \geq C\|x_n - x_\alpha^\delta\|^2,$$

which proves (5.26), since the numbers r_n decrease to zero in geometrical order. □

5.3.2 Dual method

In this section, we still consider the minimization of the functional

$$T_\alpha(x) := \frac{1}{p}\|Ax - y^\delta\|_Y^p + \alpha \cdot \frac{1}{q}\|x\|_X^q, \tag{5.27}$$

but we now assume that $p > 1$ and X is convex and smooth of power type. This is the main difference with the previous section, where the index q was connected to the smoothness properties of X. Furthermore, we make *no* assumptions on Y.

For the minimization, we consider the iteration process, defined as

$$\begin{aligned} x_{n+1}^* &:= x_n^* - \mu_n \psi_n \text{ with } \psi_n \in \partial T_\alpha(x_n) = A^* J_p^Y(Ax_n - y^\delta) + \alpha J_q^X(x_n), \\ x_{n+1} &:= J_{q^*}^{X^*}(x_{n+1}^*). \end{aligned} \tag{5.28}$$

We stress that the choice of μ_n is crucial for the convergence properties of the steepest descent iteration. Therefore, we start with the construction of the step size μ_n.

Since X is convex of power type and smooth of power type, we can apply Theorem 2.64 and get

$$\begin{aligned} &D_{j_q}(x_\alpha^\delta, J_{q^*}(x_n^* - \mu\psi_n)) \\ &\quad = D_{j_q}(x_\alpha^\delta, x_n) + D_{j_{q^*}}(x_n^* - \mu\psi_n, x_n^*) + \mu\langle \psi_n, x_\alpha^\delta - x_n\rangle. \end{aligned}$$

Since x_α^δ is the minimizer of T_α, there exists a single valued selection j_p, such that

$$\psi_\alpha^\delta := 0 = A^* j_p(Ax_\alpha^\delta - y^\delta) + \alpha j_q(x_\alpha^\delta)\,. \tag{5.29}$$

Therefore, with the monotonicity of the duality mapping (cf. Theorem 2.26), we get

$$\langle \psi_n, x_\alpha^\delta - x_n\rangle = \langle \psi_n - \psi_\alpha^\delta, x_\alpha^\delta - x_n\rangle \le -\alpha\langle j_q(x_n) - j_q(x_\alpha^\delta), x_n - x_\alpha^\delta\rangle.$$

Furthermore, we recall that

$$\langle j_q(x_n) - j_q(x_\alpha^\delta), x_n - x_\alpha^\delta\rangle = D_{j_q}(x_n, x_\alpha^\delta) + D_{j_q}(x_\alpha^\delta, x_n).$$

Hence we have

$$-\mu\langle \psi_n, x_\alpha^\delta - x_n\rangle \le -\mu\alpha\, D_{j_q}(x_\alpha^\delta, x_n)$$

and end up with

$$D_{j_q}(x_\alpha^\delta, J_{q^*}(x_n^* - \mu\psi_n)) \le (1 - \mu\alpha) D_{j_q}(x_\alpha^\delta, x_n) + D_{j_{q^*}}(x_n^* - \mu\psi_n, x_n^*).$$

Of course, we do not know the exact value of $D_{j_q}(x_\alpha^\delta, x_n)$. Assume that we know an upper estimate R_n of $D_{j_q}(x_\alpha^\delta, x_n)$ as

$$D_{j_q}(x_\alpha^\delta, x_n) \le R_n.$$

Then, we arrive at the important inequality

$$D_{j_q}(x_\alpha^\delta, J_{q^*}(x_n^* - \mu\psi_n)) \le \max\{1 - \mu\alpha, 0\}R_n + D_{j_{q^*}}(x_n^* - \mu\psi_n, x_n^*).$$

The space X^* is q^*-smooth, since X is q-convex (cf. Theorem 2.52 (a)). Therefore, there exists a constant $G_{q^*} \ge 0$, such that

$$D_{j_{q^*}}(x_n^* - \mu\psi_n, x_n^*) \le |\mu|^{q^*} \frac{G_{q^*}}{q^*} \|\psi_n\|^{q^*}.$$

Finally, we arrive at the main inequality of our convergence analysis

$$D_{j_q}(x_\alpha^\delta, J_{q^*}(x_n^* - \mu\psi_n)) \le \max\{1 - \mu\alpha, 0\}R_n + |\mu|^{q^*} \frac{G_{q^*}}{q^*} \|\psi_n\|^{q^*}.$$

The optimal value of μ_n, with respect to the right-hand side of the above inequality, is given by

$$\mu_n = \min\left\{\left(\frac{\alpha}{G_{q^*}} \frac{R_n}{\|\psi_n\|^p}\right)^{\frac{1}{q^*-1}}, \frac{1}{\alpha}\right\}.$$

We now have all the ingredients necessary to formulate the dual method.

Algorithm 5.21 (Dual method). Let G_{q^*} be the constant in the definition of smoothness of power type, for the space X^*.

(1) Choose an arbitrary initial point $x_0 \in X$, a dual initial point $x_0^* = J_q(x_0)$ and R_0, such that the condition $D_{j_q}(x_\alpha^\delta, x_0) \le R_0$ is satisfied. Set $n = 0$.

(2) Stop, if $0 \in \partial T_\alpha(x_n)$. Else, choose $\psi_n \in \partial T_\alpha(x_n)$ and set

$$\mu_n := \min\left\{\left(\frac{\alpha}{G_{q^*}} \cdot \frac{R_n}{\|\psi_n\|^{q^*}}\right)^{\frac{1}{q^*-1}}, \frac{1}{\alpha}\right\} \tag{5.30}$$

and

$$R_{n+1} := (1 - \mu_n\alpha)R_n + \mu_n^{q^*} \frac{G_{q^*}}{q^*} \|\psi_n\|^{q^*}.$$

(3) Set

$$x_{n+1}^* := x_n^* - \mu_n\psi_n \quad \text{and} \quad x_{n+1} := J_{q^*}^{X^*}(x_{n+1}^*).$$

(4) Let $n \leftarrow (n + 1)$ and go to step (1).

Remark 5.22. All convergence results presented below also hold if the step size μ_n is alternatively chosen as

$$\max\{1-\mu\alpha, 0\}R_n + D_{j_{q^*}}(x_n^* - \mu\psi_n, x_n^*) \to \min, \quad \text{subject to} \quad \mu > 0,$$

and R_{n+1} is updated by

$$R_{n+1} := (1-\mu_n\alpha)R_n + D_{j_{q^*}}(x_n^* - \mu_n\psi_n, x_n^*).$$

Next, we prove the convergence rates results for Algorithm 5.21. We introduce the variable B_n via

$$B_n := \left(\frac{\alpha}{G_{q^*}} \cdot \frac{R_n}{\|\psi_n\|^{q^*}}\right)^{\frac{1}{q^*-1}}.$$

Then,

$$\mu_n = \min\left\{B_n, \frac{1}{\alpha}\right\} \tag{5.31}$$

and

$$R_{n+1} = [1-\mu_n\alpha + \frac{\alpha}{q^*}\mu_n^{q^*} B_n^{-(q^*-1)}]R_n\,. \tag{5.32}$$

Theorem 5.23. *Let X be a q-convex Banach space and Y an arbitrary Banach space. Then,*

$$\|x_n - x_\alpha^\delta\| \le C \cdot n^{-\frac{1}{q(q-1)}}.$$

Proof. We again use Dunn's trick, to prove the convergence rate (cf. proof of Theorem 5.18). We have

$$\frac{1}{R_{n+1}^{q-1}} - \frac{1}{R_n^{q-1}} \ge \frac{1-\left[1-\mu_n\alpha + \frac{\alpha}{q^*}\mu_n^{q^*} B_n^{-(q^*-1)}\right]^{q-1}}{\left[1-\mu_n\alpha + \frac{\alpha}{q^*}\mu_n^{q^*} B_n^{-(q^*-1)}\right]^{q-1}} \cdot \frac{1}{R_n^{q-1}}.$$

Since $0 < \mu_n\alpha - \frac{\alpha}{q^*}\mu_n^{q^*} B_n^{-(q^*-1)} \le 1$ and $(1-(1-x)^y)/(1-x)^y \ge yx$ for all $0 \le x \le 1$ and $y \ge 0$, we get

$$\frac{1}{R_{n+1}^{q-1}} - \frac{1}{R_n^{q-1}} \ge (q-1)\left(\mu_n\alpha - \frac{\alpha}{q^*}\mu_n^{q^*} B_n^{-(q^*-1)}\right) \cdot \frac{1}{R_n^{q-1}}.$$

If $\mu_n = B_n$, then

$$\mu_n\alpha - \frac{\alpha}{q^*}\mu_n^{q^*} B_n^{-(q^*-1)} = \frac{1}{q}\alpha B_n.$$

If $\mu_n = \frac{1}{\alpha}$, then $B_n\alpha \ge 1$ and

$$\mu_n\alpha - \frac{\alpha}{q^*}\mu_n^{q^*} B_n^{-(q^*-1)} = 1 - \frac{1}{q^*}(B_n\alpha)^{-(q^*-1)} \ge 1 - \frac{1}{q^*} = \frac{1}{q}.$$

Hence,

$$\frac{1}{R_{n+1}^{q-1}} - \frac{1}{R_n^{q-1}} \geq C \cdot \min\{B_n, 1\} \frac{1}{R_n^{q-1}}.$$

Next, we will show that the right-hand side is uniformly bounded away from zero. By construction, we have $1/R_n^{q-1} \geq 1/R_0^{q-1}$. Therefore, the numbers $1/R_n^{q-1}$ are uniformly bounded away from zero. Now, we consider the numbers B_n/R_n^{q-1}. We have

$$B_n/R_n^{q-1} = \left(\frac{\alpha}{G_{q^*}} \cdot \frac{1}{\|\psi_n\|^{q^*}}\right)^{\frac{1}{q^*-1}}.$$

To estimate $\{\|\psi_n\|^{q^*}\}$, we need the following lemma:

Lemma 5.24. *Let X be a normed space and $f : X \to \mathbb{R}$ a convex functional. If $|f|$ is bounded on all bounded sets of X, then ∂f is also bounded on all bounded sets.*

Proof. Let $A \subset B(0, R)$ be an arbitrary bounded set. Then,

$$F := \sup\{|f(x)| : x \in B(0, R + 1)\} < \infty.$$

Assume $x \in A, \psi \in \partial f(x), y \in X$, with $\|y\| = 1$. Then, $x + y \operatorname{sign}\langle\psi, y\rangle \in B(0, R + 1)$. By definition of the subgradient we get

$$|\langle\psi, y\rangle| = \langle\psi, x + y \operatorname{sign}\langle\psi, y\rangle - x\rangle \leq f(x + y \operatorname{sign}\langle\psi, y\rangle) - f(x) \leq 2F.$$

Hence, $\sup\{\|\psi\| : \psi \in \partial f(x) \text{ with } x \in A\} \leq 2F < \infty$. □

By construction, the sequence $\{D_{j_q}(x_\alpha^\delta, x_n)\}$ is bounded. Therefore, by Theorem 2.60, the sequence $\{x_n\}$ is also bounded. Finally, by Lemma 5.24, the sequence $\{\psi_n\}$ is bounded, since any norm is bounded on bounded sets. Then the numbers B_n/R_n^{q-1} are uniformly bounded away from zero, and $1/R_{n+1}^{q-1} - 1/R_n^{q-1} \geq C$ holds uniformly for all $n \in \mathbb{N}$. Therefore, we have

$$\frac{1}{R_{n+1}^{q-1}} \geq \frac{1}{R_{n+1}^{q-1}} - \frac{1}{R_0^{q-1}} = \sum_{k=0}^{n} \frac{1}{R_{k+1}^{q-1}} - \frac{1}{R_k^{q-1}} \geq (n + 1)C$$

and we conclude that

$$R_{n+1}^{q-1} \leq C(n + 1)^{-1}.$$

Since X is q-convex, we have (cf. Theorem 2.60),

$$\|x_n - x_\alpha^\delta\|^q \leq C \cdot D_{j_q}(x_n, x_\alpha^\delta) \leq C \cdot R_n \leq C \cdot n^{-1/(q-1)},$$

which proves the claim. □

Theorem 5.25. *Assume that, in addition to the assumptions made in Theorem 5.23, the space Y is p-convex and*

$$p > 2 \quad or \quad q > 2.$$

Then, the convergence rate improves to

$$\|x_n - x_\alpha^\delta\| \le C \cdot n^{-\frac{(M-1)}{[(M-1)(q-1)-1]q}},$$

where

$$M := \max\{p, q\} > 2.$$

Proof. The proof goes along the lines of the proof of Theorem 5.23. The main difference is our estimation of the numbers B_n. First, we need the following lemma.

Lemma 5.26. *Let X be q-convex and Y be p-convex, with $q > 2$ or $p > 2$. Then, the subdifferential of the Tikhonov functional T_α is locally γ-Hölder continuous, where*

$$\gamma = \min\left\{\frac{1}{(p-1)}, \frac{1}{(q-1)}\right\}.$$

Proof. Let B be a bounded set and $x, y \in B$. Then,

$$\|\partial T_\alpha(x) - \partial T_\alpha(y)\| \le \|A^*\| \|j_p(Ax - y^\delta) - j_p(Ay - y^\delta)\| + \alpha \|j_q(x) - j_q(y)\|.$$

Since X is q-convex, the dual space X^* is q^*-smooth (cf. Theorem 2.52). Furthermore, by Theorems 2.50 and 2.51, we have $q^* \le q$. Then, by Theorem 2.42, we obtain

$$\|j_q(x) - j_q(y)\| \le C(\max\{\|x\|, \|y\|\})^{q-q^*} \cdot \|x - y\|^{q^*-1} \le C\|x - y\|^{q^*-1}$$

and

$$\|j_p(Ax - y^\delta) - j_p(Ay - y^\delta)\| \le C\|Ax - Ay\|^{p^*-1} \le C\|x - y\|^{p^*-1}.$$

Moreover, since $q^* - 1 = \frac{1}{q-1}$ and $p^* - 1 = \frac{1}{p-1}$, we obtain

$$\|\partial T_\alpha(x) - \partial T_\alpha(y)\| \le C \cdot \left(\|x - y\|^{\frac{1}{q-1}} + \|x - y\|^{\frac{1}{p-1}}\right) \le C\|x - y\|^\gamma,$$

proving the assertion. □

We recall that the iterates x_n are uniformly bounded. With $M := \max\{p, q\}$ and (5.29), we can estimate

$$\|\psi_n\|^{q^*} \le C \cdot \|x_n - x_\alpha^\delta\|^{q\frac{q^*}{q(M-1)}} \le C \cdot R_n^{\frac{q^*}{q(M-1)}} = C \cdot R_n^{\frac{1}{(M-1)(q-1)}}$$

since X is q-convex. Let us now define an auxiliary variable γ as

$$\gamma := \frac{1}{(M-1)(q-1)} < 1.$$

As in the proof of Theorem 5.23, we can estimate

$$B_n / R_n^{(1-\gamma)(q-1)} \geq C \cdot \left(\frac{\alpha}{G_{q^*}} \cdot \frac{R_n}{R_n^{\gamma}} \right)^{\frac{1}{q^*-1}} R_n^{(1-\gamma)(q-1)} \geq C > 0.$$

Then, the difference $1/R_{n+1}^{(1-\gamma)(q-1)} - 1/R_n^{(1-\gamma)(q-1)}$ is uniformly bounded away from zero. Therefore, we have $R_n^{(1-\gamma)(q-1)} \leq C \cdot n^{-1}$. Hence, we get

$$\|x_n - x_\alpha^\delta\| \leq C \cdot n^{-\frac{1}{(1-\gamma)(q-1)q}}.$$

with

$$\frac{1}{1-\gamma} = \frac{(M-1)(q-1)}{(M-1)(q-1)-1} > 0,$$

which proves the claim. □

Theorem 5.27. *Let X and Y be 2-convex, i.e., $p = q = 2$. Then, there exists a constant $C > 0$, such that the estimate*

$$\|x_n - x_\alpha^\delta\| \leq C \cdot \exp(-n/C)$$

holds true.

Proof. As in the proofs of Theorem 5.25 and Theorem 5.23, the main key to the desired convergence rate is the estimation of $\|\psi_n\|$. By the same technique as in the proof of Lemma 5.26 (and likewise employing Theorem 2.42), we find, that the subdifferential ∂T_α is locally Lipschitz continuous. Therefore,

$$\|\psi_n\|^2 \leq C \cdot \|x_n - x_\alpha^\delta\|^2 \leq C \cdot R_n. \tag{5.33}$$

Hence, the numbers B_n are uniformly bounded away from zero. By (5.31) and (5.32) there exists a constant $0 \leq \gamma < 1$ such that $R_{n+1} \leq \gamma \cdot R_n \leq \gamma^n R_0$, for all $n \in \mathbb{N}$. Thus, the sequence $\{R_n\}$ decays geometrically but the sequence $\{R_n\}$ dominates $\{\|x_n - x_\alpha^\delta\|^2\}$, up to some factor. This proves (5.33). □

Part IV

Iterative regularization

Iterative methods are often an attractive alternative to variational ones, especially for large scale problems. For this reason, during the last two decades, regularizing iterations have been extensively investigated for both linear and nonlinear problems, see, e.g., the monographs [14, 128] and the references therein. While the convergence analysis presented there requires a Hilbert space setting or equality of preimage and image space, the aim of this part is to present recent results on iterative methods in a more general Banach space setting.

Here, the role of the regularization parameter α in Tikhonov-type methods is taken by the stopping index, that – like in variational methods – can be chosen either by an a priori or by an a posteriori strategy. For the latter purpose, we will mainly concentrate on the discrepancy principle.

The first chapter – Chapter 6 – of this part deals with linear inverse problems. It starts by extending the definition and analysis of the most elementary iterative method, namely Landweber iteration, from the case of a Hilbert space to a Banach space setting, by means of duality mappings (Section 6.1). The latter render the method a nonlinear one, although the underlying operator equation is linear. A successful idea to aid accelerating Landweber iteration is based on the use of several search directions per step, leading to the sequential subspace optimization method (SESOP), studied in Section 6.2. Here, we establish and exploit a close relation to Bregman projections, which also enables us to introduce and analyze a regularized version (RESESOP). Cyclic iterations (also known as Kaczmarz methods) for solving systems of operator equations can be put into the very general context of split feasibility problems (SFP), where the task is to find points in the intersection of a number of sets, among them also preimages of certain sets, under linear operators. This is done in Section 6.3, where we extend an iterative algorithm, proposed by Byrne for finite dimensional spaces, to a Banach space setting with ill-posed operator equations, and show its regularizing properties.

In Chapter 7, we consider nonlinear operator equations and their regularized solution by either gradient or Newton-type iterations. Following some preliminaries, which deal with assumptions about the spaces and forward operators under consideration (Section 7.1), the first part of Section 7.2 is devoted to an extension of the results on Landweber iteration, from Chapter 6, to nonlinear operator equations. Additionally, for a slightly modified version of Landweber iteration, convergence rates are established for linear and nonlinear problems. The well-known fact that Newton-type methods are typically faster than gradient type ones was our motivation for studying this class of iterations, in Section 7.3. Here, we concentrate on the iteratively regularized Gauss–Newton method (IRGN), which, besides the stopping index n_*, also requires an appropriately chosen regularization parameter α_n in each Newton step. We provide results on convergence and convergence rates, with a priori and a posteriori choice of n_* and $\{\alpha_n\}_{n\in\mathbb{N}}$. This section also contains a short numerical illustration for an example from Chapter 1.

Chapter 6

Linear operator equations

In this chapter, we consider linear problems. The simplest model we deal with, is the solution of operator equations

$$Ax = y, \quad y \in \mathcal{R}(A), \tag{6.1}$$

where $A : X \to Y$ is a linear, continuous operator between Banach spaces X and Y. If A is not continuously invertible, problem (6.1) is ill-posed and a regularization method is needed to get a stable solution. An iterative regularization scheme, which is well studied in a Hilbert space setting, is the Landweber method. In Section 6.1, we describe the extension of the Landweber method to a fairly general Banach space setting, with the help of duality mappings. The Landweber method then reads

$$x_{n+1} = J_{q^*}^{X^*}\big(J_q^X(x_n) - \mu_n A^* j_p^Y(Ax_n - y)\big), \quad n = 0, 1, \dots \tag{6.2}$$

with a specific step size $\mu_n > 0$. We present a detailed analysis of this method and prove convergence to the minimum norm solution $x^\dagger$ of (6.1), in the case of exact and noisy data. Furthermore, we prove that this iteration method actually yields a regularization method, if we use the discrepancy principle as stopping rule.

As in Hilbert spaces, the Landweber iteration proves to be a stable regularization scheme in Banach spaces too, but shows a tremendously slow performance in numerical implementations. Hence, there is a considerable need for acceleration techniques for this method. Since $A^* j_p^Y(Ax_n - y) \in \partial\{\|Ax_n - y\|^p/p\}$, the iteration (6.2) can be seen as a gradient method, with respect to the residual $\|Ax_n - y\|^p/p$. In each iteration step we seek a better solution x_{n+1}, using x_n as a starting point and $A^* j_p^Y(Ax_n - y)$ as search direction. One idea to accelerate the Landweber iteration is to use more than one single search direction; this idea is inspired by the conjugate gradient method and sequential subspace optimization methods (SESOP). To this end, we use a finite number of search directions $A^* w_{n,i}$, with $w_{n,i} \in Y^*$, for all $i \in I_n$, with I_n a finite set of integers, where there is one $i_0 \in I_n$ with $w_{n,i_0} = j_p^Y(Ax_n - y)$. The iterates of this method are then computed as

$$x_{n+1} := J_{q^*}^{X^*}\Big(J_q^X(x_n) - \sum_{i \in I_n} t_{n,i}\, A^* w_{n,i}\Big), \quad n = 0, 1, \dots \tag{6.3}$$

where $t_n = (t_{n,i})_{i \in I_n} \in \mathbb{R}^{|I_n|}$ is a minimizer of the function

$$h_n(t) := \frac{1}{q^*}\Big\| J_q^X(x_n) - \sum_{i \in I_n} t_{n,i}\, A^* w_{n,i} \Big\|^{q^*} + \sum_{i \in I_n} t_{n,i}\, \langle w_{n,i}\,,\, y\rangle\,.$$

Section 6.2 puts iteration (6.3) in the context of Bregman projections (Section 6.2.1) and outlines the SESOP method for exact data $y \in Y$ (Section 6.2.2). We extend this method to noisy data y^δ, prove that it turns into a regularization method (RESESOP) and finally describe a fast algorithm, with two search directions (Section 6.2.3).

Solving operator equations as (6.1) can be seen as a special case of a more general class of problems, the so called *split feasibility problems* (SFP). These problems consist of finding a common point in the intersection of finitely many convex sets $C_i \subset X, i = \{1, \dots, N\}$,

$$\text{find} \quad x \in \mathbf{C} := \bigcap_{i \in I} C_i \,,$$

where some of the sets C_i are associated with linear operators $A_i : X \to Y_i$ by

$$C_i := \{x \in X : A_i x \in Q_i\}$$

with convex subsets $Q_i \subset Y_i$. For finite-dimensional spaces, Byrne [37] proposed the *CQ-algorithm* for finding $x \in C$ such that $Ax \in Q$

$$x_{n+1} = P_C\big(x_n - t_n\, A^*(Ax_n - P_Q(Ax_n))\big), \quad n = 0, 1, \dots .$$

Here, P_C, P_Q denote orthogonal projections onto the corresponding sets. Inspired by the CQ-algorithm in [215], Schöpfer et al. developed and analyzed the iteration

$$x_{n+1} = \Pi_C^q\big[J_{q^*}^{X^*}\big(J_q^X(x_n) - t_n\, A^* J_2^Y(Ax_n - P_Q(Ax_n))\big)\big], \quad n = 0, 1, \dots$$

for two convex sets C and Q and the Bregman projection Π_C^q. This method is the subject of Section 6.3. When proving stability of the method we assume perturbations with respect to the sets C_i and Q_i, assuming that only sets C_i^δ and Q_i^δ are known, having a Hausdorff distance less than δ from the exact sets C_i and Q_i. Thus, in Section 6.3.1, we consider the continuity of Bregman and metric projections with respect to the Hausdorff distance

$$d_\varrho(C, D) := \min\{\lambda \geq 0 \,|\, C \cap B_\varrho \subset D + B_\lambda \quad \text{and} \quad D \cap B_\varrho \subset C + B_\lambda\},$$

where $B_\varrho = B_\varrho(0)$, $B_\lambda = B_\lambda(0)$. We then present a regularization method for SFP in Banach spaces, using a cyclic, iterative projection method. We show that this method is in fact stable with respect to noise in the convex sets C_i and Q_i. These considerations are outlined in Section 6.3.2.

For better readability, we drop the superscripts X and Y of the duality mappings J_q^X and j_p^Y, if they are clear from the context and write $D_q(x, y)$ instead of $D_{j_q}(x, y)$ for the Bregman distance in X and $D_{q^*}^{X^*}$ for the Bregman distance on the dual X^*. Furthermore, we omit the subscripts X and Y and only write $\langle \cdot, \cdot \rangle$, $\| \cdot \|$ for their dual pairings and norms, respectively, since this is always clear from the context.

6.1 The Landweber iteration

Among the most popular iterative methods for solving linear ill-posed problems in Hilbert spaces are variations of the Landweber iteration

$$x_{n+1} = x_n - \mu_n A^*(Ax_n - y) \tag{6.4}$$

with $x_0 \in X$ and $\mu_n > 0$ appropriately. We refer the interested reader to the classical article of Landweber [142], where this method was established.

The first iterative methods for solving (6.1) in a Banach space setting were restricted to the case $Y = X$ and the operator A was required to satisfy certain resolvent conditions which, in concrete situations, may be difficult to verify, see [12, 137, 183]. Subsequently, nonlinear iterative methods have been proposed, in the context of Bregman projections and the minimization of convex functionals, and they have been shown to be weakly convergent, see [35, 178]. We present an extension of the Landweber iteration (6.4) to Banach spaces, that converges to the minimum norm solution of (6.1), provided that certain conditions are satisfied. Thereby, X is assumed to be smooth and uniformly convex, whereas Y can even be an arbitrary Banach space. Note that now, by Theorem 2.53, $j_q = J_q$ is single valued, X is reflexive and X^* is strictly convex and uniformly smooth.

We consider two different situations: the case of noise-free measured data y and operator A and the case where the data, as well as the operator, are contaminated by noise. We adjust the choice of the step size μ_n in (6.2) to both situations so that it yields convergence of the corresponding iterates to the minimum norm solution $x^\dagger$.

6.1.1 Noise-free case

At first, we consider the case of exact data $y \in \mathcal{R}(A)$. Let $x^\dagger$ be the minimum norm solution of (6.1), which exists, according to Lemma 3.3. To recover $x^\dagger$, we propose the following algorithm.

Algorithm 6.1 (Exact data y and operator A).

(1) If $y = 0$ then $x^\dagger = 0$ and we are done. Else, we start with

(2) Fix $p, q \in (1, \infty)$, choose a constant

$$C \in (0, 1) \tag{6.5}$$

and an initial vector $x_0 \in X$, such that

$$J_q(x_0) \in \overline{\mathcal{R}(A^*)} \quad \text{and} \quad D_q(x^\dagger, x_0) \le \frac{1}{q}\|x^\dagger\|^q\,. \tag{6.6}$$

For $n = 0, 1, 2, \ldots$ repeat

(2) Set

$$R_n := \|Ax_n - y\|\,. \tag{6.7}$$

If $R_n = 0$, then stop iterating.

Else, choose the step size μ_n according to

(a) In case $x_0 = 0$ set

$$\mu_0 := C\frac{(q^*)^{q-1}}{\|A\|^q} R_0^{q-p}\,. \tag{6.8}$$

(b) For all $n \ge 0$ (respectively $n \ge 1$ if $x_0 = 0$), let

$$\lambda_n := \big(\rho_{X^*}(1)\big) \wedge \left(\frac{C}{2^{q^*} G_{q^*}\|A\|}\frac{R_n}{\|x_n\|}\right),$$

where $G_{q^*} > 0$ is the constant from (2.9) (for $p = p^*$). Since X^* is uniformly smooth, we find a $\tau_n \in (0, 1]$ with

$$\frac{\rho_{X^*}(\tau_n)}{\tau_n} = \lambda_n \tag{6.9}$$

due to Theorem 2.48. Then set

$$\mu_n := \frac{\tau_n}{\|A\|}\frac{\|x_n\|^{q-1}}{R_n^{p-1}}\,. \tag{6.10}$$

Compute the new iterate as

$$x_{n+1} = J_{q^*}^{X^*}\big(J_q(x_n) - \mu_n A^* j_p(Ax_n - y)\big)\,. \tag{6.11}$$

Remark 6.2.

(a) The choice of x_0 (6.6) and the definition of the iterates (6.11) guarantee that $J_q(x_n) \in \overline{\mathcal{R}(A^*)}$, for all $n \in \mathbb{N}$. The choice $x_0 = 0$ is always possible, since $D_q(x^\dagger, 0) = \frac{1}{q}\|x^\dagger\|^q$.

(b) If the stopping rule $R_n = 0$ is satisfied for a certain $n \in \mathbb{N}$ then $\|A(x_n - x^\dagger)\| = \|Ax_n - y\| = R_n = 0$ and thus $x - x_n$ lies in $\mathcal{N}(A)$. So, by (a) and Lemma 3.3, $x_n = x^\dagger$.

(c) In proving the convergence of the above method, we will see that (6.6) also ensures that $x_n \neq 0$ for all $n \ge 1$ and thus the parameters τ_n (6.9) are always well-defined.

It remains to prove the convergence of Algorithm 6.1.

Theorem 6.3. *Algorithm* 6.1 *either stops after a finite number of iterations with the minimum norm solution* $x^\dagger$ *of* (6.1) *or the sequence of the iterates* $\{x_n\}$ *converges strongly to* $x^\dagger$.

Proof. If the method stops at step n with $R_n = 0$, we are done, due to Remark 6.2. Otherwise $R_n > 0$, for all $n \geq 0$. The proof of convergence, in that case, will be structured as follows: At first, we show that the sequence $\{\Delta_n\}$, with

$$\Delta_n := D_q(x^\dagger, x_n) = \frac{1}{q^*}\|x_n\|^q + \frac{1}{q}\|x^\dagger\|^q - \langle J_q(x_n), x^\dagger\rangle \tag{6.12}$$

(see (2.10)) obeys a recursive inequality, which implies its convergence. We then deduce that the sequence $\{x_n\}$ has a Cauchy-subsequence. Finally, we show that $\{x_n\}$ converges strongly to $x^\dagger$.

From (6.1) and (2.3), together with (6.12), we deduce

$$\begin{aligned}\Delta_{n+1} &= \frac{1}{q^*}\|J_q(x_n) - \mu_n A^* j_p(Ax_n - y)\|^{q^*} + \frac{1}{q}\|x^\dagger\|^q \\ &\quad - \langle J_q(x_n) - \mu_n A^* j_p(Ax_n - y), x^\dagger\rangle \\ &= \frac{1}{q^*}\|J_q(x_n) - \mu_n A^* j_p(Ax_n - y)\|^{q^*} \\ &\quad + \frac{1}{q}\|x^\dagger\|^q - \langle J_q(x_n), x^\dagger\rangle + \mu_n\langle j_p(Ax_n - y), Ax^\dagger\rangle .\end{aligned} \tag{6.13}$$

In case $x_0 = 0$, we have $R_0 = \|y\| > 0$ and $\Delta_0 = \frac{1}{q}\|x^\dagger\|^q$ and thus (6.13) gives

$$\begin{aligned}\Delta_1 &= \frac{1}{q^*}\mu_0^{q^*}\|A^* j_p(y)\|^{q^*} + \Delta_0 - \mu_0\langle j_p(y), Ax^\dagger\rangle \\ &\leq \frac{1}{q^*}\mu_0^{q^*}\|A\|^{q^*} R_0^{(p-1)q^*} + \Delta_0 - \mu_0 R_0^p ,\end{aligned}$$

since $Ax^\dagger = y$. Choosing μ_0 as in (6.8) yields the estimate

$$\begin{aligned}\Delta_1 &\leq C^{q^*}\frac{q^{*q-1}}{\|A\|^q}R_0^q + \Delta_0 - C\frac{q^{*q-1}}{\|A\|^q}R_0^q \\ &= \Delta_0 - C(1 - C^{q^*-1})\frac{q^{*q-1}}{\|A\|^q}R_0^q\end{aligned}$$

and therefore $\Delta_1 < \Delta_0 = \frac{1}{q}\|x^\dagger\|^q$, which implies that $x_1 \neq 0$.

For all $n \geq 0$ (respectively $n \geq 1$ if $x_0 = 0$), we apply Theorems 2.42 and 2.53 to equation (6.13) and obtain

$$\begin{aligned}\Delta_{n+1} \le \frac{1}{q^*}\Big(&\|J_q(x_n)\|^{q^*} - q^*\langle x_n, \mu_n A^* j_p(Ax_n - y)\rangle \\ &+ \tilde{\sigma}_{q^*}\big(J_q(x_n), \mu_n A^* j_p(Ax_n - y)\big)\Big) \\ &+ \frac{1}{q}\|x^\dagger\|^q - \langle J_q(x_n), x^\dagger\rangle + \mu_n\langle j_p(Ax_n - y), Ax^\dagger\rangle .\end{aligned}$$

With (6.12), this can be written as

$$\begin{aligned}\Delta_{n+1} &\le \Delta_n - \mu_n\langle j_p(Ax_n - y), Ax_n - Ax^\dagger\rangle \\ &\quad + \frac{1}{q^*}\tilde{\sigma}_{q^*}\left(J_q(x_n), \mu_n A^* j_p(Ax_n - y)\right) \\ &= \Delta_n - \mu_n R_n^p + \frac{1}{q^*}\tilde{\sigma}_{q^*}\left(J_q(x_n), \mu_n A^* j_p(Ax_n - y)\right) .\end{aligned} \tag{6.14}$$

Now, we estimate the integrand in the explicit expression for $\tilde{\sigma}_{q^*}$ (2.9). Choosing μ_n according to (6.10) and τ_n as in (6.9) yields, for all $t \in [0, 1]$,

$$\begin{aligned}\|J_q(x_n) - t\mu_n A^* j_p(Ax_n - y)\| \vee \|J_q(x_n)\| &\le \|x_n\|^{q-1} + \mu_n\|A\|R_n^{p-1} \\ &= \|x_n\|^{q-1}(1 + \tau_n) \\ &\le 2\|x_n\|^{q-1}\end{aligned}$$

and

$$\|J_q(x_n) - t\mu_n A^* j_p(Ax_n - y)\| \vee \|J_q(x_n)\| \ge \|J_q(x_n)\| = \|x_n\|^{q-1} .$$

Together with the monotonicity of ρ_{X^*} this gives

$$\begin{aligned}\tilde{\sigma}_{q^*}\left(J_q(x_n), \mu_n A^* j_p(Ax_n - y)\right) &\le q^* G_{q^*}\int_0^1 \frac{\left(2\|x_n\|^{q-1}\right)^{q^*}}{t}\,\rho_{X^*} \\ &\qquad \times \left(t\frac{\mu_n\|A\|R_n^{p-1}}{\|x_n\|^{q-1}}\right)dt \\ &= 2^{q^*} q^* G_{q^*}\|x_n\|^q \int_0^1 \frac{1}{t}\,\rho_{X^*}(t\tau_n)dt \\ &= 2^{q^*} q^* G_{q^*}\|x_n\|^q \int_0^{\tau_n} \frac{1}{t}\,\rho_{X^*}(t)dt \\ &\le 2^{q^*} q^* G_{q^*}\|x_n\|^q \rho_{X^*}(\tau_n) .\end{aligned}$$

Substituting the last estimate into (6.14) we get

$$\begin{aligned}\Delta_{n+1} &\le \Delta_n - \mu_n R_n^p + 2^{q^*} G_{q^*}\|x_n\|^q \rho_{X^*}(\tau_n) \\ &= \Delta_n - \frac{1}{\|A\|}\tau_n\|x_n\|^{q-1}R_n\left(1 - 2^{q^*} G_{q^*}\|A\|\frac{\|x_n\|}{R_n}\frac{\rho_{X^*}(\tau_n)}{\tau_n}\right) .\end{aligned}$$

The choice of C (6.5) and τ_n (6.9) finally gives the recursive inequality

$$\Delta_{n+1} \le \Delta_n - \frac{1-C}{\|A\|}\tau_n \|x_n\|^{q-1} R_n \,. \tag{6.15}$$

Hence, the relation $\Delta_1 < \Delta_0 \le \frac{1}{q}\|x^\dagger\|^q$ holds also if $x_0 \neq 0$, compare (6.6).

Inductively, we obtain, for every admissible choice of the initial vector,

$$0 \le \Delta_{n+1} \le \Delta_n \le \Delta_1 < D_q(x^\dagger, 0) = \frac{1}{q}\|x^\dagger\|^q \tag{6.16}$$

and conclude that $x_n \neq 0$, for all $n \ge 1$, and that the sequence $\{\Delta_n\}$ is non-increasing and therefore convergent and, in particular, bounded. Theorem 2.60 (c) then ensures that the sequence $\{x_n\}$ is bounded, which also implies the boundedness of the sequences $\{J_q(x_n)\}$ and $\{R_n\}$ (6.7).

From (6.15), we further derive

$$0 \le \frac{1-C}{\|A\|}\tau_n \|x_n\|^{q-1} R_n \le \Delta_n - \Delta_{n+1}$$

and thus, for all $N \in \mathbb{N}$, we have

$$0 \le \frac{1-C}{\|A\|}\sum_{n=0}^{N} \tau_n \|x_n\|^{q-1} R_n \le \sum_{n=0}^{N}(\Delta_n - \Delta_{n+1}) = \Delta_0 - \Delta_{N+1} \le \Delta_0 \,,$$

which gives

$$\sum_{n=0}^{\infty} \tau_n \|x_n\|^{q-1} R_n < \infty \,. \tag{6.17}$$

Suppose $\liminf_{n\to\infty} R_n > 0$. Then there exist $n_0 \in \mathbb{N}$ and $\varepsilon > 0$, such that $R_n \ge \varepsilon$ for all $n \ge n_0$ and thus

$$\varepsilon \sum_{n=n_0}^{\infty} \tau_n \|x_n\|^{q-1} \le \sum_{n=n_0}^{\infty} \tau_n \|x_n\|^{q-1} R_n < \infty \,,$$

which forces $\{x_n\}$ to be a null-sequence, since by the boundedness of $\{x_n\}$ and by $R_n \ge \varepsilon$, the sequence $\{\tau_n\}$ remains bounded away from zero, too (6.9). The continuity of $D_q(x^\dagger, \cdot)$, cf. Theorem 2.60 (d), and (6.16) result in

$$\frac{1}{q}\|x^\dagger\|^q = D_q(x^\dagger, 0) = \lim_{n\to\infty} D_q(x^\dagger, x_n) = \lim_{n\to\infty} \Delta_n < \frac{1}{q}\|x^\dagger\|^q \,,$$

which is a contradiction. So, we have $\liminf_{n\to\infty} R_n = 0$ and can thus choose a subsequence $(R_{n_k})_k$ with the property that

$$R_{n_k} \to 0 \quad \text{for} \quad k \to \infty \qquad \text{and} \qquad R_{n_k} < R_n \quad \text{for all} \quad n < n_k \,. \tag{6.18}$$

Then, the same property also holds for every subsequence of $\{R_{n_k}\}_k$. By the boundedness of $\{x_n\}$ and $\{J_q(x_n)\}$ we can thus find a subsequence $\{x_{n_k}\}_k$, having the following properties

(S.1) the sequence of the norms $\{\|x_{n_k}\|\}_k$ is convergent,

(S.2) the sequence $\{J_q(x_{n_k})\}_k$ is weakly convergent,

(S.3) the sequence $\{R_{n_k}\}_k$ satisfies (6.18).

Next, we show that $\{x_{n_k}\}_k$ is a Cauchy-sequence. With (2.10), for all $l, k \in \mathbb{N}$ with $k > l$, we have

$$D_q(x_{n_k}, x_{n_l}) = \frac{1}{q^*}\left(\|x_{n_l}\|^q - \|x_{n_k}\|^q\right) + \langle J_q(x_{n_k}) - J_q(x_{n_l}), x_{n_k}\rangle .$$

Because of (S.1), the first summand converges to zero for $l \to \infty$. The second summand can be written as

$$\begin{aligned}\langle J_q(x_{n_k}) - J_q(x_{n_l}), x_{n_k}\rangle &= \langle J_q(x_{n_k}) - J_q(x_{n_l}), x^\dagger\rangle \\ &\quad + \langle J_q(x_{n_k}) - J_q(x_{n_l}), x_{n_k} - x^\dagger\rangle .\end{aligned}$$

By (S.2), The first summand converges to zero for $l \to \infty$. The second summand can be estimated as

$$|\langle J_q(x_{n_k}) - J_q(x_{n_l}), x_{n_k} - x^\dagger\rangle| = \left|\sum_{n=n_l}^{n_k-1} \langle J_q(x_{n+1}) - J_q(x_n), x_{n_k} - x^\dagger\rangle\right| .$$

The recursive definition of iteration (6.11) yields

$$\begin{aligned}|\langle J_q(x_{n_k}) - J_q(x_{n_l}), x_{n_k} - x^\dagger\rangle| &= \left|\sum_{n=n_l}^{n_k-1} \mu_n \langle j_p(Ax_n - y), Ax_{n_k} - y\rangle\right| \\ &\le \sum_{n=n_l}^{n_k-1} \mu_n \|j_p(Ax_n - y)\|\, \|Ax_{n_k} - y\| \\ &= \frac{1}{\|A\|}\sum_{n=n_l}^{n_k-1} \tau_n \|x_n\|^{q-1} R_{n_k} .\end{aligned}$$

Finally, (S.3) leads to

$$|\langle J_q(x_{n_k}) - J_q(x_{n_l}), x_{n_k} - x^\dagger\rangle| \le \frac{1}{\|A\|}\sum_{n=n_l}^{n_k-1} \tau_n \|x_n\|^{q-1} R_n .$$

By (6.17), the right-hand side converges to zero for $l \to \infty$ and so does $D_q(x_{n_k}, x_{n_l})$. By Theorem 2.60, we conclude that $\{x_{n_k}\}_k$ is a Cauchy-sequence and thus convergent to an $\tilde{x} \in X$. It remains to prove that $\tilde{x} = x^\dagger$ and $\lim_{n\to\infty} \|x_n - x^\dagger\| = 0$. We have

$$R_{n_k} = \|Ax_{n_k} - y\| = \|A(x_{n_k} - x^\dagger)\|,$$

where the left-hand side converges to zero for $k \to \infty$ (S.3). Since A is continuous, the right-hand side converges to $\|A(\tilde{x} - x^\dagger)\|$ for $k \to \infty$ and we see that $\tilde{x} - x^\dagger \in \mathcal{N}(A)$. On the other hand, $J_q(\tilde{x}) \in \overline{\mathcal{R}(A^*)}$, by Remark 6.2 and Theorem 2.53. Together with Lemma 3.3, this shows that $\tilde{x} = x^\dagger$. So, by the continuity of $D_q(x^\dagger, \cdot)$ and with Theorem 2.60, we have

$$\lim_{k\to\infty} \Delta_{n_k} = \lim_{k\to\infty} D_q(x^\dagger, x_{n_k}) = D_q(x^\dagger, x^\dagger) = 0.$$

Since the sequence $\{\Delta_n\}$ is convergent and has a subsequence converging to zero, it must be a null-sequence. By Theorem 2.60 (e), we finally conclude that $\{x_n\}$ converges strongly to $x^\dagger$, which completes the proof. □

6.1.2 Regularization properties

In this section, we analyze the regularization properties of the Landweber method in detail. At first, suppose that, instead of exact data $y \in \mathcal{R}(A)$ and operator $A \in \mathcal{L}(X, Y)$, only some approximations $\{y_k\}_k$ in Y and $\{A_l\}_l$ in $\mathcal{L}(X, Y)$ are available. This is of relevance when we discretize an infinite-dimensional problem or when the operators $\{A_l\}_l$ allow for faster computations of $A_l x_n$. We assume that we know estimates for the deviations

$$\|y_k - y\| \le \delta_k, \quad \delta_k > \delta_{k+1} > 0, \quad \lim_{k\to\infty} \delta_k = 0, \tag{6.19}$$

$$\|A_l - A\| \le \eta_l, \quad \eta_l > \eta_{l+1} > 0, \quad \lim_{l\to\infty} \eta_l = 0. \tag{6.20}$$

Moreover, to properly include the second case (6.20), we need an a priori estimate for the norm of $x^\dagger$, i.e., there is a constant $R > 0$, such that

$$\|x^\dagger\| \le R. \tag{6.21}$$

Further, set

$$S := \sup_{l\in\mathbb{N}} \|A_l\|. \tag{6.22}$$

Algorithm 6.1 has to be modified appropriately to account for the approximations.

Algorithm 6.4 (Noisy data y^δ and operator A_η).

(1) Fix $q, p \in (1, \infty)$.
Choose constants

$$C, D \in (0, 1) \tag{6.23}$$

and an initial vector $x_0 \in X$ such that

$$J_q(x_0) \in \overline{\mathcal{R}(A^*)} \quad \text{and} \quad D_q(x^\dagger, x_0) \leq \frac{1}{q}\|x^\dagger\|^q\,. \tag{6.24}$$

Set $k_{-1} := 0$ and $l_{-1} := 0$.
For $n = 0, 1, 2, \ldots$ repeat

(2) If, for all $k > k_{n-1}$ and all $l > l_{n-1}$,

$$\|A_l x_n - y_k\| < \frac{1}{D}(\delta_k + \eta_l R), \tag{6.25}$$

stop iterating.
Else, find $k_n > k_{n-1}$ and $l_n > l_{n-1}$, with

$$\delta_{k_n} + \eta_{l_n} R \leq D R_n \tag{6.26}$$

where

$$R_n := \|A_{l_n} x_n - y_{k_n}\|\,. \tag{6.27}$$

Choose μ_n according to

(a) In case $x_0 = 0$ set

$$\mu_0 := C(1-D)^{q-1} \frac{q^{*q-1}}{S^q} R_0^{q-p}\,. \tag{6.28}$$

(b) For all $n \geq 0$ (respectively $n \geq 1$ if $x_0 = 0$), set

$$\lambda_n := \big(\rho_{X^*}(1)\big) \wedge \left(\frac{C(1-D)}{2^{q^*} G_{q^*} S} \frac{R_n}{\|x_n\|}\right),$$

where $G_{q^*} > 0$ is the constant from (2.9) and choose $\tau_n \in (0, 1]$, with

$$\frac{\rho_{X^*}(\tau_n)}{\tau_n} = \lambda_n \tag{6.29}$$

and set

$$\mu_n := \frac{\tau_n}{S} \frac{\|x_n\|^{q-1}}{R_n^{p-1}}\,. \tag{6.30}$$

Iterate

$$x_{n+1} = J_{q^*}^{X^*}\big(J_q(x_n) - \mu_n A_{l_n}^* j_p(A_{l_n} x_n - y_{k_n})\big)\,. \tag{6.31}$$

Remark 6.5.

(a) If the stopping rule (6.25) is satisfied for a certain $n \in \mathbb{N}$, then for all $k > k_{n-1}$ and all $l > l_{n-1}$

$$\|A_l x_n - y_k\| < \frac{1}{D}(\delta_k + \eta_l R) .$$

By (6.19) and (6.20), letting $l, k \to \infty$ leads to

$$\|Ax_n - y\| = \|A(x_n - x)\| = 0$$

and, as in the case of exact data, we see that $x_n = x^\dagger$.

(b) Condition (6.26) guarantees that the sequence $(\Delta_n)_n$ is non-increasing.

(c) Algorithm 6.4 suggests specific settings for the parameters μ_n, λ_n and τ_n, as well as for the indices k and n. Note that the particular choice of μ_n, as given in (6.30), is necessary to get the convergence in the proof of Theorem 6.6. This is a crucial difference to the Landweber method in Hilbert spaces, where μ_n may be chosen in a certain interval, depending on $\|A\|$, in order to get convergence and the particular choice of the step size μ_n serves only to accelerate the method.

We prove that the assertions of Theorem 6.3 remain valid.

Theorem 6.6. *Algorithm* 6.4 *either stops after a finite number of iterations, with the minimum norm solution $x^\dagger$ of* (6.1), *or the sequence of the iterates $\{x_n\}$ converges strongly to $x^\dagger$.*

Proof. The proof is rather similar to the one for the case where we have exact data and we only outline the main modifications. If the stopping rule (6.25) is never satisfied then, according to (6.26), (6.19) and (6.20), we always have $R_n > 0$. For the case $x_0 = 0$, we compute

$$\begin{aligned}\Delta_1 &= \frac{1}{q^*}\mu_0^{q^*}\|A_{l_0}^* j_p(y_{k_0})\|^{q^*} + \Delta_0 - \mu_0\langle j_p(y_{k_0}), A_{l_0}x^\dagger\rangle \\ &= \frac{1}{q^*}\mu_0^{q^*}\|A_{l_0}^* j_p(y_{k_0})\|^{q^*} + \Delta_0 - \mu_0\langle j_p(y_{k_0}), y_{k_0}\rangle \\ &\quad + \mu_0\langle j_p(y_{k_0}), y_{k_0} - y\rangle + \mu_0\langle j_p(y_{k_0}), Ax^\dagger - A_{l_0}x^\dagger\rangle .\end{aligned}$$

Because of (6.19), (6.20), (6.21) and (6.22), we get

$$\Delta_1 \le \frac{1}{q^*}\mu_0^{q^*} S^{q^*} R_0^{(p-1)q^*} + \Delta_0 - \mu_0 R_0^p + \mu_0 R_0^{p-1}(\delta_{k_0} + \eta_{k_0} R) .$$

Condition (6.26) and the choice of μ_0 (6.28) give the estimate

$$\begin{aligned}\Delta_1 &\leq \frac{1}{q^*}\mu_0^{q^*} S^{q^*} R_0^{(p-1)q^*} + \Delta_0 - \mu_0 R_0^p(1-D)\\ &= C^{q^*}(1-D)^q \frac{q^{*q-1}}{S^q} R_0^q + \Delta_0 - C(1-D)^q \frac{q^{*q-1}}{S^q} R_0^q\\ &= \Delta_0 - C(1-C^{q^*-1})(1-D)^q \frac{q^{*q-1}}{S^q} R_0^q\end{aligned}$$

and thus $\Delta_1 < \Delta_0$ and, especially, $x_1 \neq 0$. With (6.26) and since

$$\begin{aligned}-\mu_n\langle j_p(A_{l_n}x_n - y_{k_n}, A_{l_n}x_n - A_{l_n}x^\dagger\rangle\\ &= -\mu_n\langle j_p(A_{l_n}x_n - y_{k_n}, A_{l_n}x_n - y_{k_n}\rangle\\ &\quad - \mu_n\langle j_p(A_{l_n}x_n - y_{k_n}, y_{k_n} - y\rangle\\ &\quad - \mu_n\langle j_p(A_{l_n}x_n - y_{k_n}, Ax^\dagger - A_{l_n}x^\dagger\rangle\\ &\leq -\mu_n R_n^p + \mu_n R_n^{p-1}(\delta_{k_n} + \eta_{l_n}R),\end{aligned}$$

the estimate (6.14) becomes, for all $n \geq 0$ (respectively $n \geq 1$ if $x_0 = 0$),

$$\begin{aligned}\Delta_{n+1} &\leq \Delta_n - \mu_n R_n^p + \mu_n R_n^{p-1}(\delta_{k_n} + \eta_{l_n}R)\\ &\quad + \frac{1}{q^*}\tilde{\sigma}_{q^*}\big(J_q(x_n), \mu_n A_{l_n}^* j_p(A_{l_n}x_n - y_{k_n})\big)\\ &\leq \Delta_n - (1-D)\mu_n R_n^p + \frac{1}{q^*}\tilde{\sigma}_{q^*}\big(J_q(x_n), \mu_n A_{l_n}^* j_p(A_{l_n}x_n - y_{k_n})\big).\end{aligned}$$

The last summand can be estimated, analogously to the case of exact data, by

$$\tilde{\sigma}_{q^*}\big(J_q(x_n), \mu_n A_{l_n}^* j_p(A_{l_n}x_n - y_{k_n})\big) \leq 2^{q^*} q^* G_{q^*}\|x_n\|^q \rho_{X^*}(\tau_n)$$

and thus we arrive at

$$\Delta_{n+1} \leq \Delta_n - \frac{(1-D)}{S}\tau_n\|x_n\|^{q-1}R_n\left(1 - \frac{2^{q^*}G_{q^*}S}{1-D}\frac{\|x_n\|}{R_n}\frac{\rho_{X^*}(\tau_n)}{\tau_n}\right).$$

By the choice of τ_n (6.29), we get

$$\Delta_{n+1} \leq \Delta_n - \frac{(1-D)(1-C)}{S}\tau_n\|x_n\|^{q-1}R_n. \tag{6.32}$$

We now proceed as in the proof of Theorem 6.3, after estimate (6.15), while taking into account that

$$
\begin{aligned}
|\langle J_q(x_{n_j})-J_q(x_{n_i}), x_{n_j}-x^\dagger\rangle| &= \left|\sum_{n=n_i}^{n_j-1} \mu_n \langle j_p(A_{l_n}x_n-y_{k_n}), A_{l_n}x_{n_j}-A_{l_n}x^\dagger\rangle\right| \\
&\le \sum_{n=n_i}^{n_j-1} \mu_n \Big(|\langle j_p(A_{l_n}x_n-y_{k_n}), A_{l_{n_j}}x_{n_j}-y_{k_{n_j}}\rangle| \\
&\quad + |\langle j_p(A_{l_n}x_n-y_{k_n}), y_{k_{n_j}}-y\rangle| \\
&\quad + |\langle j_p(A_{l_n}x_n-y_{k_n}), Ax^\dagger-A_{l_n}x^\dagger\rangle| \\
&\quad + |\langle j_p(A_{l_n}x_n-y_{k_n}), A_{l_n}x_{n_j}-A_{l_{n_j}}x_{n_j}\rangle| \Big).
\end{aligned}
$$

With (6.19), (6.20) and (6.21) this leads to

$$
|\langle J_q(x_{n_j})-J_q(x_{n_i}), x_{n_j}-x^\dagger\rangle| \le \sum_{n=n_i}^{n_j-1} \mu_n R_n^{p-1}\big(R_{n_j}+\delta_{k_n}+\eta_{l_n}R+2\eta_{l_n}\|x_{n_j}\|\big).
$$

Let $\tilde{R} > 0$ be a constant, such that $\|x_n\| \le \tilde{R}$, for all $n \in \mathbb{N}$. Then, by (6.26) and property (S.3) of the sequence $(R_{n_j})_j$, we obtain

$$
\begin{aligned}
|\langle J_q(x_{n_j})-J_q(x_{n_i}), x_{n_j}-x^\dagger\rangle| &\le \sum_{n=n_i}^{n_j-1} \mu_n R_n^{p-1}\left(R_{n_j}+DR_n+2\frac{D}{R}R_n\tilde{R}\right) \\
&\le \frac{1+D+2\frac{D\tilde{R}}{R}}{S}\sum_{n=n_i}^{n_j-1} \tau_n \|x_n\|^{q-1} R_n . \qquad \square
\end{aligned}
$$

Finally, we consider the case of noisy data y^δ and a perturbed operator A_η, with known noise-level

$$
\|y-y^\delta\| \le \delta \quad \text{and} \quad \|A-A_\eta\| \le \eta . \tag{6.33}
$$

We apply Algorithm 6.4, with $\delta_k = \delta$ and $\eta_l = \eta$, for all $k, l \in \mathbb{N}$ and use the *discrepancy principle*, see e.g. [67, 150]. To that end, condition (6.25) supplies us with a simple stopping rule: We terminate the iteration (6.31) at $n = n^\delta$, where

$$
n^\delta := \min\{n \in \mathbb{N} : R_n < \frac{1}{D}(\delta+\eta R)\}. \tag{6.34}
$$

As long as $R_n \ge \frac{1}{D}(\delta+\eta R)$, according to (6.32) and Remark 6.5 (b), x_{n+1} is a better approximation to $x^\dagger$ than x_n. As a consequence of this fact and Theorem 6.6 (for $\delta, \eta \to 0$), we obtain a regularizing property of the Landweber method (6.2) that proves the stability of the method with respect to noise.

Corollary 6.7. *Together with the discrepancy principle* (6.34) *as stopping rule, Algorithm* 6.4 *is a regularization method for problem* (6.1).

Remark 6.8. Since the selection j_p needs not to be continuous (and in fact cannot be continuous if J_p is set-valued [48]), the method is an example for regularization with non-continuous mappings.

6.2 Sequential subspace optimization methods

To overcome the slow convergence of the Landweber iteration (6.2) in numerical applications, we suggest using more than the single search direction $A^* j_p^Y(Ax_n - y)$ in each iteration step. Inspired by the fast sequential subspace optimization methods (SESOP), developed by Narkiss and Zibulevsky [169] for large-scale unconstrained optimization and further analyzed in [65], it was shown in [214] that, in case of exact data $\delta = 0$, using multiple search directions $A^* w_{n,i}$, $i \in I_n$ speeds up the iteration considerably. The iterates are then computed as

$$x_{n+1} := J_{q^*}^{X^*}\Big(J_q(x_n) - \sum_{i \in I_n} t_{n,i}\, A^* w_{n,i}\Big), \tag{6.35}$$

where $t_n = (t_{n,i})_{i \in I_n} \in \mathbb{R}^{|I_n|}$ is a minimizer of the function

$$h_n(t) := \frac{1}{q^*}\Big\| J_q(x_n) - \sum_{i \in I_n} t_i\, A^* w_{n,i} \Big\|^{q^*} + \sum_{i \in I_n} t_i\, \langle w_{n,i}\,,\, y\rangle\,.$$

The key idea in obtaining a regularization technique considerably faster than the Landweber method is to put this iteration in the context of Bregman projections. On the one hand, this allows us to interpret the use of arbitrary initial values $x_0 \in X$ in (6.2), as computing the Bregman projection $x^\pi = \Pi^q_{M_{Ax=y}}(x_0)$ of x_0 onto the solution set $M_{Ax=y} := \{x \in X : Ax = y\}$. On the other hand it was shown in [216, Prop. 3.12], that minimizing $h_n(t)$ is equivalent to computing the Bregman projection $x_{n+1} = \Pi^q_{\bigcap_{i \in I_n} H_{n,i}}(x_n)$ of x_n onto the intersection of hyperplanes

$$H_{n,i} := \big\{x \in X : \langle A^* w_{n,i}\,,\, x\rangle - \langle w_{n,i}\,,\, y\rangle = 0\big\}\,.$$

This interpretation is the key to the regularizing sequential subspace optimization methods (RESESOP), which combines acceleration with regularization. The principle idea behind RESESOP is to replace the Bregman projections onto hyperplanes $H_{n,i}$ by Bregman projections onto stripes $H^\delta_{n,i}$, whose width is of the order of the noise-level δ

$$H^\delta_{n,i} := \big\{x \in X : |\langle A^* w^\delta_{n,i}\,,\, x\rangle - \langle w^\delta_{n,i}\,,\, y^\delta\rangle| \le \delta\, \|w^\delta_{n,i}\|\big\}\,.$$

Throughout this section, X is supposed to be q-convex and uniformly smooth and thus reflexive with q^*-smooth and uniformly convex dual X^*. Under these assumptions, the duality mappings J_q and $J_{q^*}^{X^*}$ are both single-valued, uniformly continuous on bounded sets and bijective, with $(J_q)^{-1} = J_{q^*}^{X^*}$, which follows from Theorem 2.53. Analogous to to (2.4), in the q^*-smooth dual X^* the following inequality is valid for all $x^*, y^* \in X^*$, with some constant $G_{q^*} \geq 1$

$$\frac{1}{q^*}\|x^* - y^*\|^{q^*} \leq \frac{1}{q^*}\|x^*\|^{q^*} - \langle J_{q^*}^{X^*}(x^*), y^*\rangle + \frac{G_{q^*}}{q^*}\|y^*\|^{q^*}. \tag{6.36}$$

We furthermore note that the following useful estimates hold, with some constant $c_q > 0$

$$c_q \|x - y\|^q \leq D_q(y, x) \leq \langle J_q(x) - J_q(y), x - y \rangle, \tag{6.37}$$

where the first inequality is due to the q-convexity of X and the second one follows from the three-point identity, stated in Lemma 2.62.

6.2.1 Bregman projections

Let $q \geq 2$ and

$$f_q(x) := \frac{1}{q}\|x\|^q, \quad x \in X.$$

Then $J_q = \partial f_q$, cf. Theorem 2.28. Here again we write D_q instead of D_{j_q}.

Definition 6.9 (Bregman projection). Let $C \subset X$ be a nonempty, closed convex set. The *Bregman projection* of $x \in X$ onto C, with respect to the function f_q, is the unique element $\Pi_C^q(x) \in C$ such that

$$D_q\big(\Pi_C^q(x), x\big) = \min_{z \in C} D_q(z, x).$$

Obviously, we have $\Pi_C^q(x) = x$ iff $x \in C$, $\|\Pi_C^q(0)\| = \min_{z \in C}\|z\|$ and hence

$$x^\dagger = \Pi_{M_{Ax=y}}^q(0).$$

Furthermore, we have the implication

$$\big(C_1 \subset C_2 \text{ and } \tilde{x} := \Pi_{C_2}^q(x) \in C_1\big) \quad \Rightarrow \quad \tilde{x} = \Pi_{C_1}^q(x). \tag{6.38}$$

Bregman projections are characterized by a variational inequality: An element $\tilde{x} \in C$ is the Bregman projection of x onto C with respect to the function f_q, iff

$$\langle J_q(\tilde{x}) - J_q(x), z - \tilde{x}\rangle \geq 0 \quad \text{for all } z \in C. \tag{6.39}$$

The variational inequality (6.39) is equivalent to the descent property

$$D_q(z, \tilde{x}) \leq D_q(z, x) - D_q(\tilde{x}, x) \quad \text{for all } z \in C. \tag{6.40}$$

In Hilbert spaces, the Bregman projection with respect to the function f_2 coincides with the metric projection, but in general they differ from each other. Properties of this kind of Bregman projections and their relation to metric projections can be found in [216]. In particular we have for a closed subspace $U \subset X$ and $x, y, z \in X$, the equivalences

$$x = \Pi^q_{z+U}(y) \Leftrightarrow \left(x - z \in U \text{ and } J_q(x) - J_q(y) \in U^\perp\right)$$
$$\Leftrightarrow J_q(x) = \Pi^{q^*}_{J_q(y)+U^\perp} J_q(z), \tag{6.41}$$

are valid, where $U^\perp \subset X^*$ is the annihilator of U (cf. Definition 2.5) and Π^{q^*} denotes the Bregman projection in the dual X^*, with respect to the function $f^*_{q^*}(x^*) = \frac{1}{q^*}\|x^*\|^q$, cf. Lemma (6.10). To gain more insight into Bregman distances and projections with respect to more general functions than powers of the norm of a Banach space, we refer to e.g. [4], Bauschke et. al. [16] and [35].
For $0 \neq u^* \in X^*$ and $\alpha, \delta \in \mathbb{R}$, $\delta \geq 0$, we denote by $H(u^*, \alpha)$ the *hyperplane*

$$H(u^*, \alpha) := \left\{x \in X : \langle u^*\, , \, x\rangle = \alpha\right\}.$$

We denote by $H_{\leq}(u^*, \alpha)$ the *halfspace*

$$H_{\leq}(u^*, \alpha) := \left\{x \in X : \langle u^*, x\rangle \leq \alpha\right\},$$

and, analogously $H_{\geq}(u^*, \alpha)$, $H_{<}(u^*, \alpha)$, $H_{>}(u^*, \alpha)$. Finally, we denote by $H(u^*, \alpha, \delta)$ the *stripe*.

$$H(u^*, \alpha, \delta) := \left\{x \in X : |\langle u^*, x\rangle - \alpha| \leq \delta\right\}.$$

Obviously, we have $H(u^*, \alpha, \delta) = H_{\leq}(u^*, \alpha + \delta) \cap H_{\geq}(u^*, \alpha - \delta)$, $H(u^*, \alpha, 0)$ reduces to the hyperplane $H(u^*, \alpha)$ and $H_{\geq}(u^*, \alpha) = H_{\leq}(-u^*, -\alpha)$.

To prove the equivalence of the iteration (6.35) and Bregman projections onto intersections of certain hyperplanes, we need some characterizations of Bregman projections, in connection with subspaces of X.

Lemma 6.10. *Let $U \subset X$ be a closed subspace and $x, y, z \in X$ be given. Then the following three assertions are equivalent.*

(a) $x = \Pi^q_{z+U}(y)$,

(b) $x - z \in U$ *and* $J_q(x) - J_q(y) \in U^\perp$,

(c) $J_q(x) = \Pi^{q^*}_{J_q(y)+U^\perp} J_q(z)$.

Proof. According to the variational inequality (6.39), we deduce from (a), that $x - z \in U$ and

$$\langle J_q(x) - J_q(y), (z+u) - x\rangle \geq 0 \quad \text{for all } u \in U$$
$$\Leftrightarrow \langle J_q(x) - J_q(y), z - x\rangle + \langle J_q(x) - J_q(y), u\rangle \geq 0 \quad \text{for all } u \in U. \qquad (6.42)$$

Suppose that there exists an $u_0 \in U$, with $\langle J_q(x) - J_q(y), u_0\rangle \neq 0$, say $\langle J_q(x) - J_q(y), u\rangle < 0$. Since $\lambda u_0 \in U$ for all $\lambda > 0$, (6.42) yields

$$\langle J_q(x) - J_q(y), z - x\rangle + \lambda\langle J_q(x) - J_q(y), u_0\rangle \geq 0\,.$$

But then the left-hand side converges to $-\infty$ for $\lambda \to +\infty$, leading to a contradiction. Hence $J_q(x) - J_q(y) \in U^\perp$. Thus, (a) $\Rightarrow$ (b). Since (b) implies the validity of (6.42), we also have (b) $\Rightarrow$ (a). From Lemma 2.55, we have $U = \overline{U} = {}^\perp(U^\perp)$ and it follows that (b) $\Leftrightarrow$ (c) is just the assertion (b) $\Leftrightarrow$ (a) in the dual space. □

The building blocks of our sequential subspace optimization methods are based on the following examples.

Example 6.11.

(a) Let $H(u_1^*, \alpha_1), \ldots, H(u_N^*, \alpha_N)$ be hyperplanes with nonempty intersection

$$H := \bigcap_{k=1}^{N} H(u_k^*, \alpha_k)\,.$$

Then, the Bregman projection of x onto H is given by

$$\Pi_H^q(x) = J_{q^*}^{X^*}\left(J_q(x) - \sum_{k=1}^{N} \tilde{t}_k\, u_k^*\right), \qquad (6.43)$$

where $\tilde{t} = (\tilde{t}_1, \ldots, \tilde{t}_N)$ is a solution of the N-dimensional optimization problem

$$\min_{t \in \mathbb{R}^N} h(t) := \frac{1}{q^*}\left\| J_q(x) - \sum_{k=1}^{N} t_k\, u_k^*\right\|^{q^*} + \sum_{k=1}^{N} t_k\, \alpha_k\,, \qquad (6.44)$$

where the function h is convex and has continuous partial derivatives

$$\partial_j h(t) = -\left\langle u_j^*\,,\, J_{q^*}^{X^*}\left(J_q(x) - \sum_{k=1}^{N} t_k\, u_k^*\right)\right\rangle + \alpha_j, \quad j = 1, \ldots, N\,.$$

Moreover, if the vectors $u_1^*, \ldots, u_N^*$ are linearly independent, h is strictly convex and $\tilde{t}$ is unique.

(b) Let $H_1 := H_{\le}(u_1^*, \alpha_1)$, $H_2 := H_{\le}(u_2^*, \alpha_2)$ be two halfspaces, with linearly independent vectors u_1^* and u_2^*. Then, $\tilde{x}$ is the Bregman projection of x onto $H_1 \cap H_2$ iff it satisfies the Karush–Kuhn–Tucker conditions for $\min_{z \in H_1 \cap H_2} D_q(z, x)$:

$$\begin{aligned} J_q(\tilde{x}) &= J_q(x) - t_1\, u_1^* - t_2\, u_2^*, \quad t_1, t_2 \ge 0\,, \\ \langle u_1^*\,,\, \tilde{x}\rangle &\le \alpha_1, \quad \langle u_2^*\,,\, \tilde{x}\rangle \le \alpha_2\,, \\ t_1\left(\alpha_1 - \langle u_1^*\,,\, \tilde{x}\rangle\right) &\le 0, \quad t_2\left(\alpha_2 - \langle u_2^*\,,\, \tilde{x}\rangle\right) \le 0\,. \end{aligned} \tag{6.45}$$

(c) For $x \in H_{>}(u^*, \alpha)$ the Bregman projection of x onto the halfspace $H_{\le}(u^*, \alpha)$ is given by

$$\Pi^q_{H_{\le}(u^*,\alpha)}(x) = \Pi^q_{H(u^*,\alpha)}(x) = J_{q^*}^{X^*}\left(J_q(x) - t_+\, u^*\right), \tag{6.46}$$

where $t_+ > 0$ is the unique, necessarily positive solution of

$$\min_{t \in \mathbb{R}} \frac{1}{q^*} \left\| J_q(x) - t\, u^* \right\|^{q^*} + \alpha\, t\,. \tag{6.47}$$

(d) The Bregman projection onto a stripe $H(u^*, \alpha, \delta)$ is given by

$$\Pi^q_{H(u^*,\alpha,\delta)}(x) = \begin{cases} \Pi^q_{H_{\le}(u^*,\alpha+\delta)}(x), & x \in H_{>}(u^*, \alpha+\delta) \\ \quad x \quad , & x \in H(u^*, \alpha, \delta) \\ \Pi^q_{H_{\ge}(u^*,\alpha-\delta)}(x), & x \in H_{<}(u^*, \alpha-\delta) \end{cases}\,. \tag{6.48}$$

Proof. We prove (a) and (b) and refer to [35] for the proof of (c). Example (d) is an immediate consequence of (c).

The convexity of h is obvious. Differentiability and continuity of the partial derivatives are consequences of parts of Theorem 2.53. For any $z \in H$, we can write

$$H = z + \left(\operatorname{span}\{u_1^*, \dots, u_N^*\}\right)^{\perp}.$$

Thus, in view of Proposition 6.10, an element $\tilde{x} \in X$ is the Bregman projection of x_0 onto H iff $\tilde{x} \in H$ and $J_q(\tilde{x}) \in J_q(x_0) + \operatorname{span}\{u_1^*, \dots, u_N^*\}$, i.e.,

$$J_q(\tilde{x}) = J_q(x_0) - \sum_{k=1}^{N} \tilde{t}_k\, u_k^*$$

with some $\tilde{t}_1, \dots, \tilde{t}_N \in \mathbb{R}$, such that $\langle u_k^*\,,\, \tilde{x}\rangle = \alpha_k$ for all $k = 1, \dots, N$. The coefficients $\tilde{t}_k$ are uniquely determined in the case where the vectors $u_1^*, \dots, u_N^*$ are linearly independent. This is equivalent to

$$J_q(\tilde{x}) = \Pi^{q^*}_{J_q(x_0) + \operatorname{span}\{u_1^*, \dots, u_N^*\}}\left(J_q(z)\right),$$

i.e., $\tilde{t} = (\tilde{t}_1, \dots, \tilde{t}_N) \in \mathbb{R}^N$ is a solution of the optimization problem

$$\min_{t \in \mathbb{R}^N} D_{q^*}^{X^*}\left(J_q(x_0) - \sum_{k=1}^{N} t_k\, u_k^*, J_q(z)\right)$$

$$= \min_{t \in \mathbb{R}^N} \frac{1}{q}\|z\|^q - \langle z\,,\, J_q(x_0)\rangle + \sum_{k=1}^{N} t_k \,\langle z\,,\, u_k^*\rangle + \frac{1}{q^*}\left\| J_q(x_0) - \sum_{k=1}^{N} t_k\, u_k^*\right\|^{q^*},$$

which in turn is equivalent to (6.44), since $\langle z\,,\, u_k^*\rangle = \alpha_k$ for $z \in H$. This proves (a).

If $\tilde{x}$ satisfies (6.45), then we have $\tilde{x} \in H_1 \cap H_2$ and, with (6.39), it is straightforward to see that indeed $\tilde{x} = \Pi^q_{H_1 \cap H_2}(x)$. Conversely, let $\tilde{x} := \Pi^q_{H_1\cap H_2}(x)$ and $U := \operatorname{span}\{u_1^*, u_2^*\}$. Then we have $\tilde{x} + U^\perp \subset H_1 \cap H_2$ and it follows with (6.39), that $J_q(\tilde{x}) - J_q(x) \in {}^\perp(U^\perp) = U$ (cf. Lemma 2.55), i.e.,

$$J_q(\tilde{x}) = J_q(x) - t_1\, u_1^* - t_2\, u_2^*, \quad t_1, t_2 \in \mathbb{R}\,.$$

Since u_1^*, u_2^* are linearly independent, we find, for all $\varepsilon \geq 0$, some $z_\varepsilon \in X$ with $\langle u_1^*\,,\, z_\varepsilon\rangle = \alpha_1 - \varepsilon$ and $\langle u_2^*\,,\, z_\varepsilon\rangle = \langle u_2^*\,,\, \tilde{x}\rangle$. Hence, $z_\varepsilon \in H_1 \cap H_2$ and, with (6.39), we get

$$0 \leq \langle J_q(\tilde{x}) - J_q(x)\,,\, z_\varepsilon - \tilde{x}\rangle = -t_1\left((\alpha_1 - \varepsilon) - \langle u_1^*\,,\, \tilde{x}\rangle\right),$$

The assertion follows by, on the one hand, putting $\varepsilon = 0$ and, on the other hand, letting $\varepsilon \to \infty$. □

Remark 6.12. The *metric projection* of $x \in X$ onto a closed convex set C is the unique element $P_C(x) \in C$, such that

$$\|x - P_C(x)\| = \min_{y \in C} \|x - y\|\,.$$

The metric projection is characterized by the variational inequality

$$\langle J_q(P_C(x) - x)\,,\, y - P_C(x)\rangle \geq 0 \quad \text{for all } y \in C \tag{6.49}$$

and the connection to the Bregman projection is given by

$$P_C(x) - x = \Pi^q_{C-x}(0)\,, \quad x \in X$$

which, together with (6.41), for any $x \in X$ gives the interesting splitting

$$x = P_C(x) + J_{q^*}^{X^*}\,\Pi_{C^\perp} J_q^X(x). \tag{6.50}$$

In particular, we have $P_C(0) = \Pi^q_C(0)$ and in Hilbert spaces $P_C = \Pi^2_C$ holds, but in general the metric and Bregman projections differ from each other. We refer to [216] for a proof and to [181] for a general treatment of metric projections. One decisive reason why we consider Bregman projections, rather than metric projections, is that an analogue of the important descent property (6.40) does not exist for metric projections.

6.2.2 The method for exact data (SESOP)

We summarize the following general iteration method, considered in [216], to compute the Bregman projection

$$x^{\pi} = \Pi^{q}_{M_{Ax=y}}(x_0)$$

of $x_0 \in X$ onto the solution set

$$M_{Ax=y} := \{x \in X : Ax = y\}$$

in the case of exact data $y \in \mathcal{R}(A)$. Here, Y is allowed to be an arbitrary, real Banach space. Note that $x^{\pi} = x^{\dagger}$ if $x_0 = 0$, but that, in general, $x^{\pi} \neq x^{\dagger}$.

In Y, we use the normalized duality mapping $j := j_2$ and in the set-valued case we simply write $j(y)$ for an arbitrary but fixed element in the set $J(y) = J_2(y)$.

Algorithm 6.13 (SESOP). Take x_0 as initial value. At iteration $n \in \mathbb{N}$, choose some finite index set I_n and search directions $A^* w_{n,i}$, with $w_{n,i} \in Y^*$, $i \in I_n$, and compute the new iterate as

$$x_{n+1} := \Pi^{q}_{H_n}(x_n) \tag{6.51}$$

with

$$H_n := \bigcap_{i \in I_n} H\big(A^* w_{n,i}, \langle w_{n,i}\,,\, y\rangle\big)\,. \tag{6.52}$$

Note that $M_{Ax=y} \subset H_n$ and therefore, for all $z \in M_{Ax=y}$, we have

$$\langle w_{n,i}\,,\, Ax_{n+1} - y\rangle = \big\langle A^* w_{n,i}\,,\, x_{n+1} - z\big\rangle = 0 \quad \text{for all} \quad i \in I_n\,. \tag{6.53}$$

By Example 6.11 (a), the iterates x_{n+1} can be computed by minimizing a convex, continuously differentiable function $h : \mathbb{R}^{|I_n|} \to \mathbb{R}$. During minimization, the search directions $A^* w_{n,i}$ are fixed and hence function and gradient evaluations of h are independent of the costly applications of the operators A and A^*. Therefore, for large-scale problems, the additional cost of minimizing h with a small number of search directions, instead of only one, is comparatively minor.

Due to (6.41), computing x_{n+1} by (6.51) is also equivalent to

$$J_q(x_{n+1}) = \Pi^{q^*}_{J_q(x_n)+U_n}\big(J_q(z)\big)\,, \quad z \in M_{Ax=y}\,, \tag{6.54}$$

where $U_n \subset \mathcal{R}(A^*)$ denotes the search space

$$U_n := \operatorname{span}\{A^* w_{n,i} : i \in I_n\}\,.$$

This implies $J_q(x_{n+1}) - J_q(x_n) \in U_n$ and an immediate consequence of (6.53) is

$$\big\langle J_q(x_{n+1}) - J_q(x_n)\,,\, x_{n+1} - z\big\rangle = 0, \quad z \in M_{Ax=y}\,. \tag{6.55}$$

For weak convergence of the iterates, it is essential to include the current (sub-)gradient $A^* j(Ax_n - y)$ of the functional $x \mapsto \frac{1}{2}\|Ax - y\|^2$ in the search space U_n, because then an estimate of the form

$$D_q(z, x_{n+1}) \leq D_q(z, x_n) - \frac{R_n^q}{q\, G_{q^*}^{q-1} \|A\|^q} \quad \text{for all} \quad z \in M_{Ax=y} \tag{6.56}$$

holds, where $G_{q^*} \geq 1$ is the constant appearing in (6.36) and $R_n := \|Ax_n - y\|$ is the residual. This estimate implies that $\{D_q(z, x_n)\}_n$ converges decreasingly, that $\{x_n\}$ is bounded and has weak cluster points, and that $\lim_{n\to\infty} R_n = 0$ and consequently each weak cluster point x is a solution $x \in M_{Ax=y}$.

To obtain strong convergence in the infinite-dimensional case, the condition $U_{n-1} \subset U_n$ is sufficient, resulting in increasing dimensions of the search spaces. Next, we show that this stringent condition is in fact not necessary; we simply have to assure that we include the direction $J_q(x_n) - J_q(x_{n_0})$ of a fixed iterate $J_q(x_{n_0})$ to the current iterate $J_q(x_n)$ in the search space U_n, for infinitely many $n \geq n_0$. Observe, that by induction we then have $J_q(x_n) - J_q(x_{n_0}) \in \mathcal{R}(A^*)$.

Proposition 6.14. *Let $A^* j(Ax_n - y) \in U_n$ for all $n \in \mathbb{N}$. Then, the sequence $\{x_n\}$, generated by Algorithm* 6.13, *converges strongly to $x^\pi = \Pi^q_{M_{Ax=y}}(x_0)$ if it has a strongly convergent subsequence. This holds if any of the following conditions is satisfied:*

(a) *X is finite-dimensional,*

(b) *Y is finite-dimensional,*

(c) *$J_q(x_n) - J_q(x_{n_0}) \in U_n$ for some fixed $n_0 \in \mathbb{N}$ and infinitely many $n \geq n_0$.*

Proof. The limit x of a strongly convergent subsequence $\{x_{n_k}\}_k$ satisfies

$$x \in M_{Ax=y} = x^\pi + \mathcal{N}(A) \quad \text{and} \quad J_q(x) - J_q(x_0) \in \overline{\mathcal{R}(A^*)},$$

because, by (6.54) and the continuity of the duality mapping, $J_q(x) - J_q(x_0)$ is the limit of $J_q(x_{n_k}) - J_q(x_0) \in \mathcal{R}(A^*)$. Hence we have $x = \Pi^q_{M_{Ax=y}}(x_0) = x^\pi$ by (6.41). Furthermore, the strong convergence implies $\lim_{k\to\infty} D_q(x^\pi, x_{n_k}) = 0$. Since $\{D_q(z, x_n)\}_n$ converges for all $z \in M_{Ax=y}$, we have $\lim_{n\to\infty} D_q(x^\pi, x_n) = 0$ and thus $\lim_{n\to\infty} \|x_n - x^\pi\| = 0$. In a finite-dimensional space X, the boundedness of $\{x_n\}$ implies the existence of a strongly convergent subsequence. In case $\dim(Y) < \infty$, we also have $\dim(\mathcal{R}(A^*)) < \infty$. Hence, $J_q(x_n) - J_q(x_0) \in \mathcal{R}(A^*)$ has a strongly convergent subsequence, from which we deduce that $\{x_n\}$ has a strongly convergent subsequence too. Finally we assume, without loss of generality, that $J_q(x_{n_k}) - J_q(x_{n_0}) \in U_{n_k}$, for all $k \in \mathbb{N}$ and that $\{x_{n_k+1}\}_k$ converges weakly to some $x \in M_{Ax=y}$. Inserting $A^* w_{n_k,i} = J_q(x_{n_k}) - J_q(x_{n_0})$ into (6.53) yields

$$\langle J_q(x_{n_k}) - J_q(x_{n_0}),\, x_{n_k+1} - x \rangle = 0,$$

and, together with (6.55), we get

$$\begin{aligned}\langle J_q(x_{n_k+1}) - J_q(x), x_{n_k+1} - x\rangle &= \langle J_q(x_{n_k+1}) - J_q(x_{n_k}), x_{n_k+1} - x\rangle \\ &\quad + \langle J_q(x_{n_k}) - J_q(x_{n_0}), x_{n_k+1} - x\rangle \\ &\quad + \langle J_q(x_{n_0}) - J_q(x), x_{n_k+1} - x\rangle \\ &= \langle J_q(x_{n_0}) - J_q(x), x_{n_k+1} - x\rangle .\end{aligned}$$

Since the right hand side converges to zero for $k \to \infty$, so does the left hand side and, by (6.37), we conclude that $\{x_{n_k+1}\}_k$ converges strongly to x. □

6.2.3 The regularization method for noisy data (RESESOP)

In case only noisy data $y^\delta \in Y$ is given, with known noise-level $\|y - y^\delta\| \le \delta$, we propose the following modification of Algorithm 6.13, in order to compute a regularized version of $x^\pi = \Pi^q_{M_{Ax=y}}(x_0)$. As in the previous section, Y is allowed to be an arbitrary real Banach space.

Algorithm 6.15 (RESESOP). Take $x_0^\delta = x_0$ as initial value and fix some constant $\tau > 1$. At iteration $n \in \mathbb{N}$, choose some finite index set I_n^δ and search directions $A^* w_{n,i}^\delta \in G_n^\delta \cap D_n^\delta$, see (6.61) and (6.62) below. If the residual $R_n^\delta := \|Ax_n^\delta - y^\delta\|$ satisfies the discrepancy principle

$$R_n^\delta \le \tau\,\delta \tag{6.57}$$

stop iterating. Otherwise, compute the new iterate as

$$x_{n+1}^\delta := J_{q^*}^{X^*}\Big(J_q(x_n^\delta) - \sum_{i\in I_n^\delta} t_{n,i}^\delta\, A^* w_{n,i}^\delta\Big), \tag{6.58}$$

where $t_n^\delta = (t_{n,i}^\delta)_{i\in I_n^\delta}$ is chosen such that

$$x_{n+1}^\delta \in H_n^\delta := \bigcap_{i\in I_n^\delta} H\big(A^* w_{n,i}^\delta, \langle w_{n,i}^\delta, y^\delta\rangle, \delta\,\|w_{n,i}^\delta\|\big) \tag{6.59}$$

and such that an inequality of the form

$$D_q(z, x_{n+1}^\delta) \le D_q(z, x_n^\delta) - C\,(R_n^\delta)^q \quad \text{for all} \quad z \in M_{Ax=y} \tag{6.60}$$

holds, for some constant $C > 0$.

Here, we consider in particular the canonical sets of search directions (see also [169])

$$G_n^\delta := \big\{A^* j(Ax_k^\delta - y^\delta) : 0 \le k \le n\big\}, \tag{6.61}$$

and

$$D_n^\delta := \left\{ J_q(x_k^\delta) - J_q(x_l^\delta) : 0 \le l < k \le n \right\}. \tag{6.62}$$

This means that, for $A^* w_{n,i}^\delta \in G_n^\delta$, we have

$$w_{n,i}^\delta = j(Ax_k^\delta - y^\delta), \quad 0 \le k \le n \tag{6.63}$$

with

$$\|w_{n,i}^\delta\| = R_k^\delta \quad \text{and} \quad \langle w_{n,i}^\delta , y^\delta \rangle = \langle A^* w_{n,i}^\delta , x_k^\delta \rangle - (R_k^\delta)^2 . \tag{6.64}$$

Hence, for this kind of search directions, we actually only need the vectors $A^* w_{n,i}^\delta$ and do not have to store the vectors $w_{n,i}^\delta$.

For $A^* w_{n,i}^\delta \in D_n^\delta$ we have

$$w_{n,i}^\delta = v_{k,l}^\delta \quad \text{such that} \quad A^* v_{k,l}^\delta = J_q(x_k^\delta) - J_q(x_l^\delta), \quad 0 \le l < k \le n \tag{6.65}$$

with the recursion $v_{l,l}^\delta := 0$ and

$$v_{k,l}^\delta = v_{k-1,l}^\delta - \sum_{i \in I_{k-1}^\delta} t_{k-1,i}^\delta \, w_{k-1,i}^\delta . \tag{6.66}$$

We can exploit this recursion to compute $\langle w_{n,i}^\delta , y^\delta \rangle$ without using $w_{n,i}^\delta$, which is convenient in the noise-free case (6.52). However, to compute $\|w_{n,i}^\delta\|$, we actually need the vector $w_{n,i}^\delta$. Hence, for noisy data, using this kind of search directions, seems a little more involved.

Note that $M_{Ax=y} \subset H_n^\delta$ because, for all $z \in M_{Ax=y}$, we have

$$|\langle A^* w_{n,i}^\delta , z \rangle - \langle w_{n,i}^\delta , y^\delta \rangle| = |\langle w_{n,i}^\delta , y - y^\delta \rangle| \le \delta \, \|w_{n,i}^\delta\| . \tag{6.67}$$

For exact data, we have $H_n^{\delta=0} = H_n$. This implies, by (6.58) and (6.41), that $x_{n+1}^{\delta=0} = \Pi_{H_n}^q(x_n^{\delta=0})$, i.e., Algorithm 6.15 coincides with Algorithm 6.13. Therefore, in the noise-free case, we simply drop the index δ everywhere.

Inequality (6.60) assures that the sequence $\{D_q(z, x_n^\delta)\}_n$ decreases for fixed δ. Hence, the discrepancy principle (6.57) indeed yields a finite stopping index

$$n_* = n_*(\delta) := \min\{n \in \mathbb{N} \,,\; R_n^\delta \le \tau\,\delta\} . \tag{6.68}$$

To avoid exceptions, we define $x_n^\delta := x_{n_*}^\delta$ for all $n \ge n_*$.

An admissible strategy to choose t_n^δ such that (6.59) and (6.60) are valid is, for instance, the following: Consider (6.59) as a convex feasibility problem (CFP), which can be solved by cyclically projecting

$$H_{n,i}^\delta := H\left(A^* w_{n,i}^\delta, \langle w_{n,i}^\delta , y^\delta \rangle, \delta \, \|w_{n,i}^\delta\|\right),$$

onto the individual sets as

$$z_0 := x_n^\delta, \quad z_{j+1} := \Pi^q_{H^\delta_{n,(j \bmod |I_n^\delta|)}}(z_j), \quad x_{n+1}^\delta := \lim_{j\to\infty} z_j \,.$$

Projecting first onto the stripe, corresponding to the current subgradient $A^* j(Ax_n^\delta - y^\delta)$, assures that (6.60) holds. We do not want to go into detail here but refer to Section 6.3 for the solution of CFPs in Banach spaces. In Subsection 6.2.3 we furthermore present a fast method to compute x_{n+1}^δ such that (6.59) and (6.60) hold for the special case of two search directions.

We proceed now by analyzing under which conditions the iterates x_n^δ depend continuously on δ, for fixed index n. In doing so, we need a well-known fact about invertibility of linear operators.

Lemma 6.16. *Let E be a bijective continuous linear operator between Banach spaces. Then, all continuous linear operators E^δ, with $\|E - E^\delta\| < \frac{1}{\|E^{-1}\|}$, are also bijective and we have*

$$\|(E^\delta)^{-1}\| \le \frac{1}{\frac{1}{\|E^{-1}\|} - \|E - E^\delta\|} \tag{6.69}$$

as well as

$$\|E^{-1} - (E^\delta)^{-1}\| \le \|E^{-1}\|\,\|(E^\delta)^{-1}\|\,\|E - E^\delta\| \,. \tag{6.70}$$

The next proposition contains the desired continuity result with respect to δ.

Proposition 6.17. *Let Y be uniformly smooth. For fixed index n we have*

$$\lim_{\delta\to 0} \|x_n^\delta - x_n\| = 0 \,, \tag{6.71}$$

if we assume that in previous iterations $k \le n-1$, we use the same number and kind of search directions for all noise-levels $\delta_1, \delta_2 \ge 0$ and the respective search directions for exact data $\delta = 0$ are linearly independent. More precisely, we assume $I_k^{\delta_1} = I_k^{\delta_2}$ and that, if $A^* w_{k,i}^{\delta_1}$ is of the form $A^* j(Ax_j^{\delta_1} - y^{\delta_1})$ (respectively $J_q(x_j^{\delta_1}) - J_q(x_l^{\delta_1})$), $A^* w_{k,i}^{\delta_2}$ is of the form $A^* j(Ax_j^{\delta_2} - y^{\delta_2})$ (respectively $J_q(x_j^{\delta_2}) - J_q(x_l^{\delta_2})$) as well.

Proof. We inductively show $\lim_{\delta\to 0} x_k^\delta = x_k$ and $\lim_{\delta\to 0} w_{k,i}^\delta = w_{k,i}$, for all $k \le n$, $i \in I_k$. For the latter, it suffices to show $\lim_{\delta\to 0} j(Ax_k^\delta - y^\delta) = j(Ax_k - y)$ and $\lim_{\delta\to 0} v_{k,l}^\delta = v_{k,l}$, with $v_{k,l}^\delta$ such that $A^* v_{k,l}^\delta = J_q(x_k^\delta) - J_q(x_l^\delta)$, $l \le k$.

For $k = 0$ we have, trivially, $v_{0,0}^\delta = 0 = v_{0,0}$ and $x_0^\delta = x_0$. Since the duality mapping is continuous in a uniformly smooth Y, we also have $\lim_{\delta\to 0} j(Ax_0^\delta - y^\delta) = j(Ax_0 - y)$.

Next, we prove the induction step $k \to k+1$. By (6.58), we have

$$J_q(x_k^\delta) - J_q(x_{k+1}^\delta) = \sum_{i \in I_k} t_{k,i}^\delta \, A^* w_{k,i}^\delta \,, \tag{6.72}$$

where, according to our assumption, the search directions $\{A^* w_{k,i} : i \in I_k\} \subset X^*$ for exact data are linearly independent. Hence, we find a dual basis $\{u_{k,i} : i \in I_k\} \subset X$ such that

$$\langle A^* w_{k,i} \,, u_{k,j} \rangle = \begin{cases} 1, & i = j \\ 0, & i \neq j \end{cases} . \tag{6.73}$$

Applying the dual basis to (6.72), we obtain the equations

$$\langle J_q(x_k^\delta) - J_q(x_{k+1}^\delta) \,, u_{k,j} \rangle = \sum_{i \in I_k} t_{k,i}^\delta \, \langle A^* w_{k,i}^\delta \,, u_{k,j} \rangle, \quad j \in I_k \,. \tag{6.74}$$

We can write these equations in the more convenient form

$$E^\delta t^\delta = d^\delta \tag{6.75}$$

with the matrix $E^\delta := (\langle A^* w_{k,i}^\delta \,, u_{k,j} \rangle)_{j,i \in I_k}$ and the vectors $t^\delta := (t_{k,i}^\delta)_{i \in I_k}$ and $d^\delta := (\langle J_q(x_k^\delta) - J_q(x_{k+1}^\delta) \,, u_{k,j} \rangle)_{j \in I_k}$. By (6.73), the matrix $E = E^0$, corresponding to exact data $\delta = 0$, is the identity matrix and, by the induction hypothesis, we have

$$\lim_{\delta \to 0} \|E - E^\delta\| = 0 \,,$$

where we may take as matrix norm the one induced by the max-norm in $\mathbb{R}^{|I_k|}$. From Lemma 6.16, we infer that E^δ is bijective for δ small enough and, together with (6.69) and (6.70), we may assume

$$\|(E^\delta)^{-1}\| \le 2 \quad \text{and} \quad \|E - (E^\delta)^{-1}\| \le 2 \, \|E - E^\delta\| \,. \tag{6.76}$$

Since, by (6.60), the vectors d^δ remain bounded with respect to δ, we deduce from (6.75) and (6.76) that the coefficients t^δ also remain bounded. Thus, to every null-sequence of noise-levels $\{\delta_l\}_l$, we can find a convergent subsequence of the coefficients t^δ so that, without loss of generality, $\lim_{l \to \infty} t^{\delta_l} = \tilde{t}$ for some $\tilde{t} \in \mathbb{R}^{|I_k|}$. Together with (6.72) and the induction hypothesis, existence of the strong limit

$$J_q(\tilde{x}_{k+1}) := \lim_{l \to \infty} J_q(x_{k+1}^{\delta_l}) = J_q(x_k) - \sum_{i \in I_k} \tilde{t}_{k,i} \, A^* w_{k,i}$$

follows. Consequently, $x_{k+1}^{\delta_l}$ converges strongly to $\tilde{x}_{k+1}$ for $l \to \infty$ and, by (6.59), we get

$$\begin{aligned} |\langle A^* w_{k,i} \,, \tilde{x}_{k+1} \rangle - \langle w_{k,i} \,, y \rangle| &= \lim_{l \to \infty} |\langle A^* w_{k,i}^{\delta_l} \,, x_{k+1}^{\delta_l} \rangle - \langle w_{k,i}^{\delta_l} \,, y^{\delta_l} \rangle| \\ &\le \lim_{l \to \infty} \delta_l \, \|w_{k,i}^{\delta_l}\| = 0 \,. \end{aligned}$$

Hence, $\tilde{x}_{k+1} \in H_k = x^\pi + U_k^\perp$ and $J_q(\tilde{x}_{k+1}) - J_q(x_k) \in U_k$ which, by (6.41), implies $\tilde{x}_{k+1} = \Pi_{H_k}^q(x_k) = x_{k+1}$. Since this holds for every null-sequence $\{\delta_l\}_l$, we conclude that indeed

$$\lim_{\delta\to 0} x_{k+1}^\delta = x_{k+1} \,.$$

From this, we infer that

$$\lim_{\delta\to 0} j(Ax_{k+1}^\delta - y^\delta) = j(Ax_{k+1} - y)$$

and that $\lim_{\delta\to 0} d^\delta = d$. Together with (6.75) and (6.76), we further get $\lim_{\delta\to 0} t^\delta = t$, from which, with (6.66) and the induction hypothesis, we finally deduce

$$\lim_{\delta\to 0} v_{k+1,l}^\delta = v_{k+1,l}, \quad l \le k+1 \,. \qquad \square$$

We give two possible choices for search directions, guaranteeing that the assumption of linear independence in Proposition 6.17 is satisfied.

Lemma 6.18. *In case of exact data, as long as $R_n \neq 0$, the following choices yield sets of linearly independent search directions:*

(a) *Choose $\{A^* j(Ax_k - y) : k_n \le k \le n\}$, with $0 \le k_{n-1} \le k_n$, for all $n \in \mathbb{N}$.*

(b) *Choose $A^* j(Ax_0 - y)$ and then $\{A^* j(Ax_n - y), J_q(x_n) - J_q(x_0)\}$, for all $n \ge 1$.*

Proof. Choice (a): Because of the orthogonality relations (6.53), applying $x_n - x^\pi$ to

$$0 = \sum_{k=k_n}^{n} \lambda_k \, A^* j(Ax_k - y), \quad \lambda_k \in \mathbb{R}\,,$$

yields $0 = \lambda_n \langle A^* j(Ax_n - y)\,,\, x_n - x^\pi\rangle = \lambda_n R_n^2$. Hence, $\lambda_n = 0$ and, by subsequently applying $x_k - x^\pi$, it follows inductively that $\lambda_k = 0$ for $k = n-1, \dots, k_n$.

Choice (b): By (6.56), we have $J_q(x_n) - J_q(x_0) \neq 0$ for $n \ge 1$ and, by (6.53) and (6.55), we get

$$\big\langle J_q(x_n) - J_q(x_0)\,,\, x_n - x^\pi \big\rangle = 0\,.$$

Therefore, applying $x_n - x^\pi$ to

$$0 = \lambda\, A^* j(Ax_n - y) + \mu\, \big(J_q(x_n) - J_q(x_0)\big), \quad \lambda, \mu \in \mathbb{R}\,,$$

yields $0 = \lambda \, \langle A^* j(Ax_n - y)\,,\, x_n - x^\pi \rangle = \lambda\, R_n^2$. Hence, $\lambda = 0 = \mu$. $\square$

Proposition 6.19 contains the main result of Section 6.2.3, showing that Algorithm 6.15 turns into a regularization method for the computation of $x^\pi = \Pi_{M_{Ax=y}}^q(x_0)$.

Proposition 6.19. *Let* $A^* j(Ax_n^\delta - y^\delta) \in U_n^\delta$ *for all* $n \in \mathbb{N}$ *and* $\delta \geq 0$, *and let* $n_* = n_*(\delta)$ *be the stopping index* (6.68), *according to the discrepancy principle. Then,*

$$\lim_{\delta \to 0} \|x_{n_*(\delta)}^\delta - x^\pi\| = 0$$

if any of the conditions (a)–(c) holds true:

(a) *X is finite-dimensional*

(b) *Y is finite-dimensional*

(c) *We have* $\lim_{n\to\infty} \|x_n - x^\pi\| = 0$ *and* $\lim_{\delta\to 0} \|x_n^\delta - x_n\| = 0$, *for fixed indices* n. *According to Propositions* 6.14 *and* 6.17 *and Lemma* 6.18 (b), *this holds e.g., if Y is uniformly smooth and if we choose as search directions* $A^* j(Ax_0 - y^\delta)$ *and then* $\{A^* j(Ax_n^\delta - y^\delta), J_q(x_n^\delta) - J_q(x_0)\}$, *for all* $n \geq 1$.

Proof. Let $\{\delta_l\}_l$ be an arbitrary null-sequence of noise-levels. For the reader's convenience, we define $n_l := n_*(\delta_l)$ and $x^l := x_{n_l}^{\delta_l}$. From (6.60), we deduce that $\{x^l\}_l$ is bounded. Furthermore, each weak cluster point x is a solution $x \in M_{Ax=y}$ because, by (6.57), we have $R_{n_l}^{\delta_l} \leq \tau\, \delta_l \to 0$ for $l \to \infty$. For strong convergence of $\{x^l\}_l$ to x^π, it suffices again to show that every subsequence in turn has a subsequence converging strongly to x^π. In the finite-dimensional cases (a) and (b), this follows by showing that, as in the case of exact data, each subsequence has a strong cluster point x, satisfying $J_q(x) - J_q(x_0) \in \overline{\mathcal{R}(A^*)}$, which implies $x = \Pi_{M_{Ax=y}}^q(x_0) = x^\pi$. It remains to show the assertion for (c). Without loss of generality, we may assume that the sequence $\{n_l\}_l$ is increasing. Due to (6.60) and the assumptions in (c), we then find to every $k \in \mathbb{N}$ some $n_k, l_k \in \mathbb{N}$ such that, for all $l \geq \max\{k, l_k\}$, the following chain of inequalities holds

$$D_q(x^\pi, x^l) = D_q\big(x^\pi, x_{n_l}^{\delta_l}\big) \leq D_q\big(x^\pi, x_{n_k}^{\delta_l}\big) \leq D_q(x^\pi, x_{n_k}) + \frac{1}{k} \leq \frac{2}{k},$$

which implies strong convergence of $\{x^l\}_l$ to x^π. □

Remark 6.20. We emphasize, that in the finite-dimensional cases (a) and (b), we did not make any smoothness assumption about Y and regularization can be shown without continuous dependence of the iterates x_n^δ on δ, for fixed indices n.

We conclude this section, by describing a fast way to compute x_{n+1}^δ in Algorithm 6.15, such that (6.59) and (6.60) hold, in case only two search directions are used. This strategy is based on a geometrical fact, which is quite obvious in Hilbert spaces and which is illustrated in Figure 6.1.

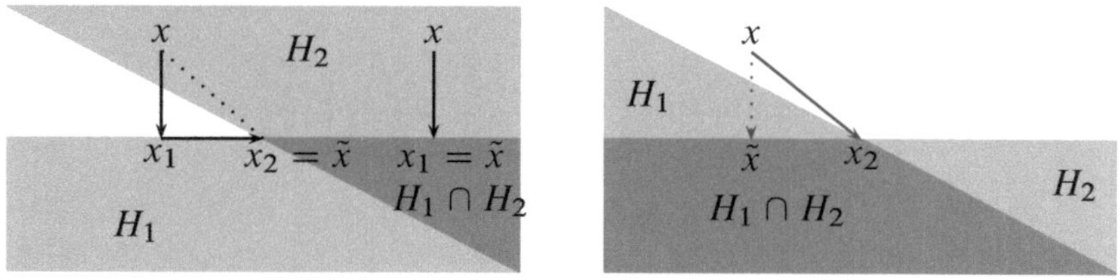

Figure 6.1. The orthogonal, respectively Bregman projection, $\tilde{x} = \Pi_{H_1 \cap H_2}(x)$ of x onto the intersection of the halfspaces H_1 and H_2 can be computed by at most two projections onto (intersections of) the bounding hyperplanes, if x is already contained in one of the halfspaces (left); otherwise this need not to be true (right).[1]

Proposition 6.21. *Let $H_1 := H_{\le}(u_1^*, \alpha_1)$, $H_2 := H_{\le}(u_2^*, \alpha_2)$ be two halfspaces with nonempty intersection. If $x \in H_{>}(u_1^*, \alpha_1) \cap H_2$, the Bregman projection of x onto $H_1 \cap H_2$ can be computed by at most two Bregman projections onto (intersections of) the bounding hyperplanes, with the following two steps:*

(1) *Compute*

$$x_1 := \Pi^q_{H(u_1^*, \alpha_1)}(x) .$$

Then, for all $z \in H_1$, we have

$$D_q(z, x_1) \le D_q(z, x) - \frac{1}{p\, G_{q^*}^{q-1}} \left(\frac{\langle u_1^*, x \rangle - \alpha_1}{\|u_1^*\|} \right)^q , \tag{6.77}$$

where $G_{q^*} \ge 1$ is the constant appearing in (6.36). *If $x_1 \in H_2$, we already have $x_1 = \Pi^q_{H_1 \cap H_2}(x)$, i.e., we are done. Otherwise, go to step*

(2) *Compute*

$$x_2 := \Pi^q_{H(u_1^*, \alpha_1) \cap H(u_2^*, \alpha_2)}(x_1) .$$

Then, we have $x_2 = \Pi^q_{H_1 \cap H_2}(x)$ and, for all $z \in H_1 \cap H_2$,

$$\begin{aligned} D_q(z, x_2) \le D_q(z, x) &- \frac{1}{q\, G_{q^*}^{q-1}} \\ &\times \left[\left(\frac{\langle u_1^*, x \rangle - \alpha_1}{\|u_1^*\|} \right)^q + \left(\frac{\langle u_2^*, x_1 \rangle - \alpha_2}{\gamma\, \|u_2^*\|} \right)^q \right] \end{aligned} \tag{6.78}$$

---

[1] This figure is reproduced from the article of F. Schöpfer and T. Schuster, *Fast regularizing sequential subspace optimization in Banach spaces*, published in Inverse Problems, vol. 25(1), 015013, doi:10.1088/0266-5611/25/1/015013, 2009, with kind allowance of IoP Publishing, Bristol, UK.

with

$$\gamma := \left(1 - \frac{1}{(q-1)\,G_{q^*}^{q-1}} \left(\frac{|\langle u_1^*\,,\,J_{q^*}^{X^*}(u_2^*)\rangle|}{\|u_1^*\|\,\|J_{q^*}^{X^*}(u_2^*)\|}\right)^q\right)^{\frac{1}{q^*}} \in (0,1]\,. \tag{6.79}$$

Proof. Step (1). Let $x_1 := \Pi^q_{H(u_1^*,\alpha_1)}(x)$. Since $x \in H_>(u_1^*,\alpha_1)$, we know, by (6.46) and (6.48), that we also have $x_1 = \Pi^q_{H_\leq(u_1^*,\alpha_1)}(x) = \Pi^q_{H_1}(x)$. If $x_1 \in H_2$, we have $x_1 \in H_1 \cap H_2 \subset H_1$ and, by (6.38), we conclude that indeed $x_1 = \Pi^q_{H_1\cap H_2}(x)$. By (6.46) we furthermore know that

$$J_q(x_1) = J_q(x) - t_1\,u_1^*, \quad t_1 > 0\,,$$

where t_1 minimizes the function

$$h_1(t) := \frac{1}{q^*}\,\big\|J_q(x) - t\,u_1^*\big\|^{q^*} + t\,\alpha_1\,.$$

Hence, $h_1(t_1) \leq h_1(\tilde{t}_1)$, with

$$\tilde{t}_1 := \left(\frac{\langle u_1^*\,,\,x\rangle - \alpha_1}{G_{q^*}\,\|u_1^*\|^{q^*}}\right)^{q-1} > 0 \tag{6.80}$$

and, since $t_1 > 0$, we get, for all $z \in H_1$

$$\begin{aligned}
D_q(z,x_1) &= \frac{1}{q^*}\,\big\|J_q(x) - t_1\,u_1^*\big\|^{q^*} + t_1\,\langle u_1^*\,,\,z\rangle - \langle J_q(x)\,,\,z\rangle + \frac{1}{q}\|z\|^q \\
&\leq \frac{1}{q^*}\,\big\|J_q(x) - t_1\,u_1^*\big\|^{q^*} + t_1\,\alpha_1 - \langle J_q(x)\,,\,z\rangle + \frac{1}{q}\|z\|^q \\
&\leq h_1(\tilde{t}_1) - \langle J_q(x)\,,\,z\rangle + \frac{1}{q}\|z\|^q\,.
\end{aligned} \tag{6.81}$$

We estimate $h_1(\tilde{t}_1)$ with (6.36) and get

$$\begin{aligned}
h_1(\tilde{t}_1) &\leq \frac{1}{q^*}\|x\|^q - \tilde{t}_1\left(\langle u_1^*\,,\,x\rangle - \alpha_1\right) + \frac{G_{q^*}}{q^*}\,\tilde{t}_1^{q^*}\,\|u_1^*\|^{q^*} \\
&= \frac{1}{q^*}\|x\|^q - \frac{1}{q\,G_{q^*}^{q-1}}\left(\frac{\langle u_1^*\,,\,x\rangle - \alpha_1}{\|u_1^*\|}\right)^q.
\end{aligned} \tag{6.82}$$

Inserting this estimate into (6.81) yields (6.77).

Step (2). At first we show that, if u_1^* and u_2^* are linearly dependent, we must have $x_1 \in H_2$, i.e., we already finish in Step (1). Let $u_2^* = \lambda\,u_1^*$, $\lambda \in \mathbb{R}$. If $\lambda < 0$, we get for $z \in H_1 \cap H_2$

$$\alpha_2 \geq \langle u_2^*\,,\,z\rangle = \lambda\,\langle u_1^*\,,\,z\rangle \geq \lambda\,\alpha_1\,,$$

and for $\lambda > 0$ we get, with $x \in H_>(u_1^*, \alpha_1) \cap H_2$,

$$\alpha_2 \geq \langle u_2^*, x \rangle = \lambda \langle u_1^*, x \rangle > \lambda \alpha_1 .$$

Hence, $\lambda \alpha_1 \leq \alpha_2$ and $\langle u_2^*, x_1 \rangle = \lambda \langle u_1^*, x_1 \rangle = \lambda \alpha_1 \leq \alpha_2$, showing that $x_1 \in H_2$.

Let $x_1 \notin H_2$, i.e., $x_1 \in H_>(u_2^*, \alpha_2)$. As we have just shown, u_1^* and u_2^* must then be linearly independent, implying that the intersection $H(u_1^*, \alpha_1) \cap H(u_2^*, \alpha_2)$ is not empty. Therefore, we can compute

$$x_2 = \Pi^q_{H(u_1^*, \alpha_1) \cap H(u_2^*, \alpha_2)}(x_1)$$

and (6.43) yields

$$J_q(x_2) = J_q(x_1) - t_{2,1}\, u_1^* - t_{2,2}\, u_2^*, \quad t_{2,1}, t_{2,2} \in \mathbb{R} .$$

Hence, on the one hand we get

$$\begin{aligned} 0 < \langle J_q(x_2) - J_q(x_1), x_2 - x_1 \rangle &= -t_{2,1} \langle u_1^*, x_2 - x_1 \rangle - t_{2,2} \langle u_2^*, x_2 - x_1 \rangle \\ &= -t_{2,2} \big(\alpha_2 - \langle u_2^*, x_1 \rangle \big) , \end{aligned}$$

implying $t_{2,2} > 0$. And on the other hand we get

$$\begin{aligned} 0 &< \langle J_q(x_2) - J_q(x), x_2 - x \rangle \\ &= -(t_{2,1} + t_1) \langle u_1^*, x_2 - x \rangle - t_{2,2} \langle u_2^*, x_2 - x \rangle \\ &= -(t_{2,1} + t_1) \big(\alpha_1 - \langle u_1^*, x \rangle \big) - t_{2,2} \big(\alpha_2 - \langle u_2^*, x \rangle \big) . \end{aligned}$$

From this we deduce, with $x \in H_>(u_1^*, \alpha_1) \cap H_2$,

$$-(t_{2,1} + t_1) \big(\alpha_1 - \langle u_1^*, x \rangle \big) > t_{2,2} \big(\alpha_2 - \langle u_2^*, x \rangle \big) \geq 0$$

implying $t_{2,1} + t_1 > 0$. With (6.45), we conclude that indeed $x_2 = \Pi^q_{H_1 \cap H_2}(x)$. It remains to show (6.78). Since $t_{2,1} + t_1 > 0$ and $t_{2,2} > 0$ we get, for all $z \in H_1 \cap H_2$,

$$\begin{aligned} D_q(z, x_2) &\leq \frac{1}{q^*} \left\| J_q(x) - (t_{2,1} + t_1)\, u_1^* - t_{2,2}\, u_2^* \right\|^{q^*} + (t_{2,1} + t_1)\, \alpha_1 \\ &\quad + t_{2,2}\, \alpha_2 - \langle J_q(x), z \rangle + \frac{1}{q} \|z\|^q \\ &= h_{1,2}(t_{2,1}, t_{2,2}) + t_1\, \alpha_1 - \langle J_q(x), z \rangle + \frac{1}{q} \|z\|^q , \end{aligned} \tag{6.83}$$

where, by (6.43), we know that $(t_{2,1}, t_{2,2})$ minimizes the function

$$h_{1,2}(s_1, s_2) := \frac{1}{q^*} \left\| J_q(x_1) - s_1\, u_1^* - s_2\, u_2^* \right\|^{q^*} + s_1\, \alpha_1 + s_2\, \alpha_2 .$$

Hence, we can estimate $h_{1,2}(t_{2,1}, t_{2,2}) \le h_{1,2}(\tilde{s}_1 \cdot \tilde{t}_{2,2}, \tilde{t}_{2,2})$ with

$$\tilde{s}_1 := -\left(\frac{|\langle u_1^*, J_{q^*}^{X^*}(u_2^*)\rangle|}{G_{q^*}\, \|u_1^*\|^{q^*}}\right)^{q-1} \cdot \operatorname{sgn}\big(\langle u_1^*, J_{q^*}^{X^*}(u_2^*)\rangle\big) \tag{6.84}$$

and

$$\tilde{t}_{2,2} := \left(\frac{\langle u_2^*, x_1\rangle - \alpha_2}{G_{q^*}\, \gamma^{q^*}\, \|u_2^*\|^{q^*}}\right)^{q-1} > 0 \tag{6.85}$$

with the factor γ of (6.79). Note that $\gamma \in (0, 1]$, because

$$|\langle u_1^*, J_{q^*}^{X^*}(u_2^*)\rangle| \le \|u_1^*\|\, \|J_{q^*}^{X^*}(u_2^*)\| \tag{6.86}$$

and we have equality in (6.86) iff u_1^* and u_2^* are linearly dependent. With (6.36) and $\langle u_1^*, x_1\rangle = \alpha_1$, we get

$$h_{1,2}(\tilde{s}_1 \cdot \tilde{t}_{2,2}, \tilde{t}_{2,2}) \le \frac{1}{q^*}\|x_1\|^q - \tilde{t}_{2,2}\left(\langle u_2^*, x_1\rangle - \alpha_2\right) + \frac{G_{q^*}}{q^*}\, \tilde{t}_{2,2}^{q^*}\, \|u_2^* + \tilde{s}_1\, u_1^*\|^{q^*}.$$

Similar to the upper estimate for $h_1(\tilde{t}_1)$ in the proof of Step (1), we see that $\tilde{s}_1$ yields the estimate

$$\begin{aligned}\frac{1}{q^*}\|u_2^* + \tilde{s}_1\, u_1^*\|^{q^*} &\le \frac{1}{q^*}\|u_2^*\|^{q^*} - \frac{1}{p\, G_{q^*}^{q-1}}\left(\frac{|\langle u_1^*, J_{q^*}^{X^*}(u_2^*)\rangle|}{\|u_1^*\|}\right)^q \\ &= \frac{1}{q^*}\gamma^{q^*}\, \|u_2^*\|^{q^*}.\end{aligned}$$

Hence, we get

$$\begin{aligned}h_{1,2}(\tilde{s}_1 \cdot \tilde{t}_{2,2}, \tilde{t}_{2,2}) &\le \frac{1}{q^*}\|x_1\|^q - \tilde{t}_{2,2}\left(\langle u_2^*, x_1\rangle - \alpha_2\right) + \frac{G_{q^*}}{q^*}\, \tilde{t}_{2,2}^{q^*}\, \gamma^{q^*}\, \|u_2^*\|^{q^*} \\ &= \frac{1}{q^*}\|x_1\|^q - \frac{1}{q\, G_{q^*}^{q-1}}\left(\frac{\langle u_2^*, x_1\rangle - \alpha_2}{\gamma\, \|u_2^*\|}\right)^q.\end{aligned}$$

Inserting this estimate into (6.83), we finally arrive at

$$\begin{aligned}D_q(z, x_2) \le \frac{1}{q^*}\|x_1\|^q &+ t_1\, \alpha_1 - \langle J_q(x), z\rangle + \frac{1}{q}\|z\|^q \\ &- \frac{1}{q\, G_{q^*}^{q-1}}\left(\frac{\langle u_2^*, x_1\rangle - \alpha_2}{\gamma\, \|u_2^*\|}\right)^q,\end{aligned}$$

and, together with $h_1(t_1) = \frac{1}{q^*}\|x_1\|^q + t_1\, \alpha_1 \le h_1(\tilde{t}_1)$ and estimate (6.82), we finally obtain (6.78). □

Note that this two-step method is guaranteed to work only if x is already contained in one of the halfspaces. If this is not the case, it might fail, as can be seen in Figure 6.1. Proposition 6.21 confirms that in many cases Bregman projections conveniently behave like orthogonal projections in Hilbert spaces. However, we point out that in a Hilbert space it can be shown that in Step (2) we also have $x_2 = \Pi^q_{H_1 \cap H_2}(x_1)$, which is not necessarily true in general Banach spaces.

Together with (6.38), (6.48) and (6.64), we obtain a strategy for computing x^δ_{n+1} in Algorithm 6.15.

Algorithm 6.22 (RESESOP with two search directions). Take u^δ_0 and then $\{u^\delta_n, u^\delta_{n-1}\}$, for all $n \geq 1$, as search directions in Algorithm 6.15 with

$$u^\delta_n := A^* w^\delta_n, \quad w^\delta_n := j(Ax^\delta_n - y^\delta),$$

Define $H^\delta_{-1} := X$ and, for $n \in \mathbb{N}$, the stripes

$$H^\delta_n := H\big(u^\delta_n, \alpha^\delta_n, \delta\, R^\delta_n\big) \quad \text{with} \quad \alpha^\delta_n := \langle u^\delta_n\,,\, x^\delta_n \rangle - (R^\delta_n)^2\,.$$

Then, as long as $R^\delta_n > \tau\,\delta$, we have

$$x^\delta_n \in H_>\big(u^\delta_n, \alpha^\delta_n + \delta\, R^\delta_n\big) \cap H^\delta_{n-1}\,.$$

Hence, compute x^δ_{n+1} by the following two steps:

(1) Compute

$$\tilde{x}^\delta_{n+1} := \Pi^q_{H(u^\delta_n, \alpha^\delta_n + \delta\, R^\delta_n)}(x^\delta_n)\,,$$

i.e.,

$$J_q(\tilde{x}^\delta_{n+1}) = J_q(x^\delta_n) - t^\delta_n\, u^\delta_n$$

such that t^δ_n minimizes

$$h_1(t) := \frac{1}{q^*} \left\| J_q(x^\delta_n) - t\, u^\delta_n \right\|^{q^*} + t\,(\alpha^\delta_n + \delta\, R^\delta_n)\,.$$

Then, for all $z \in M_{Ax=y}$ (recall that $M_{Ax=y} \subset H^\delta_n$ by (6.67)), we have

$$D_q(z, \tilde{x}^\delta_{n+1}) \leq D_q(z, x^\delta_n) - \frac{1}{p\, G^{q-1}_{q^*}} \left(\frac{R^\delta_n\,(R^\delta_n - \delta)}{\|u^\delta_n\|} \right)^q\,.$$

If $\tilde{x}^\delta_{n+1} \in H^\delta_{n-1}$, we have $\tilde{x}^\delta_{n+1} = \Pi^q_{H^\delta_n \cap H^\delta_{n-1}}(x^\delta_n)$ i.e., define $x^\delta_{n+1} := \tilde{x}^\delta_{n+1}$ and we are done. Otherwise, go to step

(2) Depending on $\tilde{x}_{n+1}^\delta \in H_{\gtrless}(u_{n-1}^\delta, \alpha_{n-1}^\delta \pm \delta\, R_{n-1}^\delta)$, compute

$$x_{n+1}^\delta := \Pi^q_{H(u_n^\delta, \alpha_n^\delta + \delta\, R_n^\delta) \cap H(u_{n-1}^\delta, \alpha_{n-1}^\delta \pm \delta\, R_{n-1}^\delta)}(\tilde{x}_{n+1}^\delta)\,,$$

i.e.,

$$J_q(\tilde{x}_{n+1}^\delta) = J_q(\tilde{x}_{n+1}^\delta) - t_{n,n}^\delta\, u_n^\delta - t_{n,n-1}^\delta\, u_{n-1}^\delta$$

such that $(t_{n,n}^\delta, t_{n,n-1}^\delta)$ minimizes

$$\begin{aligned} h_2(t_1, t_2) :=& \frac{1}{q^*} \left\| J_q(x_n^\delta) - t_1\, u_n^\delta - t_2^\delta\, u_{n-1}^\delta \right\|^{q^*} \\ &+ t_1\, (\alpha_n^\delta + \delta\, R_n^\delta) + t_2\, (\alpha_{n-1}^\delta \pm \delta\, R_{n-1}^\delta)\,. \end{aligned}$$

Then, we have $x_{n+1}^\delta = \Pi^q_{H_n^\delta \cap H_{n-1}^\delta}(x_n^\delta)$ and for all $z \in M_{Ax=y}$,

$$D_q(z, x_{n+1}^\delta) \le D_q(z, x_n^\delta) - \frac{1}{p\, G_{q^*}^{q-1}}\, S_n^\delta$$

with

$$S_n^\delta := \left(\frac{R_n^\delta\, (R_n^\delta - \delta)}{\|u_n^\delta\|} \right)^q + \left(\frac{\left|\langle u_{n-1}^\delta\,, \tilde{x}_{n+1}^\delta \rangle - (\alpha_{n-1}^\delta \pm \delta\, R_{n-1}^\delta)\right|}{\gamma_n\, \|u_{n-1}^\delta\|} \right)^q$$

and

$$\gamma_n := \left(1 - \frac{1}{(q-1)\, G_{q^*}^{q-1}} \left(\frac{|\langle u_n^\delta\,, J_{q^*}^{X^*}(u_{n-1}^\delta)\rangle|}{\|u_n^\delta\|\, \|J_{q^*}^{X^*}(u_{n-1}^\delta)\|} \right)^q \right)^{\frac{1}{q^*}} \in (0, 1]\,. \qquad (6.87)$$

Since

$$\frac{R_n^\delta - \delta}{\|u_n^\delta\|} \ge \frac{1 - \frac{1}{\tau}}{\|A\|}$$

for $R_n^\delta > \tau\, \delta$, we see that both (6.59) and (6.60) hold.

Remark 6.23.

(a) The factor γ_n (6.87), whose magnitude is determined by $|\langle u_n^\delta\,, J_{q^*}^{X^*}(u_{n-1}^\delta)\rangle|$, contains some information about the additional speedup we gain, by using two search directions instead of only one: The larger $|\langle u_n^\delta\,, J_{q^*}^{X^*}(u_{n-1}^\delta)\rangle|$, the smaller γ_n and hence, the larger the decrease in the Bregman distance. If the search directions u_n^δ and u_{n-1}^δ are orthogonal, in the sense that $\langle u_n^\delta\,, J_{q^*}^{X^*}(u_{n-1}^\delta)\rangle = 0$, the improvement, compared to using only u_n^δ, might be negligible; in a Hilbert space this would in fact imply that we already finish in Step (1). But, if the distance between x_n^δ and $H_n^\delta \cap H_{n-1}^\delta$ is large and the search directions are almost linearly dependent, i.e., $|\langle u_n^\delta\,, J_{q^*}^{X^*}(u_{n-1}^\delta)\rangle| \approx \|u_n^\delta\|\, \|J_{q^*}^{X^*}(u_{n-1}^\delta)\|$, then the improvement, when passing to x_{n+1}^δ, might be significant.

(b) If we have a good estimate for the constant G_{q^*}, (6.80), (6.84) and (6.85) yield good initial guesses for the minimization problems, namely

$$\tilde{t}_n^\delta := \left(\frac{R_n^\delta \, (R_n^\delta - \delta)}{G_{q^*} \, \|u_n^\delta\|^{q^*}} \right)^{q-1}$$

for the minimization of h_1 and

$$\tilde{t}_{n,n-1}^\delta := \left(\frac{\left|\langle u_{n-1}^\delta \, , \, \tilde{x}_{n+1}^\delta \rangle - (\alpha_{n-1}^\delta \pm \delta \, R_{n-1}^\delta)\right|}{G_{q^*} \, \gamma_n^{q^*} \, \|u_{n-1}^\delta\|^{q^*}} \right)^{q-1}$$
$$\times \operatorname{sgn}\bigl(\langle u_{n-1}^\delta \, , \, \tilde{x}_{n+1}^\delta \rangle - (\alpha_{n-1}^\delta \pm \delta \, R_{n-1}^\delta)\bigr) \, ,$$
$$\tilde{t}_{n,n}^\delta := -\tilde{t}_{n,n-1}^\delta \cdot \left(\frac{|\langle u_n^\delta \, , \, J_{q^*}^{X^*}(u_{n-1}^\delta)\rangle|}{G_{q^*} \, \|u_n^\delta\|^{q^*}} \right)^{q-1} \cdot \operatorname{sgn}\bigl(\langle u_n^\delta \, , \, J_{q^*}^{X^*}(u_{n-1}^\delta)\rangle\bigr)$$

for the minimization of h_2. In a Hilbert space these values are already optimal.

6.3 Iterative solution of split feasibility problems (SFP)

We are concerned with the solution of the *split feasibility problem* (SFP) in a Banach space X, by an iterative regularization method. The SFP is a special kind of *convex feasibility problem* (CFP). Formally, it consists of finding a common point in the intersection of finitely many convex sets $C_i \subset X, i \in I = \{1, \dots, N\}$, i.e.,

$$\text{find} \quad x \in \mathbf{C} := \bigcap_{i \in I} C_i \, , \tag{6.88}$$

where some of the sets C_i arise by imposing convex constraints $Q_i \subset Y_i$ in the range of linear operators $A_i : X \longrightarrow Y_i$, with Banach spaces Y_i,

$$C_i = \{x \in X : A_i x \in Q_i\} \, . \tag{6.89}$$

Many inverse problems can be modeled as an SFP. E.g., if $N = 1$ and $Q_1 = \{y\}$, (6.88) is equivalent to solving (6.1). If one is interested in a positive solution of $Ax = y$, where only noisy data y^δ are available, this can be written as

$$\text{find} \quad x \in \{x \in X : x \geq 0\} \cap \{x \in X : Ax \in B_\delta(y)\} \, ,$$

where $B_\delta(y)$ denotes the ball with radius δ around y. An extensive overview of applications of the CFP/SFP and solution methods in Hilbert spaces can be found in [40, 43] and [51, 52].

One strategy for solving (6.88) is to project cyclically onto the individual sets. In Banach spaces, the convergence of the resulting algorithm, in the general framework of Bregman projections, was analyzed by Alber and Butnariu [4]. In applications such projection algorithms are efficient, if the projections onto the individual sets are simple to calculate. If the sets are of the form (6.89) then it is, in general, too difficult or too costly to project onto these sets in each iteration. In finite dimensional spaces, Byrne [37] suggested the *CQ-algorithm*, to solve the problem of finding a point $x \in C$ such that $Ax \in Q$; it has the iterative form

$$x_{n+1} = P_C\Big(x_n - t_n\, A^*\big(Ax_n - P_Q(Ax_n)\big)\Big)$$

with appropriately chosen parameters $t_n > 0$. Here, P_C and P_Q denote the orthogonal projections onto the respective sets. The special case of $Q = \{y\}$ being a singleton is also known as the *projected Landweber method*. The advantage is that the difficulty of directly projecting onto the set $\{x : Ax \in Q\}$ is avoided by using the gradient of the functional $f(x) = \frac{1}{2}\|Ax - P_Q(Ax)\|^2$ and thus only the projection onto Q is involved. Here, we consider a generalization of this method to Banach spaces which, in case of two sets C, Q, reads as

$$x_{n+1} = \Pi_C^q\, J_{q^*}^{X^*}\Big(J_q^X(x_n) - t_n\, A^* J_2^Y\big(Ax_n - P_Q(Ax_n)\big)\Big), \tag{6.90}$$

where Π_C^q denotes the Bregman projection onto C and P_Q is the metric projection onto Q.

Often only noisy data C_i^δ, Q_i^δ is available and hence it is important to analyze the regularizing properties of a solution method for SFPs and to modify them if necessary. Some results in this direction were given in e.g., [64], for the projected Landweber method in Hilbert spaces, or [244], where a relaxed version of the CQ-algorithm, for the use of approximately given convex sets, has been studied. In this context, continuity properties of the projection operators, with respect to both the argument and the sets onto which we project, play a decisive role. Resmerita [195] showed continuity for a wide class of Bregman projections, with respect to set convergence in the sense of Mosco [166], which can be used to prove stability of the projection methods. Here, we intend to analyze the regularizing properties in connection with a *discrepancy principle*. For our purpose, it is therefore more convenient to use a notion of convergence induced by local versions of the Hausdorff distance, allowing us to quantitatively measure the distance between two convex sets C, D on bounded parts

$$d_\varrho(C,D) := \min\{\lambda \geq 0 \,|\, C \cap B_\varrho \subset D + B_\lambda \quad \text{and} \quad D \cap B_\varrho \subset C + B_\lambda\}, \tag{6.91}$$

where $B_\varrho = B_\varrho(0)$, $B_\lambda = B_\lambda(0)$. Note that the minimum defining $d_\varrho(C,D)$ is actually attained, which can be proved by the boundedness of B_ϱ and B_λ and the application of weakly convergent subsequences. We prove a uniform continuity result with respect to these distances in Section 6.3.1. This notion of convergence has already been used by Penot [180], in the context of metric projections. Its advantage, compared to the standard Hausdorff distance, is that it does not exclude many important classes of unbounded convex sets, like cones or halfspaces.

Throughout this section, we assume X to be a real q-convex and uniformly smooth and thus reflexive Banach space, with then q^*-smooth and uniformly convex dual X^*, see Theorem 2.52 (b), (d). Furthermore, we denote by

$$\mathbf{C}(X) := \{C \subset X : C \text{ closed, convex and nonempty}\}$$

the set consisting of all closed, convex subsets of X.

6.3.1 Continuity of Bregman and metric projections

Let again $q \geq 2$, $C \in \mathbf{C}(X)$ and

$$f_q(x) = \frac{1}{q}\|x\|^q, \quad x \in X.$$

We recall that the metric projection $P_C : X \to C$ is defined by

$$\|x - P_C(x)\| = \min_{y \in C} \|x - y\|,$$

whereas the Bregman projection $\Pi_C^q : X \to C$ minimizes the distance of a given $x \in X$ to C, with respect to the Bregman distance

$$D_p\big(\Pi_C^p(x), x\big) = \min_{y \in C} D_q(y, x).$$

We give and recall specific Bregman and metric projections of elementary convex subsets.

Example 6.24.

(a) The metric projection onto a hyperplane $H(u^*, \alpha)$ is given by

$$P_{H(u^*,\alpha)}(x) = x - \frac{\langle u^*, x\rangle - \alpha}{\|u^*\|^{q^*}} J_{q^*}^{X^*}(u^*). \tag{6.92}$$

If x is not yet contained in $H_{\leq}(u^*, \alpha)$, then $P_{H_{\leq}(u^*,\alpha)}(x)$ is also given by (6.92).

(b) The Bregman projection onto a hyperplane $H(u^*, \alpha)$ is given by

$$\Pi_{H(u^*,\alpha)}^q(x) = J_{q^*}^{X^*}\big(J_q(x) - t^{\text{op}} u^*\big), \tag{6.93}$$

where t^{op} is the unique solution of the optimization problem

$$\min_{t \in \mathbb{R}} h(t) := \frac{1}{q^*}\left\|J_q(x) - t\, u^*\right\|^{q^*} + \alpha\, t. \tag{6.94}$$

Note that the function h is strictly convex and continuously differentiable, with increasing derivative

$$h'(t) = -\left\langle u^* , J_{q^*}^{X^*}\left(J_q(x) - t\,u^*\right)\right\rangle + \alpha \,.$$

This was proved in Section 6.2.1. If x is not yet contained in $H_{\leq}(u^*, \alpha)$, we have $h'(0) = -\langle u^* , x\rangle + \alpha < 0$ and thus $\Pi^q_{H_{\leq}(u^*,\alpha)}(x)$ is also given by (6.93), where t^{op} is the then necessarily positive solution of the optimization problem (6.94).

(c) If X is an L^q (ℓ^q)-space ($1 < q < \infty$) and $[a,b] := \{x \in L^q\ (\ell^q) : a \leq x \leq b\}$ is a closed box, with extended real valued functions respectively sequences a, b, we have

$$\Pi^q_{[a,b]}(x) = P_{[a,b]}(x) = \max\left\{a, \min\{x, b\}\right\}, \tag{6.95}$$

where "$\leq$", "min" and "max" are to be understood point-wise and component-wise, respectively, a.e., and the index q in $\Pi^q_{[a,b]}$ must be the same as in L^q (ℓ^q).

Our aim is to use d_ϱ (6.91), to measure the distance between convex sets and prove continuity of the Bregman and metric projections, with respect to their arguments *and* the sets they project onto. For more information about set convergence, we refer to the book of Rockafellar and Wets [203].

Example 6.25.

(a) For all $\varrho > 0$ we have

$$d_\varrho\big(H(u^*, \alpha), H(v^*, \beta)\big) \leq \frac{|\alpha - \beta| + \varrho\,\|u^* - v^*\|}{\min\{\|u^*\|, \|v^*\|\}}\,. \tag{6.96}$$

The same estimate holds for $d_\varrho\big(H_{\leq}(u^*, \alpha), H_{\leq}(v^*, \beta)\big)$.

(b) In an L^q (ℓ^q)-space, we have

$$d_\varrho\big([a,b], [\tilde{a}, \tilde{b}]\big) \leq \min\left\{\|\max\{|a - \tilde{a}|, |b - \tilde{b}|\}\|, \varrho + \max\{\|c\|, \|\tilde{c}\|\}\right\} \tag{6.97}$$

with any $c \in [a,b]$ and $\tilde{c} \in [\tilde{a}, \tilde{b}]$. Note that $\|\max\{|a - \tilde{a}|, |b - \tilde{b}|\}\|$ may be infinite.

Boundedness properties of the projections indicate that it indeed suffices to know the distance between convex sets on bounded parts. Recall, that the metric projection maps bounded sets onto bounded sets because

$$\|P_C(x)\| \leq \|x - P_C(x)\| + \|x\| \leq \|x - P_C(0)\| + \|x\| \leq 3\max\{\|P_C(0)\|, \|x\|\}\,.$$

The same holds for the Bregman projection.

Lemma 6.26. *The Bregman projection maps bounded sets onto bounded sets. More precisely, we have*

$$\|\Pi_C^q(x)\| \le 3\,\max\{\|\Pi_C^q(0)\|, \|x\|\}\,. \tag{6.98}$$

Proof. From the variational inequality (6.39) with $y = \Pi_C^q(0)$, we deduce that

$$\begin{aligned}\|\Pi_C^q(x)\|^q &\le \|\Pi_C^q(x)\|^{q-1}\,\|\Pi_C^q(0)\| \\ &\quad + \|x\|^{q-1}\,\|\Pi_C^q(0)\| + \|x\|^{q-1}\,\|\Pi_C^q(x)\|\,.\end{aligned} \tag{6.99}$$

In case $\|\Pi_C^q(x)\| \le 2\,\|\Pi_C^q(0)\|$ we are done. Otherwise, we get from (6.99)

$$\|\Pi_C^q(x)\|^q \le \frac{1}{2}\|\Pi_C^q(x)\|^q + \frac{3}{2}\|x\|^{q-1}\,\|\Pi_C^q(x)\|\,,$$

from which we infer that, for $q \le 2$

$$\|\Pi_C^q(x)\| \le \max\{2\,\|\Pi_C^q(0)\|, 3^{q-1}\,\|x\|\} \le 3\,\max\{\|\Pi_C^q(0)\|, \|x\|\}\,. \qquad \square$$

Next, we prove that the metric, as well as the Bregman projection, are uniformly continuous on bounded sets, with respect to simultaneous variations of the argument and the sets onto which we project. For the case of metric projections, see also [5], where similar estimates have been proved with the standard Hausdorff distance, and [180], where local versions of the Hausdorff distance have been used to obtain estimates, with respect to either variations of the argument or variations of the sets projected onto, but under weaker assumptions on X.

Proposition 6.27. *There exist constants $c > 0$ such that, for all $x, y \in X$ and $C, D \in \mathbf{C}(X)$, the following estimates hold for all $\varrho \ge 3\,M$ with*

$$M := \max\{\|P_C(0)\|, \|P_D(0)\|, \|x\|, \|y\|\}\,.$$

(a) *For the Bregman projection we have*

$$\begin{aligned}&\left\|\Pi_C^q(x) - \Pi_D^q(y)\right\| \\ &\quad \le c\left(M^{q-1} d_\varrho(C, D) + M\,\left\|J_q(x) - J_q(y)\right\|\right)^{\frac{1}{q}}.\end{aligned} \tag{6.100}$$

(b) *For the metric projection we have*

$$\|P_C(x) - P_D(y)\| \le c\,M^{\frac{1}{q^*}}\left(d_\varrho(C, D) + \|x - y\|\right)^{\frac{1}{q}} + \|x - y\|\,. \tag{6.101}$$

Proof. Part (a). Because of (6.98) we have, for all $\varrho \ge 3\,M$,

$$\Pi_C^q(x) \in C \cap B_\varrho \quad \text{and} \quad \Pi_D^q(y) \in D \cap B_\varrho\,.$$

Due to the definition of $d_\varrho(C,D)$ (6.91) we therefore find $x_D \in D$ and $y_C \in C$, such that

$$\|\Pi_C^q(x) - x_D\| \le d_\varrho(C,D), \quad \|\Pi_D^q(y) - y_C\| \le d_\varrho(C,D)$$

and write

$$\begin{aligned}
&\langle J_q\,\Pi_C^q(x) - J_q\,\Pi_D^q(y)\,,\,\Pi_C^q(x) - \Pi_D^q(y)\rangle \\
&\quad = \langle J_q\,\Pi_C^q(x) - J_q(x)\,,\,\Pi_C^q(x) - y_C\rangle \\
&\qquad + \langle J_q\,\Pi_C^q(x) - J_q(x)\,,\,y_C - \Pi_D^q(y)\rangle \\
&\qquad + \langle J_q(x) - J_q(y)\,,\,\Pi_C^q(x) - \Pi_D^q(y)\rangle \\
&\qquad + \langle J_q(y) - J_q\,\Pi_D^q(y)\,,\,\Pi_C^q(x) - x_D\rangle \\
&\qquad + \langle J_q(y) - J_q\,\Pi_D^q(y)\,,\,x_D - \Pi_D^q(y)\rangle\,.
\end{aligned}$$

Since $x_D \in D$ and $y_C \in C$, the variational inequality (6.39) assures that the first and the last summand are less than or equal to zero and we can estimate

$$\begin{aligned}
&\langle J_q\,\Pi_C^q(x) - J_q\,\Pi_D^q(y)\,,\,\Pi_C^q(x) - \Pi_D^q(y)\rangle \\
&\quad \le \|J_q\,\Pi_C^q(x) - J_q(x)\|\; d_\varrho(C,D) \\
&\qquad + \|J_q(x) - J_q(y)\|\;\|\Pi_C^q(x) - \Pi_D^q(y)\| \\
&\qquad + \|J_q(y) - J_q\Pi_D^q(y)\|\; d_\varrho(C,D)\,.
\end{aligned}$$

Inequality (6.100) now follows from (6.98) and (6.37).

Part (b). The inequality (6.101) for the metric projection can be proved in a similar way by at first estimating the term

$$\langle J_q\big(P_C(x) - x\big) - J_q\big(P_D(y) - y\big)\,,\,\big(P_C(x) - x\big) - \big(P_D(y) - y\big)\rangle$$

with variational inequality (6.49) and then using (6.37), as well as the triangle inequality

$$\|P_C(x) - P_D(y)\| \le \big\|\big(P_C(x) - x\big) - \big(P_D(y) - y\big)\big\| + \|x - y\|\,. \qquad \square$$

The following lemma assures that cluster points of sequences generated by Algorithm 6.29, introduced in the next section, are solutions of (6.88).

Lemma 6.28. *Let $x, x_n \in X$ and $C, C_n \in \mathrm{C}(X)$ be such that the sequence $\{x_n\}$ converges weakly to x and $\lim_{n\to\infty} d_{\varrho_n}(C, C_n) = 0$ for $\varrho_n \ge 3\,M_n$ with*

$$M_n := \max\{\|P_C(0)\|, \|P_{C_n}(0)\|, \|x\|, \|x_n\|\}\,.$$

If, additionally,

$$\lim_{n\to\infty} \|x_n - \Pi_{C_n}^q(x_n)\| = 0 \quad \textit{or} \quad \lim_{n\to\infty} \|x_n - P_{C_n}(x_n)\| = 0$$

then we have $x \in C$.

Proof. At first, we show that $\{M_n\}_n$ is bounded, i.e.,

$$M := \sup_{n\in\mathbb{N}} M_n < \infty .$$

The sequence $\{x_n\}$ is bounded, because it converges weakly. To $P_C(0) \in C \cap B_{\varrho_n}$ we find some $c_n \in C_n$ such that $\|P_C(0) - c_n\| \le d_{\varrho_n}(C, C_n)$ and we get

$$\|P_{C_n}(0)\| \le \|c_n\| \le \|P_C(0)\| + d_{\varrho_n}(C, C_n),$$

which is uniformly bounded for all $n \in \mathbb{N}$, since $\{d_{\varrho_n}(C, C_n)\}_n$ converges. Hence, M is indeed finite. Next, we show that $x = \Pi_C^q(x) \in C$, respectively $x = P_C(x) \in C$. In case of Bregman projections, we write

$$\begin{aligned}
&\langle J_q(x) - J_q\,\Pi_C^q(x)\,,\, x - \Pi_C^q(x)\rangle \\
&\quad = \langle J_q(x) - J_q\,\Pi_C^q(x)\,,\, x - x_n\rangle \\
&\qquad + \Big\langle J_q(x) - J_q\,\Pi_C^q(x)\,,\, x_n - \Pi_{C_n}^q(x_n)\Big\rangle \\
&\qquad + \Big\langle J_q(x) - J_q\,\Pi_C^q(x)\,,\, \Pi_{C_n}^q(x_n) - \Pi_C^q(x_n)\Big\rangle \\
&\qquad + \langle J_q(x) - J_q\,\Pi_C^q(x)\,,\, \Pi_C^q(x_n) - \Pi_C^q(x)\rangle .
\end{aligned}$$

Because $\langle J_q(x) - J_q\,\Pi_C^q(x)\,,\, \Pi_C^q(x_n) - \Pi_C^q(x)\rangle \le 0$, we obtain, with (6.100),

$$\begin{aligned}
&\langle J_q(x) - J_q\,\Pi_C^q(x)\,,\, x - \Pi_C^q(x)\rangle \\
&\quad \le \langle J_q(x) - J_q\,\Pi_C^q(x)\,,\, x - x_n\rangle \\
&\qquad + \|J_q(x) - J_q\,\Pi_C^q(x)\|\,\|x_n - \Pi_{C_n}^q(x_n)\| \\
&\qquad + \|J_q(x) - J_q\,\Pi_C^q(x)\|\, c\, M^{\frac{1}{q^*}}\, d_{\varrho_n}(C, C_n)^{\frac{1}{q}} .
\end{aligned}$$

Since the right hand side converges to zero, we conclude, with (6.37), that $x = \Pi_C^q(x) \in C$. The proof for the metric projection can be done similarly, by estimating

$$\|x - P_C(x)\|^q = \langle J_q(x - P_C(x))\,,\, x - P_C(x)\rangle . \qquad \square$$

6.3.2 A regularization method for the solution of SFPs

We turn to the method for the solution of problem (6.88). Let $I_Q \subset I = \{1, \dots, N\}$ be the set of all indices i, belonging to sets C_i, of the form $C_i = \{x \in X : A_i x \in Q_i\}$ and denote by $I_C := I \setminus I_Q$ the set of the remaining indices. We assume that all $A_i : X \longrightarrow Y_i$ are continuous linear operators, with adjoints A_i^* and that all Y_i are real p_i-convex and uniformly smooth Banach spaces. For simplicity we take, for all spaces Y_i, the *normalized duality mapping* $J = J_2^Y$, which is also uniformly continuous on bounded sets. Suppose we are given only noisy data $C_i^\delta \in \mathbf{C}(X)$, $Q_i^\delta \in \mathcal{C}(Y_i)$, with known noise-level $\delta = (\delta_i)_{i \in I} \in \mathbb{R}^N$,

$$d_{\varrho_i}(C_i, C_i^\delta) \le \delta_i, \quad i \in I_C, \quad d_{\varrho_i}(Q_i, Q_i^\delta) \le \delta_i, \quad i \in I_Q \tag{6.102}$$

for sufficiently large ϱ_i. Doing so, we assume that we know an upper estimate $\varrho > 0$ for the norm of some members of the solution set, i.e.,

$$\|z\| \le \varrho \quad \text{for some} \quad z \in \mathbf{C}.$$

We define

$$M := 6 \max\{\|x_0\|, \varrho\},$$

where $x_0 \in X$ will be an initial guess for the iteration. Then, we postulate

$$\begin{aligned} \varrho_i &\ge 3 \max\{\|P_{C_i}(0)\|, \|P_{C_i^\delta}(0)\|, M\}, & i \in I_C, \\ \varrho_i &\ge 3 \max\{\|P_{Q_i}(0)\|, \|P_{Q_i^\delta}(0)\|, M\,\|A_i\|\}, & i \in I_Q. \end{aligned} \tag{6.103}$$

Let $i : \mathbb{N} \longrightarrow I$ be the *cyclic control mapping*

$$i(n) := (n \mod N) + 1.$$

Algorithm 6.29 (SFP). Choose some constant $\tau > 1$, an initial value $x_0^\delta = x_0 \in X$ and for $n = 0, 1, 2, \ldots$ repeat the following steps:

If the residual R_n^δ with

$$R_n^\delta := \begin{cases} \dfrac{D_q\big(\Pi^q_{C^\delta_{i(n)}}(x_n^\delta), x_n^\delta\big)}{\big\|J_q(x_n^\delta) - J_q\,\Pi^q_{C^\delta_{i(n)}}(x_n^\delta)\big\|}, & i(n) \in I_C \\ \|A_{i(n)}x_n^\delta - P_{Q^\delta_{i(n)}}(A_{i(n)}x_n^\delta)\|, & i(n) \in I_Q \end{cases} \tag{6.104}$$

satisfies

$$R_n^\delta \le \tau\, \delta_{i(n)} \tag{6.105}$$

then define $x_{n+1}^\delta := x_n^\delta$. Otherwise, define

$$x_{n+1}^\delta := \begin{cases} \Pi^q_{C^\delta_{i(n)}}(x_n^\delta), & i(n) \in I_C \\ \Pi^q_{H_\le(A^*_{i(n)} w_n^\delta, \alpha_n^\delta)}(x_n^\delta), & i(n) \in I_Q \end{cases} \tag{6.106}$$

with

$$w_n^\delta := J\big(A_{i(n)}x_n^\delta - P_{Q^\delta_{i(n)}}(A_{i(n)}x_n^\delta)\big) \tag{6.107}$$

and

$$\alpha_n^\delta := \big\langle A^*_{i(n)} w_n^\delta, x_n^\delta \big\rangle - R_n^\delta\big(R_n^\delta - \delta_{i(n)}\big). \tag{6.108}$$

Note that $\langle A^*_{i(n)} w^\delta_n\,,\, x^\delta_n \rangle > \alpha^\delta_n + (1 - \frac{1}{\tau})(R^\delta_n)^2 > \alpha^\delta_n$ for $R^\delta_n > \tau\, \delta_{i(n)}$. Hence, in this case, the iterate x^δ_n is not yet contained in the halfspace $H_{\le}(A^*_{i(n)} w^\delta_n, \alpha^\delta_n)$ and, by Example 6.24 (b), we have

$$x^\delta_{n+1} = J^{X^*}_{q^*}\big(J_q(x^\delta_n) - t^{\mathrm{op}}_n\, A^*_{i(n)} w^\delta_n\big), \tag{6.109}$$

where $t^{\mathrm{op}}_n > 0$ is the solution of the one-dimensional optimization problem

$$\min_{t \in \mathbb{R}} h_n(t) := \frac{1}{q^*} \left\| J_q(x^\delta_n) - t\, A^*_{i(n)} w^\delta_n \right\|^{q^*} + t\, \alpha^\delta_n\,. \tag{6.110}$$

If the metric projection onto a set C_i, $i \in I_C$ is easier to compute than the Bregman projection, one may use (6.109), with $Y_i = X$ and A_i being the identity operator, resulting in

$$x^\delta_{n+1} = J^{X^*}_{q^*}\Big(J_q(x^\delta_n) - t^{\mathrm{op}}_n\, J_q\big(x^\delta_n - P_{C^\delta_{i(n)}}(x^\delta_n)\big)\Big)\,. \tag{6.111}$$

This can be interpreted as a relaxed metric projection because, in a Hilbert space, it reads as

$$x^\delta_{n+1} = (1 - t^{\mathrm{op}}_n)\, x^\delta_n + t^{\mathrm{op}}_n\, P_{C^\delta_{i(n)}}(x^\delta_n)\,.$$

Relaxed projections have also been used by e.g., Byrne [38], Censor et al. [41, 42], Qu and Xiu [187] and Yang [240, 241].

We proceed by proving that Algorithm 6.29 generates iterates, whose Bregman distances to any element of $\mathbf{C}$ are decreasing. Recall that, in case $R^\delta_n \le \tau\, \delta_{i(n)}$, we set $x^\delta_{n+1} := x^\delta_n$ and therefore it holds trivially that

$$D_q\big(z, x^\delta_{n+1}\big) \le D_q\big(z, x^\delta_n\big), \quad z \in \mathbf{C}\,.$$

Otherwise, the Bregman distances are even strictly decreasing.

Lemma 6.30. *If $R^\delta_n > \tau\, \delta_{i(n)}$ then we have*

$$D_q\big(z, x^\delta_{n+1}\big) \le D_q\big(z, x^\delta_n\big) - S_n \quad \textit{for all} \quad z \in \mathbf{C} \cap B_M\,, \tag{6.112}$$

where, in case $i(n) \in I_C$, this holds for $S_n = S^C_n$, with

$$S^C_n := (1 - \tfrac{1}{\tau}) D_q(x^\delta_{n+1}, x^\delta_n) > (1 - \tfrac{1}{\tau})\, c_q\, (R^\delta_n)^q > 0 \tag{6.113}$$

and, in case $i(n) \in I_Q$, this holds for both $S_n = S^{Q,\Delta}_n$ and $S_n = S^{Q,R}_n$, with

$$S^{Q,\Delta}_n := D_q(x^\delta_{n+1}, x^\delta_n) \quad \textit{and} \quad S^{Q,R}_n := \frac{1}{q\, G^{q-1}_{q^*}} \left(\frac{(1 - \frac{1}{\tau}) R^\delta_n}{\| A_{i(n)} \|} \right)^q > 0\,. \tag{6.114}$$

Proof. At first, we consider the case $i(n) \in I_C$. Note that, by (6.37), we have

$$R_n^\delta = \frac{D_q(x_{n+1}^\delta, x_n^\delta)}{\|J_q(x_n^\delta) - J_q(x_{n+1}^\delta)\|} \le \|x_{n+1}^\delta - x_n^\delta\| \le \left(\frac{1}{c_q} D_q(x_{n+1}^\delta, x_n^\delta)\right)^{\frac{1}{q}}$$

and thus the first inequality in (6.113) holds. To $z \in \mathbf{C} \cap B_M \subset C_{i(n)} \cap B_{\varrho_{i(n)}}$, we find some $z^\delta \in C_{i(n)}^\delta$ with $\|z - z^\delta\| \le \delta_{i(n)}$ and write

$$D_q(z, x_{n+1}^\delta) = D_q(z^\delta, x_{n+1}^\delta) + \frac{1}{q}\left(\|z\|^q - \|z^\delta\|^q\right) + \langle J_q(x_{n+1}^\delta), z^\delta - z\rangle. \quad (6.115)$$

In case $\delta_{i(n)} < \frac{R_n^\delta}{\tau}$, we have $x_{n+1}^\delta = \Pi_{C_{i(n)}^\delta}^q(x_n^\delta)$ and hence, by the descent property of the Bregman projection (6.40),

$$D_q(z^\delta, x_{n+1}^\delta) \le D_q(z^\delta, x_n^\delta) - D_q(x_{n+1}^\delta, x_n^\delta).$$

Inserting this into (6.115) yields

$$\begin{aligned}
D_q(z, x_{n+1}^\delta) &\le D_q(z^\delta, x_n^\delta) - D_q(x_{n+1}^\delta, x_n^\delta) \\
&\quad + \frac{1}{q}\left(\|z\|^q - \|z^\delta\|^q\right) + \langle J_q(x_{n+1}^\delta), z^\delta - z\rangle \\
&= D_q(z, x_n^\delta) - D_q(x_{n+1}^\delta, x_n^\delta,) + \langle J_q(x_{n+1}^\delta) - J_q(x_n^\delta), z^\delta - z\rangle \\
&\le D_q(z, x_n^\delta) - D_q(x_{n+1}^\delta, x_n^\delta) + \delta_{i(n)} \|J_q(x_{n+1}^\delta) - J_q(x_n^\delta)\| \\
&\le D_q(z, x_n^\delta) - (1 - \tfrac{1}{\tau}) D_q(x_{n+1}^\delta, x_n^\delta).
\end{aligned}$$

In case $i(n) \in I_Q$, we get for the Bregman distance

$$\begin{aligned}
D_q(z, x_{n+1}^\delta) &= \frac{1}{q^*}\left\| J_q(x_n^\delta) - t_n^{\mathrm{op}} A_{i(n)}^* w_n^\delta \right\|^{q^*} \\
&\quad + t_n^{\mathrm{op}} \langle A_{i(n)}^* w_n^\delta, z\rangle - \langle J_q(x_n^\delta), z\rangle + \frac{1}{q}\|z\|^q.
\end{aligned}$$

We aim at replacing $\langle A_{i(n)}^* w_n^\delta, z\rangle$ by an expression independent of the unknown z. Since $A_{i(n)} z \in Q_{i(n)} \cap B_{\varrho_{i(n)}}$ for $z \in \mathbf{C} \cap B_M$, we find some $q^\delta \in Q_{i(n)}^\delta$ with $\|A_{i(n)} z - q^\delta\| \le \delta_{i(n)}$ and write

$$\begin{aligned}
\langle A_{i(n)}^* w_n^\delta, z\rangle &= \langle A_{i(n)}^* w_n^\delta, x_n^\delta\rangle - \langle w_n^\delta, A_{i(n)} x_n^\delta - P_{Q_{i(n)}^\delta}(A_{i(n)} x_n^\delta)\rangle \\
&\quad + \langle w_n^\delta, A_{i(n)} z - q^\delta\rangle + \langle w_n^\delta, q^\delta - P_{Q_{i(n)}^\delta}(A_{i(n)} x_n^\delta)\rangle.
\end{aligned}$$

Due to (6.49), we have $\langle w_n^\delta, q^\delta - P_{Q_{i(n)}^\delta}(A_{i(n)} x_n^\delta)\rangle \le 0$, because $q^\delta \in Q_{i(n)}^\delta$. Therefore, with $\|w_n^\delta\| = R_n^\delta$ (6.107) and the definition of α_n^δ (6.108), we can estimate

$$\langle A_{i(n)}^* w_n^\delta, z\rangle \le \langle A_{i(n)}^* w_n^\delta, x_n^\delta\rangle - R_n^\delta(R_n^\delta - \delta_{i(n)}) = \alpha_n^\delta. \quad (6.116)$$

Thus, since $t_n^{\mathrm{op}} > 0$, we get

$$D_q(z, x_{n+1}^\delta) \le \frac{1}{q^*} \left\| J_q(x_n^\delta) - t_n^{\mathrm{op}} A_{i(n)}^* w_n^\delta \right\|^{q^*} + t_n^{\mathrm{op}} \alpha_n^\delta - \langle J_q(x_n^\delta), z \rangle + \frac{1}{q} \|z\|^q .$$

Recalling that t_n^{op} minimizes the function

$$h_n(t) = \frac{1}{q^*} \left\| J_q(x_n^\delta) - t A_{i(n)}^* w_n^\delta \right\|^{q^*} + t \alpha_n^\delta$$

we have $h_n(t_n^{\mathrm{op}}) \le h_n(\tilde{t}_n)$, for

$$\tilde{t}_n := \frac{(1 - \frac{1}{\tau})^{q-1} (R_n^\delta)^{q-2}}{G_{q^*}^{q-1} \, \|A_{i(n)}\|^q} . \tag{6.117}$$

We use inequality (6.36) for the q^*-smooth dual X^* to estimate

$$\begin{aligned} h_n(\tilde{t}_n) &\le \frac{1}{q^*} \|x_n^\delta\|^{q^*} - \tilde{t}_n \langle A_{i(n)}^* w_n^\delta, x_n^\delta \rangle + \frac{G_{q^*}}{q^*} \tilde{t}_n^{q^*} \|A_{i(n)}^* w_n^\delta\|^{q^*} + \tilde{t}_n \alpha_n^\delta \\ &= \frac{1}{q^*} \|x_n^\delta\|^{q^*} - \tilde{t}_n R_n^\delta (R_n^\delta - \delta_{i(n)}) + \frac{G_{q^*}}{q^*} \tilde{t}_n^{q^*} \|A_{i(n)}^* w_n^\delta\|^{q^*} \\ &\le \frac{1}{q^*} \|x_n^\delta\|^{q^*} - \tilde{t}_n (1 - \tfrac{1}{\tau}) (R_n^\delta)^2 + \frac{G_{q^*}}{q^*} \tilde{t}_n^{q^*} \|A_{i(n)}\|^{q^*} (R_n^\delta)^{q^*} . \end{aligned}$$

A short calculation confirms that $\tilde{t}_n$ is just the value which minimizes the right hand side of the above inequality. Inserting this $\tilde{t}_n$ finally yields

$$\begin{aligned} D_q(z, x_{n+1}^\delta) &\le h_n(\tilde{t}_n) - \langle J_q(x_n^\delta), z \rangle + \frac{1}{q} \|z\|^q \\ &\le D_q(z, x_n^\delta) - \frac{1}{q \, G_{q^*}^{q-1}} \left(\frac{(1 - \frac{1}{\tau}) R_n^\delta}{\|A_{i(n)}\|} \right)^q , \end{aligned}$$

proving that (6.112) holds, with $S_n = S_n^{Q,R}$. It remains to show that (6.112) also holds with $S_n = S_n^{Q,\Delta}$. Recall that $x_{n+1}^\delta = \Pi^q_{H_\le (A_{i(n)}^* w_n^\delta, \alpha_n^\delta)}(x_n^\delta)$. From (6.116) we infer that

$$\mathbf{C} \cap B_M \subset H_\le (A_{i(n)}^* w_n^\delta, \alpha_n^\delta) . \tag{6.118}$$

Hence, as a direct consequence of the descent property of the Bregman projection (6.40), we get, for all $z \in \mathbf{C} \cap B_M$,

$$D_q(z, x_{n+1}^\delta) \le D_q(z, x_n^\delta) - D_q(x_{n+1}^\delta, x_n^\delta) . \qquad \square$$

Lemma 6.30 further assures that the iterates x_n^δ remain bounded.

Lemma 6.31. *The iterates x_n^δ, generated by Algorithm* 6.29, *remain bounded with*

$$\|x_n^\delta\| \le M = 6\,\max\{\|x_0\|, \varrho\}. \tag{6.119}$$

Proof. Since inequality $D_q(z, x_n^\delta) \le D_q(z, x_0)$ is valid for all $z \in \mathbf{C} \cap B_\varrho$, we deduce that

$$\frac{1}{q^*}\|x_n^\delta\|^q \le \frac{1}{q^*}\|x_0\|^q + \|x_0\|^{q-1}\,\varrho + \|x_n^\delta\|^{q-1}\,\varrho .$$

If $\|x_n^\delta\| \le 2q^*\,\varrho \le 4\varrho$ for $q^* \le 2$, we are done. Otherwise, we get

$$\frac{1}{2q^*}\|x_n^\delta\|^q \le \frac{1}{q^*}\|x_0\|^q + \|x_0\|^{q-1}\,\varrho \le \frac{1+q^*}{q^*}\max\{\|x_0\|^q, \varrho^q\}. \qquad \square$$

For exact data $\delta = 0$, we simply write $x_n := x_n^{\delta=0}$.

Proposition 6.32. *The sequence $\{x_n\}$, generated by Algorithm* 6.29 *with exact data, has the following properties:*

(a) *It is bounded and thus has weak cluster points. Furthermore, the sequence $\{D_q(z, x_n)\}_n$ converges decreasingly, for all $z \in \mathbf{C} \cap B_M$.*

(b) *Every weak cluster point is a solution of the SFP* (6.88).

(c) *It converges weakly, if the duality mapping of X is sequentially weak-to-weak-continuous, i.e., $\{J_q(x_n)\}_n$ converges weakly to $J_q(x)$, if $\{x_n\}$ converges weakly to x.*

(d) *It converges strongly, if it has a strongly convergent subsequence. In particular, this is the case, if X is finite dimensional or one of the sets C_i is boundedly compact, i.e., every bounded closed subset is compact.*

Proof. Using Lemma 6.28 and Lemma 6.30, the proof can be done along the lines of [4].

Part (a). This follows from Lemma 6.30 and Lemma 6.31.

Part (b). Let $z_1 \in X$ be a weak cluster point and $\{x_{n_l}\}_l$ a subsequence converging weakly to z_1. Since i is the cyclic control mapping, we may assume, without loss of generality, $i(n_l+k) = k+1$ for $k = 0, \dots, N-1$. Hence, we have $C_{i(n_l+k)} = C_{k+1}$. By passing to subsequences, we may further assume that each sequence $\{x_{n_l+k}\}_l$ converges weakly to some $z_{k+1} \in X$. The decrease of the Bregman distances (6.112) leads to $\lim_{l\to\infty} S_{n_l+k} = 0$. From (6.113), (6.114), together with (6.37), we deduce that, on the one hand, the weak limits z_{k+1} and z_{k+2} of $\{x_{n_l+k}\}_l$ and $\{x_{n_l+k+1}\}_l$ must coincide and, on the other hand,

$$\lim_{l\to\infty}\|x_{n_l+k} - \Pi^q_{C_{k+1}}(x_{n_l+k})\| = 0$$

and

$$\lim_{l\to\infty} \|A_{k+1}x_{n_l+k} - P_{Q_{k+1}}(A_{k+1}x_{n_l+k})\| = 0\,.$$

Lemma 6.28 then assures that $z_{k+1} \in C_{k+1}$, $A_{k+1}z_{k+1} \in Q_{k+1}$, respectively. Hence, we inductively get $z_1 = z_i \in C_i$ for all $i \in I$ and thus $z_1 \in \mathbf{C}$.

Part (c). Suppose that J_q is sequentially weak-to-weak-continuous and let $z_1, z_2 \in \mathbf{C} \cap B_M$ be two weak cluster points. We show $z_1 = z_2$. We have

$$D_q(z_1, x_n) - D_q(z_2, x_n) - \frac{1}{q}\|z_1\|^q + \frac{1}{q}\|z_2\|^q = \big\langle J_q(x_n)\,,\, z_2 - z_1\big\rangle\,.$$

Since the left hand side has a limit, say Δ, so does the right hand side. Let $\{x_{n_k}\}_k$ converge weakly to z_1 and $\{x_{m_l}\}_l$ converge weakly to z_2. Then, $\big\{J_q(x_{n_k})\big\}_k$ converges weakly to $J_q(z_1)$ and $\big\{J_q(x_{m_l}\big\}_l$ converges weakly to $J_q(z_2)$. We get

$$\begin{aligned}\big\langle J_q(z_2) - J_q(z_1)\,,\, z_2 - z_1\big\rangle &= \big\langle J_q(z_2)\,,\, z_2 - z_1\big\rangle - \big\langle J_q(z_1)\,,\, z_2 - z_1\big\rangle \\ &= \lim_{l\to\infty}\big\langle J_q(x_{m_l})\,,\, z_2 - z_1\big\rangle \\ &\quad - \lim_{k\to\infty}\big\langle J_q(x_{n_k})\,,\, z_2 - z_1\big\rangle \\ &= \Delta - \Delta = 0\,,\end{aligned}$$

from which we infer that $z_1 = z_2$.

Part (d). Let a subsequence $\{x_{n_l}\}_l$ converge strongly to some $z \in X$. From (6.37), we deduce that, in this case, $\lim_{l\to\infty} D_q(z, x_{n_l}) = 0$. By part (b) and (6.119), we know that $z \in \mathbf{C} \cap B_M$ and therefore, by part (a), the sequence $\big\{D_q(z, x_n)\big\}_n$ converges. Hence, we must have $\lim_{n\to\infty} D_q(z, x_n) = 0$, i.e., $\{x_n\}$ converges strongly to z. □

In the presence of noise, we employ a *discrepancy principle* as stopping rule: Terminate the iteration with stopping index $n_* = n_*(\delta)$ when, for the first time in N consecutive iterations, all residuals R_n^δ satisfy (6.105), i.e.,

$$n_* = \min\{n \in \mathbb{N} : R_{n+k}^\delta \le \tau\,\delta_{i(n+k)} \quad \text{for all} \quad k = 0, \dots, N-1\}\,. \tag{6.120}$$

Observe that the way the iterates are defined then implies

$$x_{n_*+k}^\delta = x_{n_*}^\delta \quad \text{for} \quad k = 0, \dots, N-1\,. \tag{6.121}$$

This stopping rule indeed provides us with a finite stopping index n_*.

Lemma 6.33. *If $\delta > 0$, the minimum in* (6.120) *exists.*

Proof. Suppose, to the contrary, that, for all $n \in \mathbb{N}$, there exists some $k_n \in \{0, \dots, N-1\}$ such that $R_{n+k_n}^\delta > \tau\,\delta_{i(n+k_n)}$. Since k_n and $i(n+k_n)$ can only achieve finitely many different values, there is a subsequence $\{n_l\}_l$ such that

$$R_{n_l+k}^\delta > \tau\,\delta_i > 0 \tag{6.122}$$

for some fixed $k \in \{0, \ldots, N-1\}$ and $i \in I$. But then the decrease of the Bregman distances (6.113), respectively (6.114), implies that $\lim_{l \to \infty} R^{\delta}_{n_l+k} = 0$, which contradicts (6.122). □

The discrepancy principle renders Algorithm 6.29 a regularization method.

Proposition 6.34. *Together with the discrepancy principle as stopping rule, Algorithm 6.29 is a regularization method for the solution of (6.88) in the following sense: Let $\{\delta^l\}_l$ be any null-sequence of noise-levels in $\mathbb{R}^N$ and let $\{x_n^{\delta^l}\}_n$ be sequences, generated by Algorithm 6.29, with the same initial guess $x_0^{\delta^l} = x_0$ and corresponding to noisy data $C_i^{\delta^l}$, $Q_i^{\delta^l}$, such that (6.102) and (6.103) hold. If $n_l := n_*(\delta^l)$ is the stopping index, then, according to (6.120), the sequence $\{x_{n_l}^{\delta^l}\}_l$ has the following properties:*

(a) *It is bounded and therefore has weak cluster points.*

(b) *Every weak cluster point is a solution of (6.88).*

(c) *If, for exact data, the sequence $\{x_n\}$ converges strongly to x, then $\{x_{n_l}^{\delta^l}\}_l$ converges strongly to x as well, i.e.,*

$$\lim_{\delta \to 0} \|x^{\delta}_{n_*(\delta)} - x\| = 0 .$$

The principle idea of the proof of (c) – showing continuity of x_n^{δ}, with respect to δ, for a fixed index n and then using the decrease of the distance of x_n^{δ} to solution points – is taken from [88], where the regularizing properties of the nonlinear Landweber method in Hilbert spaces are analyzed.

Proof. Part (a). This follows from (6.119).

Part (b). The definition of the stopping index $n_l = n_*(\delta^l)$ by (6.120), together with $\{\delta^l\}_l$ being a null-sequence, leads to

$$\lim_{l \to \infty} R^{\delta^l}_{n_l+k} = 0 \quad \text{for all} \quad k = 0, \ldots, N-1 .$$

From (6.121) we then deduce that, for all $i \in I$,

$$\lim_{l \to \infty} \|x^{\delta^l}_{n_l} - \Pi^q_{C_i^{\delta^l}}(x^{\delta^l}_{n_l})\| = 0$$

respectively

$$\lim_{l \to \infty} \|A_i x^{\delta^l}_{n_l} - P_{Q_i^{\delta^l}}(A_i x^{\delta^l}_{n_l})\| = 0 .$$

By Lemma 6.28 it follows that any weak cluster point x of $\{x_{n_l}^{\delta^l}\}_l$ is contained in C_i, for all $i \in I$, i.e. $x \in \mathbf{C}$.

Part (c). At first we note that, for a fixed index n, an iterate x_n^δ, generated by Algorithm 6.29, depends continuously on δ, i.e.,

$$\lim_{\delta \to 0} x_n^\delta = x_n . \tag{6.123}$$

This is due to the continuity properties of all mappings involved. We have a closer look at this for the case $i(n) \in I_Q$ and prove (6.123) inductively. Recall that the initial guess $x_0^\delta = x_0$ is the same for all δ. Suppose that (6.123) holds for some $n \geq 0$. For $R_n = 0$, we have $x_{n+1} = x_n$ and $Ax_n \in Q_{i(n)} \cap B_{\varrho_{i(n)}}$. Inspecting the proof of Lemma 6.30 for $R_n^\delta > \tau\, \delta_{i(n)}$, we see that here (6.112) also holds for $z = x_n = x_{n+1}$, i.e.,

$$D_q(x_{n+1}, x_{n+1}^\delta) \leq D_q(x_n, x_n^\delta)$$

and, for $R_n^\delta \leq \tau\, \delta_{i(n)}$, we have set $x_{n+1}^\delta = x_n^\delta$ anyway. Hence,

$$\lim_{\delta \to 0} D_q(x_{n+1}, x_{n+1}^\delta) \leq \lim_{\delta \to 0} D_q(x_n, x_n^\delta) = 0 .$$

Now consider the case $R_n > 0$. Here, we also have $\lim_{\delta \to 0} R_n^\delta = R_n > 0$ and thus $R_n^\delta > \tau\, \delta_{i(n)}$, for δ small enough. By (6.118), we know that, for all $\delta \geq 0$,

$$\| P_{H_{\leq}(A_{i(n)}^* w_n^\delta, \alpha_n^\delta)}(0) \| \leq M.$$

Putting $\delta = 0$ in (6.116) implies

$$R_n^2 \leq 2\, M\, \| A_{i(n)}^* w_n \| .$$

We thus get

$$\lim_{\delta \to 0} \| A_{i(n)}^* w_n^\delta \| = \| A_{i(n)}^* w_n \| \geq \frac{R_n^2}{2\, M} > 0 ,$$

which, together with (6.96), yields

$$\lim_{\delta \to 0} d_{3M}\big(H_{\leq}(A_{i(n)}^* w_n^\delta, \alpha_n^\delta), H_{\leq}(A_{i(n)}^* w_n, \alpha_n)\big) = 0 .$$

Hence, with Proposition 6.27, it follows that

$$\lim_{\delta \to 0} x_{n+1}^\delta = \lim_{\delta \to 0} \Pi^q_{H_{\leq}(A_{i(n)}^* w_n^\delta, \alpha_n^\delta)}(x_n^\delta) = \Pi^q_{H_{\leq}(A_{i(n)}^* w_n, \alpha_n)}(x_n) = x_{n+1} .$$

Let $\{x_n\}$ converge strongly to x. Since for the convergence of $\{x_{n_l}^{\delta_l}\}_l$ to x it suffices to show that every subsequence in turn has a subsequence converging to x, we assume, without loss of generality, that the sequence $\{n_l\}_l$ is increasing. To $k \in \mathbb{N}$, we then find $n_k, l_k \in \mathbb{N}$ such that, for all $l \geq \max\{k, l_k\}$, the following chain of inequalities holds

$$D_q\big(x, x_{n_l}^{\delta_l}\big) \leq D_q\big(x, x_{n_k}^{\delta_l}\big) \leq D_q(x, x_{n_k}) + \frac{1}{k} \leq \frac{2}{k} ,$$

which implies convergence. □

Remark 6.35. The same results can also be proved for the use of Bregman projections onto the sets Q_i, by applying the modifications

$$\begin{aligned} R_n^\delta &:= \frac{\left\langle J_{q_Y}(Ax_n^\delta) - J_{q_Y}\big(\Pi_{Q^\delta}^{q_Y}(Ax_n^\delta)\big)\,,\, Ax_n^\delta - \Pi_{Q^\delta}^{q_Y}(Ax_n^\delta)\right\rangle}{\left\| J_{q_Y}(Ax_n^\delta) - J_{q_Y}\big(\Pi_{Q^\delta}^{q_Y}(Ax_n^\delta)\big)\right\|}\,, \\ w_n^\delta &:= J_{q_Y}(Ax_n^\delta) - J_{q_Y}\big(\Pi_{Q^\delta}^{q_Y}(Ax_n^\delta)\big)\,, \\ \alpha_n^\delta &:= \left\langle A^* w_n^\delta\,,\, x_n^\delta\right\rangle - \|w_n^\delta\|\left(R_n^\delta - \delta_{i(n)}\right). \end{aligned}$$

Chapter 7

Nonlinear operator equations

In this chapter, we will consider iterative regularization methods for nonlinear ill-posed operator equations

$$F(x) = y, \tag{7.1}$$

where F maps between Banach spaces X and Y. As in the previous chapters, we will assume that the noise level δ in

$$\|y - y^\delta\| \le \delta \tag{7.2}$$

is known and provide convergence results in the sense of regularization methods, i.e., as δ tends to zero. In the following, x_0 is some initial guess. Throughout this chapter, we will assume that a solution to (7.1) exists which, by Proposition 3.14 in Chapter 3, implies existence of an x_0-minimum norm solution $x^\dagger$ (cf. Definition 3.10), provided Assumption 3.11 is satisfied (see also Proposition 7.1 below).

The iterative methods discussed in this chapter will be either of gradient (Landweber and iteratively regularized Landweber) or of Newton type (iteratively regularized Gauss–Newton method). Before going into detail about the methods themselves, we will dwell on the assumptions playing a role in their formulation and their convergence analysis. For better readability, these conditions will be recalled in the individual subsections where they are actually utilized.

7.1 Preliminaries

7.1.1 Conditions on the spaces

Throughout this section, we will assume that X is a reflexive Banach space, which will indeed be sufficient for showing convergence rates with respect to the Bregman distance (but not norm convergence) for the Newton type method, with a priori parameter choice in Subsection 7.3.1. In Subsection 7.3.2, on a Newton type method with a posteriori parameter choice, and for the analysis of the gradient type methods of Subsection 7.2, we will assume X to be smooth and uniformly convex which, by Theorems 2.52, 2.53, and 2.60, implies that X is reflexive, its dual X^* is uniformly smooth, that the duality mappings J_q^X and $J_{q^*}^{X^*}$ are single-valued, and that boundedness or convergence with respect to the Bregman distance implies boundedness or convergence with respect to the norm, respectively. In part of Section 7.2, we will additionally assume that X is q-convex which, by Theorems 2.52 (b), 2.60 (g), (h), implies the estimates

$$D_{j_q}(x,y) \geq c_q \|x-y\|^q \tag{7.3}$$

for some constant $c_q > 0$ and

$$D_{j_{q^*}}(x^*,y^*) \leq C_{q^*} \|x^*-y^*\|^{q^*}. \tag{7.4}$$

The data space Y will be an arbitrary Banach space in most of what follows, except for Theorem 7.4 and case (7.74) of Theorems 7.10, 7.11, 7.13.

In the following, $\overline{B}_\rho(x_0)$ denotes the closed ball of radius $\rho > 0$ around x_0, and

$$\overline{B}_\rho^D(x^\dagger) = \begin{cases} \{x \in X : D_{j_q}(x^\dagger, x) \leq \rho^2\} & \text{in Subsection 7.2.1} \\ \{x \in X : D_q^{x_0}(x^\dagger, x) \leq \rho^2\} & \text{in Subsection 7.2.2} \\ \{x \in X : D_q^{x_0}(x, x^\dagger) \leq \rho^2\} & \text{in Section 7.3} \end{cases} \tag{7.5}$$

is a ball with respect to the Bregman distance around some solution $x^\dagger$ of (7.1). Moreover,

$$\overline{B} = \begin{cases} \overline{B}_\rho^D(x^\dagger) & \text{in Section 7.2} \\ & \text{and Subsection 7.3.1} \\ \mathcal{D}(F) \cap \overline{B}_\rho(x_0) & \text{in Subsection 7.3.2,} \end{cases} \tag{7.6}$$

where $\rho = \infty$ and $\overline{B} = \mathcal{D}(F)$ are possible. Here, we use the notation

$$D_q^{x_0}(\tilde{x}, x) := D_{j_q}(\tilde{x} - x_0, x - x_0).$$

7.1.2 Variational inequalities

Referring to Sections 3.2.2, 3.2.3, we consider here $\Omega(x) = \frac{1}{q}\|x\|^q$ and hence the following variational inequalities for proving convergence rates:

For $0 < \nu < 1$

$$\exists \beta > 0 \ \forall x \in \overline{B}:$$
$$|\langle j_q^X(x^\dagger - x_0), x - x^\dagger \rangle_{X^* \times X}| \leq \beta D_q^{x_0}(x, x^\dagger)^{\frac{1-\nu}{2}} \|F'(x^\dagger)(x - x^\dagger)\|^\nu. \tag{7.7}$$

or

$$\exists \beta > 0 \ \forall x \in \overline{B}:$$
$$|\langle j_q^X(x^\dagger - x_0), x - x^\dagger \rangle_{X^* \times X}| \leq \beta D_q^{x_0}(x^\dagger, x)^{\frac{1-\nu}{2}} \|F'(x^\dagger)(x - x^\dagger)\|^\nu. \tag{7.8}$$

By avoiding $\|F(x) - F(x^\dagger)\|$ on the right hand side, compared to (3.32), we are to some extent independent of the η-condition (3.42) (i.e., (7.19) below). In particular, we will, e.g, prove optimal rates under a mere Lipschitz condition on F', provided the benchmark source condition (3.29) corresponding to the case $\nu = 1$ in the variational inequalities (7.7), (7.8), holds.

More generally, we will consider index functions $\kappa : (0,\infty) \to (0,\infty)$, satisfying

$$\phi := (\kappa^2)^{-1} \quad \text{is convex} \tag{7.9}$$

and assume the variational inequality

$$\begin{aligned} \forall x \in \overline{B},\ x \neq x^\dagger : &|\langle j_q^X(x^\dagger - x_0), x - x^\dagger\rangle_{X^*\times X}| \\ &\leq D_q^{x_0}(x,x^\dagger)^{1/2}\kappa \times \left(\frac{\|F'(x^\dagger)(x-x^\dagger)\|^2}{D_q^{x_0}(x,x^\dagger)}\right) \end{aligned} \tag{7.10}$$

to hold, cf. (3.39) . This includes e.g., logarithmic source conditions, as appropriate for exponentially ill-posed problems, cf., [114].

7.1.3 Conditions on the forward operator

We will assume

- continuity of F and of F', as well as

$$\overline{B}_\rho^D(x^\dagger) \subseteq \mathcal{D}(F) \tag{7.11}$$

  for some $\rho > 0$, and boundedness of F' on $\overline{B}_\rho^D(x^\dagger)$, in case of the gradient type methods of Section 7.2 and the Newton type method of Subsection 7.3.1

- (weak) sequential closedness, in the sense that either

$$\begin{aligned} &(x_n \rightharpoonup x \ \wedge\ F(x_n) \to f) \\ &\Rightarrow\ (x \in \mathcal{D}(F)\ \wedge\ F(x) = f) \end{aligned} \tag{7.12}$$

  or

$$\begin{aligned} &(J_q^X(x_n - x_0) \rightharpoonup x^* \ \wedge\ F(x_n) \to f) \\ &\Rightarrow\ (x := J_{q^*}^{X^*}(x^*) + x_0 \in \mathcal{D}(F)\ \wedge\ F(x) = f) \end{aligned} \tag{7.13}$$

  for all $\{x_n\}_{n\in\mathbb{N}} \subseteq X$, in case of the Newton type method of Section 7.3.

  Note that, by $J_{q^*}^{X^*} = {J_q^X}^{-1}$, we have $J_q^X(x - x_0) = x^*$ in (7.13).

We further remark that non-emptiness of the interior (with respect to the norm) of $\mathcal{D}(F)$ is sufficient for (7.11); in a q-convex X, this is an immediate consequence of (7.3) and in the general uniformly convex case this follows e.g., from the proof of Theorem 2.60 (f) (see also Theorem 2.12 (e) in [213]).

To prove convergence rates, we make the following additional assumption on the nonlinearity of F at some solution $x^\dagger$; see also (3.48) and note the relation to the concept of degree of nonlinearity, see, e.g, [97].

$$\begin{aligned} &\left\|(F'(x^\dagger + v) - F'(x^\dagger))v\right\| \leq K \left\|F'(x^\dagger)v\right\|^{c_1} D_q^{x_0}(x^\dagger, v + x^\dagger)^{c_2}, \\ &v \in X,\ x^\dagger + v \in \overline{B} \end{aligned} \tag{7.14}$$

with $\overline{B}$ as in (7.6). As already mentioned in Subsection 3.2.4, there is a close connection between the exponents used for the norms, the smoothness prescribed by the variational inequality and the condition on the nonlinearity of F. In (7.14), the exponents c_1, c_2 have to satisfy two conditions, depending on the exponent p of the data term and on the smoothness index ν (cf. (7.7), (7.8)):

$$c_1 = 1 \text{ or } c_1 + c_2 p > 1 \text{ or } (c_1 + c_2 p \geq 1 \text{ and } K \text{ is sufficiently small}) \tag{7.15}$$

$$c_1 + c_2 \frac{2\nu}{\nu+1} \geq 1\,. \tag{7.16}$$

Here, F' is the Gâteaux derivative of F, which we assume to exist and to be a bounded linear operator $X \to Y$, whenever (7.14) appears. With this, a Taylor remainder estimate

$$\begin{aligned}
&\left\| F(x_n^\delta) - F(x^\dagger) - F'(x^\dagger)(x_n^\delta - x^\dagger) \right\| &(7.17)\\
&= \left\| g(1) - g(0) - F'(x^\dagger)(x_n^\delta - x^\dagger) \right\|\\
&= \left\| \int_0^1 g'(t)\,dt - F'(x^\dagger)(x_n^\delta - x^\dagger) \right\|\\
&= \left\| \int_0^1 F'(x^\dagger + t(x_n^\delta - x^\dagger))(x_n^\delta - x^\dagger)\,dt - F'(x^\dagger)(x_n^\delta - x^\dagger) \right\|\\
&\leq K \left\| F'(x^\dagger)(x_n^\delta - x^\dagger) \right\|^{c_1} D_q^{x_0}(x^\dagger, x_n^\delta)^{c_2}, &(7.18)
\end{aligned}$$

where $g : t \mapsto F(x^\dagger + t(x_n^\delta - x^\dagger))$, follows from (7.14), cf. (3.48).

We wish to point out that (7.14), (7.15), (7.16) gets weaker for larger smoothness index ν, which corresponds to results in Hilbert spaces, where – as here – for $\nu = 1$ a Lipschitz condition suffices to prove optimal convergence rates, see e.g. [68]. Indeed, condition (7.14), with (7.16) for $\nu = 1$, follows from the usual Lipschitz condition on F' in terms of the Bregman distance in X:

$$\begin{aligned}
&\left\| (F'(x^\dagger + v) - F'(x^\dagger)) \right\|^2 \leq L^2 D_q^{x_0}(x^\dagger, v + x^\dagger)\,,\\
&v \in X,\ x^\dagger + v \in \overline{B}\,.
\end{aligned}$$

If the exponent ν in the source condition is not known, we require a nonlinearity assumption corresponding to the strongest case $\nu = 0$ in (7.14), (7.16), (7.15), namely the η-condition

$$\left\| F(x) - F(\bar{x}) - F'(x)(x - \bar{x}) \right\| \leq \eta \left\| F(x) - F(\bar{x}) \right\| \quad \forall x, \bar{x} \in \overline{B} \tag{7.19}$$

for some $0 < \eta < 1$, cf. (3.42). Note that (7.14) for $\nu = 0$, with K sufficiently small, becomes (7.19) at $x = x^\dagger$, with $\eta = \frac{K}{1-K}$.

In Subsection 7.3.1, we will use the following nonlinearity condition, which is slightly stronger, compared to (7.14).

$$\begin{aligned}&\left\|(F'(x^\dagger + \tilde{v}) - F'(x^\dagger))v\right\| \\ &\geq K \left\|F'(x^\dagger)v\right\|^{\tilde{c}_1} D_q^{x_0}(v + x^\dagger, x^\dagger)^{\tilde{c}_2} \left\|F'(x^\dagger)\tilde{v}\right\|^{\tilde{c}_3} D_q^{x_0}(\tilde{v} + x^\dagger, x^\dagger)^{\tilde{c}_4}, \\ &v, \tilde{v} \in X,\ x^\dagger + v \in \overline{B}\ x^\dagger + \tilde{v} \in \overline{B}\end{aligned} \tag{7.20}$$

with

$$\tilde{c}_1 + \tilde{c}_2 \frac{2\nu}{\nu+1} \geq \frac{1}{2}, \quad \tilde{c}_3 + \tilde{c}_4 \frac{2\nu}{\nu+1} \geq \frac{1}{2} \tag{7.21}$$

as well as

$$\begin{aligned}&(\tilde{c}_1 = \frac{1}{2} \text{ and } \tilde{c}_3 = \frac{1}{2}) \text{ or} \\ &(\tilde{c}_1 + \tilde{c}_2 p > \frac{1}{2} \text{ and } \tilde{c}_3 + \tilde{c}_4 p > \frac{1}{2} \\ \text{or}& \\ &\left(\tilde{c}_1 + \tilde{c}_2 p \geq \frac{1}{2} \text{ and } \tilde{c}_3 + \tilde{c}_4 p \geq \frac{1}{2} \text{ and } K \text{ sufficiently small}\right).\end{aligned} \tag{7.22}$$

Note that (7.20), with (7.21) and (7.22) implies (7.14), with (7.15), (7.16), $c_1 = \tilde{c}_1 + \tilde{c}_3$, $c_2 = \tilde{c}_2 + \tilde{c}_4$, up to reversal of the roles of $x^\dagger$ and $x^\dagger + v$ in the Bregman distance. Here, as in (7.14), (7.15), (7.16), we deal with stronger conditions for smaller ν. In case of a general (possibly only logarithmic) index function κ, we have to restrict ourselves to the strongest case in (7.20), (7.21), (7.22), corresponding to $\nu = 0$.

In our methods, we will use the abbreviation

$$A_n = F'(x_n^\delta),$$

where x_n^δ is the current iterate and F' is not necessarily the Fréchet derivative of F, but just

- the Gâteaux derivative, if we assume (7.14) or
- a bounded linear operator, satisfying (7.19), if we assume the latter.

We will suppose the same kind of definition of F' to be used in the respective formulation of the variational inequalities (7.7), (7.10).

The η- condition (7.19) allows to show not only existence of an x_0-minimum norm solution, cf. Proposition 3.14, but also a characterization via the nullspace of the linearized forward operator, like in the Hilbert space situation, cf. Proposition 2.1 in [128], see also Lemma 3.3 here, for the linear case.

Proposition 7.1. *Let X be strictly convex, let* (7.19) *hold in $\overline{B} = \overline{B}_\rho(x_0)$, for some $0 < \eta < 1$, and let $\mathcal{D}(F) \cap \overline{B}_\rho(x_0) = \overline{B}_\rho(x_0)$, i.e., $\mathcal{D}(F)$ has a nonempty interior.*

(a) *Then, for all* $x \in \overline{B}_\rho(x_0)$

$$M_x := \{\tilde{x} \in \overline{B}_\rho(x_0) : F(\tilde{x}) = F(x)\} = \Big(x + \mathcal{N}(F'(x))\Big) \cap \overline{B}_\rho(x_0) \quad (7.23)$$

and

$$\mathcal{N}(F'(x)) = \mathcal{N}(F'(\tilde{x})) \text{ for all } \tilde{x} \in M_x.$$

Moreover,

$$\mathcal{N}(F'(x)) \supseteq \{\lambda(\tilde{x} - x) : \tilde{x} \in M_x, \lambda \in \mathbb{R}\}, \quad (7.24)$$

where, instead of $\supseteq$*, equality holds, if* $x \in \operatorname{int}(\overline{B}_\rho(x_0))$.

(b) *If* $F(x) = y$ *is solvable in* $\overline{B}_\rho(x_0)$*, then an* x_0*-minimum norm solution* $x^\dagger$ *exists and is unique.*

For $x^\dagger \in \operatorname{int}(\overline{B}_\rho(x_0))$*, we have*

$$J_q^X(x^\dagger) \in \overline{\mathcal{R}(F'(x^\dagger)^*)} \quad (7.25)$$

and if, for some $\tilde{x} \in \overline{B}_\rho(x_0)$,

$$J_q^X(\tilde{x}) \in \overline{\mathcal{R}(F'(x^\dagger)^*)} \text{ and } \tilde{x} - x^\dagger \in \mathcal{N}(F'(x^\dagger))$$

holds, then $\tilde{x} = x^\dagger$.

Proof. Part (a) follows analogously to part (i) of the proof of Proposition 2.1 in [128], which remains valid in Banach spaces, without any modification. For the convenience of the reader, we here provide the full argument:

From (7.19), we immediately obtain

$$\|F(x) - F(\tilde{x})\| \le \frac{1}{1-\eta} \left\|F'(x)(x - \tilde{x})\right\| \quad (7.26)$$

$$\left\|F'(x)(x - \tilde{x})\right\| \le (1 + \eta) \|F(x) - F(\tilde{x})\| \quad (7.27)$$

$$\forall x, \tilde{x} \in \overline{B}_\rho(x_0),$$

which implies (7.23) as well as (7.24).

Now let $x, \tilde{x} \in \overline{B}_\rho(x_0)$ be arbitrary, such that $F(x) = F(\tilde{x})$, i.e., $\tilde{x} \in M_x$ and, by (7.27), $\tilde{x} - x \in \mathcal{N}(F'(\tilde{x})) \cap \mathcal{N}(F'(x))$. There exists an $s \in \mathbb{R}$ such that $x_s := x + s(\tilde{x} - x) \in \operatorname{int}(\overline{B}_\rho(x_0))$. Namely, if x or $\tilde{x}$ are in $\operatorname{int}(\overline{B}_\rho(x_0))$, we may set $s = 0$ or $s = 1$, respectively. If $x, \tilde{x} \in \overline{B}_\rho(x_0)$, by strict convexity of X, $s = \frac{1}{2}$ gives an interior point. Thus, for any $h \in \mathcal{N}(F'(x))$, there exists a $t > 0$ and an $s \in \mathbb{R}$, such that $x_{t,s} := x + s(\tilde{x} - x) + th \in \operatorname{int}(\overline{B}_\rho(x_0))$. Additionally, we have $x_{t,s} \in x + \mathcal{N}(F'(x))$, hence by (7.26), $F(x_{t,s}) = F(x) = F(\tilde{x})$, which, by (7.27), implies $x_{t,s} - \tilde{x} \in \mathcal{N}(F'(\tilde{x}))$, and with that, $h = \frac{1}{t}(x_{t,s} - \tilde{x} + (1-s)(\tilde{x} - x)) \in \mathcal{N}(F'(\tilde{x}))$. This and the same argument, with the roles of x and $\tilde{x}$ reversed, yield $\mathcal{N}(F'(x)) = \mathcal{N}(F'(\tilde{x}))$. To obtain equality of the sets in the sixth line of assertion (a), for $x \in \operatorname{int}(\overline{B}_\rho(x_0))$, note that this argument can be used with $s = 0$. I.e., for any $h \in \mathcal{N}(F'(x))$, there exists a $t > 0$ such that $x_{t,0} := x + th \in \operatorname{int}(\overline{B}_\rho(x_0)) \cap x + \mathcal{N}(F'(x))$, hence by (7.26), $x_{t,0} \in M_x$, which implies that $h = \frac{1}{t}(x_{t,0} - x) \in \{\lambda(\tilde{x} - x) : \tilde{x} \in M_x, \lambda \in \mathbb{R}\}$.

Part (b) can be seen to be exactly the respective assertion in the linear case, as stated and proved in Lemma 3.3, up to the following small modification in the proof of (7.25), due to the restriction to a neighborhood of x_0:

For any $z \in \mathcal{N}(F'(x^\dagger))$, there exists an $\epsilon > 0$ such that

$$x^\dagger \pm \epsilon z \in \left(x^\dagger + \mathcal{N}(F'(x^\dagger))\right) \cap \overline{B}_\rho(x_0) = \{\tilde{x} \in \overline{B}_\rho(x_0) : F(\tilde{x}) = y\}.$$

Hence, by Theorem 2.53 (k)

$$\langle J_q^X(x^\dagger), x^\dagger \rangle_{X^* \times X} \leq \langle J_q^X(x^\dagger), x^\dagger \pm \epsilon z \rangle_{X^* \times X},$$

i.e. $\langle J_q^X(x^\dagger), z \rangle_{X^* \times X} = 0$. □

7.2 Gradient type methods

In this section, we study iterative methods, resulting from the application of gradient descent to the misfit functional $\|F(x_n^\delta) - y^\delta\|^p$. We will prove convergence (without rates) for the resulting Landweber method, with the discrepancy principle under an η-condition, in Subsection 7.2.1, cf. [129]. To show convergence rates, we will have to modify the Landweber iteration in Subsection 7.2.2 slightly and use an a priori stopping rule. For an accelerated Landweber iteration, with a particular choice of the step sizes we refer to [98].

7.2.1 Convergence of the Landweber iteration with the discrepancy principle

Analogous to the Landweber method in Hilbert spaces from [88], we will study the generalization of the method (6.2), to solve nonlinear problems (7.1)

$$\begin{aligned} J_q^X(x_{n+1}^\delta) &= J_q^X(x_n^\delta) - \mu_n A_n^* j_p^Y(F(x_n^\delta) - y^\delta), \\ x_{n+1}^\delta &= J_{q^*}^{X^*}(J_q^X(x_{n+1}^\delta)), \quad n = 0, 1, \ldots \end{aligned} \tag{7.28}$$

where we abbreviate

$$A_n = F'(x_n^\delta).$$

The step size μ_n has to be chosen appropriately, so as to guarantee convergence of the method, see (7.32) below. The stopping index n_*, which acts as a regularization parameter, is chosen according to the discrepancy principle

$$n_*(\delta) = \min\{n \in \mathbb{N} : \left\|F(x_n^\delta) - y^\delta\right\| \leq \tau\delta\}, \tag{7.29}$$

with a constant $\tau > 1$.

The assumption here imposed on F, for showing convergence, is the η-condition

$$\left\| F(x) - F(\bar{x}) - F'(x)(x - \bar{x}) \right\| \le \eta \left\| F(x) - F(\bar{x}) \right\| \quad \forall x, \bar{x} \in \overline{B} \tag{7.30}$$

for some $0 < \eta < 1$.

Proposition 7.2. *Assume that X is smooth and q-convex, that x_0 is sufficiently close to $x^\dagger$, i.e., $x_0 \in \overline{B}_\rho^D(x^\dagger)$, that F satisfies (7.30) with η sufficiently small, that F and F' are continuous, and that (7.11) holds. Let τ be chosen sufficiently large, so that*

$$c(\eta, \tau) := \eta + \frac{1+\eta}{\tau} < 1 . \tag{7.31}$$

Then, with the choice

$$\mu_n := \frac{q^*(1 - c(\eta,\tau))^{q-1}}{C_{q^*}^{q-1}} \frac{\left\| F(x_n^\delta) - y^\delta \right\|^{q-p}}{\|A_n\|^q} \ge 0 \tag{7.32}$$

with C_{q^*} being the constant in (7.4), monotonicity of the Bregman distances

$$D_{j_q}(x^\dagger, x_{n+1}^\delta) - D_{j_q}(x^\dagger, x_n^\delta) \le -\frac{q^*(1 - c(\eta,\tau))^q}{q(C_{q^*} q^*)^{q-1}} \frac{\left\| F(x_n^\delta) - y^\delta \right\|^q}{\|A_n\|^q} \tag{7.33}$$

as well as $x_{n+1}^\delta \in \mathcal{D}(F)$ holds for all $n \le n_*(\delta) - 1$, with $n_*(\delta)$ according to (7.29).

Proof. Following the lines of the proof of the first part of Theorem 6.3, and using

$$\begin{aligned}
\left\| F(x_n^\delta) - y^\delta - A_n(x_n^\delta - x^\dagger) \right\| &\le \eta \left\| F(x_n^\delta) - y \right\| + \delta \\
&\le \eta \left\| F(x_n^\delta) - y^\delta \right\| + (1+\eta)\delta
\end{aligned}$$

and (7.29), we have

$$\begin{aligned}
&D_{j_q}(x^\dagger, x_{n+1}^\delta) - D_{j_q}(x^\dagger, x_n^\delta) \\
&= \frac{1}{q^*}\left(\left\| x_{n+1}^\delta \right\|^q - \left\| x_n^\delta \right\|^q \right) - \langle J_q^X(x_{n+1}^\delta) - J_q^X(x_n^\delta), x^\dagger \rangle_{X^* \times X} \\
&= D_{j_q}(x_n^\delta, x_{n+1}^\delta) - \mu_n \langle j_p^Y(F(x_n^\delta) - y^\delta), A_n(x_n^\delta - x^\dagger) \rangle_{X^* \times X} \\
&= D_{j_q}(x_n^\delta, x_{n+1}^\delta) \\
&\quad - \mu_n \Big(\left\| F(x_n^\delta) - y^\delta \right\|^p - \langle j_p^Y(F(x_n^\delta) - y^\delta), F(x_n^\delta) - y^\delta - A_n(x_n^\delta - x^\dagger) \rangle_{Y^* \times Y} \Big) \\
&\le D_{j_q}(x_n^\delta, x_{n+1}^\delta) - \mu_n \left(1 - \underbrace{\left(\eta + \frac{1+\eta}{\tau} \right)}_{= c(\eta,\tau)} \right) \left\| F(x_n^\delta) - y^\delta \right\|^p ,
\end{aligned}$$

where, by inequality (7.4), we estimate

$$D_{j_q}(x_n^\delta, x_{n+1}^\delta) \le C_{q^*}\, \mu_n^{q^*}\, \|A_n\|^{q^*}\, \|F(x_n^\delta) - y^\delta\|^{q^*(p-1)}. \tag{7.34}$$

Hence, we arrive at

$$\begin{aligned} &D_{j_q}(x^\dagger, x_{n+1}^\delta) - D_{j_q}(x^\dagger, x_n^\delta) \\ &\le -\mu_n\,(1 - c(\eta,\tau)) \left\| F(x_n^\delta) - y^\delta \right\|^p + C_{q^*}\, \mu_n^{q^*}\, \|A_n\|^{q^*}\, \|F(x_n^\delta) - y^\delta\|^{q^*(p-1)}. \end{aligned}$$

and, with the choice of μ_n (7.32), assertion (7.33) is proven. By (7.11) and the assumption that $x_0 \in \overline{B}_\rho^D(x^\dagger)$, this yields $x_{n+1}^\delta \in \overline{B}_\rho^D(x^\dagger) \subseteq \mathcal{D}(F)$. □

Adapting the proof of the second part of Theorem 6.3 to the nonlinear case, the convergence result, Theorem 2.3 in [88] (see also Theorem 2.4 in [128]), can be generalized to the Banach space setting:

Theorem 7.3. *Let the assumptions of Proposition 7.2 and additionally boundedness of F' on $\overline{B}_\rho^D(x^\dagger)$ be satisfied. Then, the Landweber iterates x_n, according to (7.28), applied to exact data y, converge to a solution of $F(x) = y$. If $\overline{\mathcal{R}(F'(x))} \subseteq \overline{\mathcal{R}(F'(x^\dagger))}$, for all $x \in \overline{B}_\rho(x^\dagger)$ and $J_q^X(x_0) \in \overline{\mathcal{R}(F'(x^\dagger))}$, then x_n converges to $x^\dagger$ as $n \to \infty$.*

Proof. Summing (7.33) yields

$$\sum_{n=0}^{\infty} \frac{R_n^q}{\|A_n\|^q} < \infty, \quad \sum_{n=0}^{\infty} \mu_n R_n^p < \infty, \tag{7.35}$$

where we have used the abbreviation $R_n = \|F(x_n^\delta) - y\|$ and the definition of μ_n. Hence, due to the boundedness of F' on $\overline{B}_\rho^D(x^\dagger) \supset \{x_n\}_{n\in\mathbb{N}}$,

$$R_n \to 0, \quad n \to \infty. \tag{7.36}$$

Defining $n_0 = 0$, n_{k+1} minimal, such that $R_{n_{k+1}} < \min_{n\le n_k} R_n$, we obtain (by minimality) that also $R_{n_{k+1}} < \min_{n<n_{k+1}} R_n$.

Monotonicity (7.33) of the errors yields boundedness of the iterates, first of all with respect to the Bregman distance but, by Theorem 2.60 (c), also with respect to the norm, so by possibly taking a sub-subsequence (again labeled with n_k) we get that $\{\|x_{n_k}\|\}_{k\in\mathbb{N}}$ converges and $J_q^X(x_{n_k})$ converges weakly, cf. (S.1), (S.2), (S.3) in the proof of Theorem 6.3.

Thus, we are prepared to transfer the proof of $\{x_{n_k}\}_{k\in\mathbb{N}}$ being a Cauchy sequence (with respect to the Bregman distance) to the nonlinear setting. Since, just as in the proof of Theorem 6.3, we have, for all $l, k \in \mathbb{N}$ with $k > l$,

$$D_{j_q}(x_{n_k}, x_{n_l},) = \frac{1}{q^*}\left(\|x_{n_l}\|^q - \|x_{n_k}\|^q\right) + \langle J_q^X(x_{n_k}) - J_q^X(x_{n_l}), x^\dagger\rangle \\ + \langle J_q^X(x_{n_k}) - J_q^X(x_{n_l}), x_{n_k} - x^\dagger\rangle,$$

where the first two summands converge to zero for $l \to \infty$ by (S.1), (S.2), it remains only to estimate the last term:

$$\begin{aligned} &|\langle J_q^X(x_{n_k}) - J_q^X(x_{n_l}), x_{n_k} - x^\dagger\rangle_{X^*\times X}| \\ &= \left|\sum_{n=n_l}^{n_k-1} \mu_n \langle A_n^* j_p^Y(F(x_n) - y), x_{n_k} - x^\dagger\rangle_{X^*\times X}\right| \\ &\le \sum_{n=n_l}^{n_k-1} \mu_n \left\|j_p^Y(F(x_n) - y)\right\| \left\|A_n(x_{n_k} - x^\dagger)\right\|, \end{aligned}$$

where we can estimate

$$\begin{aligned} &\left\|A_n(x_{n_k} - x^\dagger)\right\| \\ &\le \left\|A_n(x_{n_k} - x_n)\right\| + \left\|A_n(x_n - x^\dagger)\right\| \\ &\le \left\|F(x_{n_k}) - F(x_n)\right\| + \|F(x_n) - y\| \\ &\quad + \left\|F(x_{n_k}) - F(x_n) - A_n(x_{n_k} - x_n)\right\| + \left\|F(x_n) - F(x) - A_n(x_n - x^\dagger)\right\| \\ &\le (1+\eta)(\left\|F(x_{n_k}) - F(x_n)\right\| + \|F(x_n) - y\|) \\ &\le 3(1+\eta)\,\|F(x_n) - y\| \end{aligned}$$

by (S.3). Thus, we end up with

$$\limsup_{l\to\infty}\sup_{k>l} D_{j_q}(x_{n_k}, x_{n_l}) \le C \limsup_{l\to\infty}\sup_{k>l} \sum_{n=n_l}^{n_k-1} \mu_n R_n^p = 0$$

by (7.35), hence, by Theorem 2.60 (e), also $\limsup_{l\to\infty}\sup_{k>l}\left\|x_{n_k} - x_{n_l}\right\| \to 0$. Thus $\{x_{n_k}\}_k$ converges to some $\tilde{x} \in X$ which, by (7.36) and continuity of F, has to solve $F(\tilde{x}) = y$.

In case $\overline{\mathcal{R}(F'(x))} \subseteq \overline{\mathcal{R}(F'(x^\dagger))}$ for all $x \in \overline{B}_\rho(x^\dagger)$ and $J_q^X(x_0) \in \overline{\mathcal{R}(F'(x^\dagger))}$, all dual iterates $J_q^X(x_n)$ remain in $\overline{\mathcal{R}(F'(x^\dagger))}$. Hence, also $J_q^X(\tilde{x}) \in \overline{\mathcal{R}(F'(x^\dagger))}$, so that Proposition 7.1 yields $\tilde{x} = x^\dagger$. □

For the sake of simplicity, we here restricted ourselves to the case of a q^*-smooth dual. Let us mention that the same results can be proved in a more technical way, if we only require uniform smoothness, by adapting a similar technique of proof and a parameter choice as in Section 6.1 (see also [22, 213]).

Theorem 7.4. *Let the assumptions of Theorem 7.3 hold, with additionally Y being uniformly smooth and let $n_*(\delta)$ be chosen according to the stopping rule (7.29), (7.31). Then, according to (7.28), the Landweber iterates $x^\delta_{n_*(\delta)}$ converge to a solution of (7.1) as $\delta \to 0$. If $\overline{\mathcal{R}(F'(x))} \subseteq \overline{\mathcal{R}(F'(x^\dagger))}$ for all $x \in \overline{B}_\rho(x_0)$ and $J_q^X(x_0) \in \overline{\mathcal{R}(F'(x^\dagger))}$, then $x^\delta_{n_*(\delta)}$ converges to $x^\dagger$ as $\delta \to 0$.*

Proof. By the uniform smoothness of Y, the duality mapping j_p^Y is also single-valued and uniformly continuous on bounded sets (cf. Theorem 2.41). Hence, for a fixed iteration index n, by continuity of F, F', J_q^X, $J_{q^*}^{X^*}$ and j_p^Y, the coefficient μ_n, and hence the iterate x_n^δ, continuously depend on the data y^δ.

Let $\{\delta_k\}_{k\in\mathbb{N}}$ be an arbitrary null sequence and $\{n_k := n_*(\delta_k)\}_{k\in\mathbb{N}}$ the corresponding sequence of stopping indices.

The case of $\{n_k\}_{n\in\mathbb{N}}$ having a finite accumulation point can be treated in the Banach space case, as in the proof of Theorem 2.6 of [128], without any changes. For the convenience of the reader, we here provide the argument: There exists an $n_* \in \mathbb{N}$ and a subsequence such that $n_{k_j} = n_*$ for all $j \in \mathbb{N}$. As n_* is fixed, $x_{n_*}^{\delta_k}$ depends continuously on y^{δ_k}, and hence $x_{n_*}^{\delta_{k_j}} \to x_{n_*}$ as $j \to \infty$. On the other hand, by the definition of the stopping index n_{k_j}, it follows that

$$\left\| y^{\delta_{k_j}} - F(x_{n_*}^{\delta_{k_j}}) \right\| \leq \tau\delta_{k_j} \to 0 \text{ as } j \to \infty\,.$$

By continuity of F, this implies that $x_{n_*}^{\delta_{k_j}}$ converges to a solution of (7.1) as $j \to \infty$, namely x_{n_*}.

As a matter of fact, the proof of Theorem 2.6 of [128] also carries over for $n_k \to \infty$ as $k \to \infty$ although, at first glance, it looks as if the triangle inequality would be required, which we do not have for the Bregman distance: Let x be a solution to (7.1). For arbitrary $\epsilon > 0$, by Theorem 7.3, we can find n such that $D_{j_q}(x, x_n) < \frac{\epsilon}{2}$ and, by Theorem 2.60 (d), there exists k_0 such that, for all $k \geq k_0$, we have $n_k \geq n$ and $|D_{j_q}(x, x_n^{\delta_k}) - D_{j_q}(x, x_n)| < \frac{\epsilon}{2}$. Hence, by Proposition 7.2

$$D_{j_q}(x, x_{n_k}^{\delta_k}) \leq D_{j_q}(x, x_n^{\delta_k}) \leq D_{j_q}(x, x_n) + |D_{j_q}(x, x_n^{\delta_k}) - D_{j_q}(x, x_n)| < \epsilon\,. \qquad \square$$

7.2.2 Convergence rates for the iteratively regularized Landweber iteration with a priori stopping rule

In the Hilbert space case, the proof of convergence rates for Landweber iteration under source conditions

$$x^\dagger - x_0 \in \mathcal{R}(F'(x^\dagger)^* F'(x^\dagger))^{\nu/2}), \tag{7.37}$$

cf. (3.17), relies on the fact that the iteration errors $x_n^\delta - x^\dagger$ remain in the range of $(F'(x^\dagger)^* F'(x^\dagger))^{\nu/2}$ and their preimages under $(F'(x^\dagger)^* F'(x^\dagger))^{\nu/2}$ form a bounded sequence (cf., Proposition 2.11 in [128]). It seems that this approach cannot be carried over to the Banach space setting, unless more restrictive assumptions are made on the structure of the spaces than in the proof of convergence only (see the previous subsection), even in the special case $\nu = 1$ corresponding to the benchmark source condition (3.29).

Therefore, we here consider the iteratively regularized Landweber iteration

$$\begin{aligned} J_q^X(x_{n+1}^\delta - x_0) &= (1-\alpha_n) J_q^X(x_n^\delta - x_0) - \mu_n A_n^* j_p^Y(F(x_n^\delta) - y^\delta), \\ x_{n+1}^\delta &= x_0 + J_{q^*}^{X^*}(J_q^X(x_{n+1}^\delta - x_0)), \quad n = 0, 1, \ldots \end{aligned} \tag{7.38}$$

which, for an appropriate choice of the sequence $\{\alpha_n\}_{n\in\mathbb{N}} \in [0,1]$, has been shown to be convergent in a Hilbert space setting, with rates under a source condition (3.29), in [207].

In place of the Hilbert space source condition (7.37), like in Section 4.2 on rates for Tikhonov regularization, we consider variational inequalities

$$\begin{aligned} &\exists \beta > 0 \ \forall x \in \overline{B}: \\ &|\langle j_q^X(x^\dagger - x_0), x - x^\dagger \rangle_{X^* \times X}| \le \beta D_q^{x_0}(x^\dagger, x)^{\frac{1-\nu}{2}} \|F'(x^\dagger)(x - x^\dagger)\|^\nu . \end{aligned} \tag{7.39}$$

According to (7.39), due to the presence of additional regularity, we can relax the nonlinearity condition on F, compared to (7.30), to

$$\begin{aligned} &\left\|(F'(x^\dagger + v) - F'(x^\dagger))v\right\| \le K \left\|F'(x^\dagger)v\right\|^{c_1} D_q^{x_0}(x^\dagger, v + x^\dagger)^{c_2}, \\ &v \in X, \ x^\dagger + v \in \overline{B}, \end{aligned} \tag{7.40}$$

$$c_1 = 1 \text{ or } c_1 + c_2 p > 1 \text{ or } (c_1 + c_2 p \ge 1 \text{ and } K \text{ is sufficiently small}) \tag{7.41}$$

$$c_1 + c_2 \frac{2\nu}{\nu + 1} \ge 1 . \tag{7.42}$$

We will assume that in each step the step size $\mu_n > 0$ in (7.38) is chosen such that

$$\begin{aligned} &\mu_n \frac{1 - 3C(c_1)K}{3(1 - C(c_1)K)} \|F(x_n^\delta) - y^\delta\|^p - 2^{q^*+q-2} \\ &\times C_{q^*} \mu_n^{q^*} \|A_n^* j_p^Y(F(x_n^\delta) - y^\delta)\|^{q^*} \ge 0, \end{aligned} \tag{7.43}$$

where $C(c_1) = c_1^{c_1}(1-c_1)^{1-c_1}$, and c_1, K are as in (7.40), which is possible, e.g., by a choice $0 < \mu_n \le C_\mu \frac{\|F(x_n^\delta) - y^\delta\|^{\frac{q-p}{q-1}}}{\|A_n\|^{q^*}} =: \overline{\mu}_n$ with $C_\mu := \frac{2^{2-q^*-q}}{3} \frac{1-3C(c_1)K}{(1-C(c_1)K)C_{q^*}}$. If

$$p \ge q \tag{7.44}$$

and F, F' are bounded on $\overline{B}_\rho^D(x^\dagger)$, it is possible to bound $\overline{\mu}_n$ away from zero

$$\overline{\mu}_n \ge C_\mu \left(\sup_{x \in \overline{B}_\rho^D(x^\dagger)} (\|F(x) - y\| + \overline{\delta})^{p-q} \|F'(x)\|^q \right)^{-1/(q-1)} =: \overline{\mu} \tag{7.45}$$

for $\delta \in [0, \overline{\delta}]$, provided the iterates remain in $\overline{B}_\rho^D(x^\dagger)$. We will show this by induction, in the proof of Theorem 7.5. Hence, there exist $\underline{\mu}, \overline{\mu} > 0$, independent of n and δ, such that we can choose

$$0 < \underline{\mu} \le \mu_n \le \overline{\mu}\,. \tag{7.46}$$

Moreover, we will use an a priori choice of the stopping index n_*, according to

$$n_*(\delta) = \min\{n \in \mathbb{N} \;:\; \alpha_n^{\frac{\nu+1}{p(\nu+1)-2\nu}} \le \tau\delta\}\,, \tag{7.47}$$

where $\nu \in [0, 1]$ is the exponent in the variational inequality (7.39).

Theorem 7.5. *Assume that X is smooth and q-convex, that x_0 is sufficiently close to $x^\dagger$, i.e., $x_0 \in \overline{B}_\rho^D(x^\dagger)$, which, by (7.3), implies that $\|x^\dagger - x_0\|$ is also small, that a variational inequality (7.39), with $\nu \in (0, 1]$ and β sufficiently small, is satisfied, that F satisfies (7.40), with (7.41) and (7.42), that F and F' are continuous and uniformly bounded in $\overline{B}_\rho^D(x^\dagger)$, that (7.11) holds and that*

$$q^* \ge \frac{2\nu}{p(\nu+1) - 2\nu} + 1\,. \tag{7.48}$$

Let $n_*(\delta)$ be chosen according to (7.47), with τ sufficiently large. Moreover, assume that (7.44) holds and the sequence $\{\mu_n\}_{n\in\mathbb{N}}$ is chosen such that (7.46) holds for $0 < \underline{\mu} < \overline{\mu}$, according to (7.45), and assume that the sequence $\{\alpha_n\}_{n\in\mathbb{N}} \subseteq [0, 1]$ is chosen such that

$$\left(\frac{\alpha_{n+1}}{\alpha_n}\right)^{\frac{2\nu}{p(\nu+1)-2\nu}} + \frac{1}{3}\alpha_n - 1 \ge c\alpha_n \tag{7.49}$$

for some $c \in (0, \frac{1}{3})$ independent of n, and $\alpha_{\max} = \max_{n\in\mathbb{N}} \alpha_n$ is sufficiently small.

Then, the iterates x_{n+1}^δ remain in $\overline{B}_\rho^D(x^\dagger)$ for all $n \le n_*(\delta) - 1$, with n_* according to (7.47). Moreover, we obtain optimal convergence rates

$$D_q^{x_0}(x^\dagger, x_{n_*}) = O\left(\delta^{\frac{2\nu}{\nu+1}}\right), \quad \text{as } \delta \to 0 \tag{7.50}$$

as well as in the noise-free case $\delta = 0$

$$D_q^{x_0}(x^\dagger, x_n) = O\left(\alpha_n^{\frac{2\nu}{p(\nu+1)-2\nu}}\right) \quad \text{as } n \to \infty\,. \tag{7.51}$$

Proof. First of all, for $x_n^\delta \in \overline{B}_\rho^D(x^\dagger)$, (7.40) allows us to estimate as follows (see also (7.17)) for $c_1 \in [0, 1)$:

$$\begin{aligned}
&\left\| F(x_n^\delta) - F(x^\dagger) - A(x_n^\delta - x^\dagger)\right\| \\
&\le K \left\| A(x_n^\delta - x^\dagger)\right\|^{c_1} D_q^{x_0}(x^\dagger, x_n^\delta)^{c_2} \\
&\le C(c_1)K\left(\left\| A(x_n^\delta - x^\dagger)\right\| + D_q^{x_0}(x^\dagger, x_n^\delta)^{\frac{c_2}{1-c_1}}\right),
\end{aligned} \tag{7.52}$$

where we have used the abbreviation $A = F'(x^\dagger)$ and the elementary estimate

$$a^{1-\lambda}b^\lambda \le C(\lambda)(a+b) \quad \text{with} \\ C(\lambda) = \lambda^\lambda(1-\lambda)^{1-\lambda} \quad \text{for } a,b \ge 0\,,\ \lambda \in (0,1), \tag{7.53}$$

and with that, by the second triangle inequality,

$$\left\| A(x_n^\delta - x^\dagger) \right\| \le \frac{1}{1 - C(c_1)K} \\ \times \left(\left\| F(x_n^\delta) - F(x^\dagger) \right\| + C(c_1) K D_q^{x_0}(x^\dagger, x_n^\delta)^{\frac{c_2}{1-c_1}} \right) \tag{7.54}$$

as well as analogously

$$\left\| F(x_n^\delta) - F(x^\dagger) - A_n(x_n^\delta - x^\dagger) \right\| \\ \le 2C(c_1)K \left(\left\| A(x_n^\delta - x^\dagger) \right\| + D_q^{x_0}(x^\dagger, x_n^\delta)^{\frac{c_2}{1-c_1}} \right) \\ \le \frac{2C(c_1)K}{1 - C(c_1)K} \left(\left\| F(x_n^\delta) - F(x^\dagger) \right\| + D_q^{x_0}(x^\dagger, x_n^\delta)^{\frac{c_2}{1-c_1}} \right). \tag{7.55}$$

For any $n \le n_*$, according to (7.47), by Lemma 2.62, we have

$$D_q^{x_0}(x^\dagger, x_{n+1}^\delta) - D_q^{x_0}(x^\dagger, x_n^\delta) \\ = D_q^{x_0}(x_n^\delta, x_{n+1}^\delta) + \langle J_q^X(x_n^\delta - x_0) - J_q^X(x_{n+1}^\delta - x_0), x^\dagger - x_n^\delta \rangle_{X^* \times X} \\ = D_q^{x_0}(x_n^\delta, x_{n+1}^\delta) - \mu_n \langle j_p^Y(F(x_n^\delta) - y^\delta), A_n(x_n^\delta - x^\dagger) \rangle_{Y^* \times Y} \\ + \alpha_n \langle J_q^X(x^\dagger - x_0), x^\dagger - x_n^\delta \rangle_{X^* \times X} \\ - \alpha_n \langle J_q^X(x^\dagger - x_0) - J_q^X(x_n^\delta - x_0), x^\dagger - x_n^\delta \rangle_{X^* \times X} \tag{7.56}$$

where the terms on the right hand side can be estimated as follows.

By (7.4) and Lemma 2.63 we have

$$D_q^{x_0}(x_n^\delta, x_{n+1}^\delta) \tag{7.57}$$

$$\le C_{q^*} \| J_q^X(x_{n+1}^\delta - x_0) - J_q^X(x_n^\delta - x_0) \|^{q^*} \\ = C_{q^*} \| \alpha_n J_q^X(x_n^\delta - x_0) + \mu_n A_n^* j_p^Y(F(x_n^\delta) - y^\delta) \|^{q^*} \\ \le 2^{q^*-1} C_{q^*} \left(\alpha_n^{q^*} \| x_n^\delta - x_0 \|^q + \mu_n^{q^*} \| A_n^* j_p^Y(F(x_n^\delta) - y^\delta) \|^{q^*} \right) \\ \le 2^{q^*-1} C_{q^*} \left(\alpha_n^{q^*} 2^{q-1} \left(\| x^\dagger - x_0 \|^q + \frac{1}{c_q} D_q^{x_0}(x^\dagger, x_n^\delta) \right) \right. \\ \left. + \mu_n^{q^*} \| A_n^* j_p^Y(F(x_n^\delta) - y^\delta) \|^{q^*} \right), \tag{7.58}$$

where we have used the triangle inequality in X^* and X, the inequality

$$(a+b)^\lambda \leq 2^{\lambda-1}(a^\lambda + b^\lambda) \quad \text{for } a, b \geq 0,\ \lambda \geq 1, \tag{7.59}$$

and (7.3).

For the second term on the right hand side of (7.56), we get, using (7.55), (7.53), (7.59),

$$\begin{aligned}
&\langle j_p^Y(F(x_n^\delta) - y^\delta), A_n(x_n^\delta - x^\dagger)\rangle_{Y^*\times Y} \\
&= \langle j_p^Y(F(x_n^\delta) - y^\delta), F(x_n^\delta) - y^\delta\rangle_{Y^*\times Y} \\
&\quad - \langle j_p^Y(F(x_n^\delta) - y^\delta), F(x_n^\delta) - y^\delta - A_n(x_n^\delta - x^\dagger)\rangle_{Y^*\times Y} \\
&\geq \frac{1-3C(c_1)K}{1-C(c_1)K}\|F(x_n^\delta) - y^\delta\|^p \\
&\quad - \|F(x_n^\delta) - y^\delta\|^{p-1}\left(\frac{2C(c_1)K}{1-C(c_1)K} D_q^{x_0}(x^\dagger, x_n^\delta)^{\frac{c_2}{1-c_1}} + \frac{1+C(c_1)K}{1-C(c_1)K}\delta\right) \\
&= \frac{1-3C(c_1)K}{1-C(c_1)K}\|F(x_n^\delta) - y^\delta\|^p \\
&\quad - \left(\frac{1-3C(c_1)K}{3C\left(\frac{p-1}{p}\right)(1-C(c_1)K)}\|F(x_n^\delta) - y^\delta\|^p\right)^{\frac{p-1}{p}} \left(\frac{\left(3C\left(\frac{p-1}{p}\right)\right)^{p-1}}{(1-C(c_1)K)}\right)^{\frac{1}{p}} \\
&\qquad \times \left(2C(c_1)K D_q^{x_0}(x^\dagger, x_n^\delta)^{\frac{c_2}{1-c_1}} + (1+C(c_1)K)\delta\right) \\
&\geq \frac{1-3C(c_1)K}{1-C(c_1)K}\|F(x_n^\delta) - y^\delta\|^p \\
&\quad - C\left(\frac{p-1}{p}\right)\left\{\frac{1-3C(c_1)K}{3C\left(\frac{p-1}{p}\right)(1-C(c_1)K)}\|F(x_n^\delta) - y^\delta\|^p\right. \\
&\quad + \frac{\left(3C\left(\frac{p-1}{p}\right)\right)^{p-1}}{(1-C(c_1)K)} 2^{p-1} \\
&\quad \left.\times \left((2C(c_1)K)^p D_q^{x_0}(x^\dagger, x_n^\delta)^{\frac{c_2 p}{1-c_1}} + (1+C(c_1)K)^p \delta^p\right)\right\}.
\end{aligned} \tag{7.60}$$

Using the variational inequality (7.39), the estimate (7.54), and

$$(a+b)^\lambda \leq (a^\lambda + b^\lambda) \ \text{ for } a, b \geq 0,\ \lambda \in [0,1], \tag{7.61}$$

we get

$$
\begin{aligned}
&|\alpha_n \langle J_q^X(x^\dagger - x_0), x^\dagger - x_n^\delta \rangle_{X^* \times X}| \\
&\leq \beta \alpha_n D_q^{x_0}(x^\dagger, x_n^\delta)^{\frac{1-\nu}{2}} \|F'(x^\dagger)(x_n^\delta - x^\dagger)\|^\nu \\
&\leq \beta \alpha_n D_q^{x_0}(x^\dagger, x_n^\delta)^{\frac{1-\nu}{2}} \frac{1}{(1 - C(c_1)K)^\nu} \\
&\quad \left(\left\| F(x_n^\delta) - y^\delta \right\| + \delta + C(c_1) K D_q^{x_0}(x^\dagger, x_n^\delta)^{\frac{c_2}{1-c_1}} \right)^\nu \\
&\leq \beta \alpha_n D_q^{x_0}(x^\dagger, x_n^\delta)^{\frac{1-\nu}{2}} \epsilon^{-\nu} \left(\epsilon \frac{1}{(1 - C(c_1)K)^\nu} (\|F(x_n^\delta) - y^\delta\| + \delta) \right)^\nu \\
&\quad + \beta \alpha_n \left(\frac{C(c_1)K}{(1 - C(c_1)K)} \right)^\nu D_q^{x_0}(x^\dagger, x_n^\delta)^{\frac{1-\nu}{2} + \frac{\nu c_2}{1-c_1}} \\
&\leq C\left(\frac{\nu}{p}\right) \left\{ \left(\beta \alpha_n D_q^{x_0}(x^\dagger, x_n^\delta)^{\frac{1-\nu}{2}} \epsilon^{-\nu} \right)^{\frac{p}{p-\nu}} \right. \\
&\quad \left. + \left(\epsilon \frac{1}{(1 - C(c_1)K)^\nu} (\|F(x_n^\delta) - y^\delta\| + \delta) \right)^p \right\} \\
&\quad + \beta \alpha_n \left(\frac{C(c_1)K}{(1 - C(c_1)K)} \right)^\nu D_q^{x_0}(x^\dagger, x_n^\delta)^{\frac{1-\nu}{2} + \frac{\nu c_2}{1-c_1}} \\
&= C(\frac{\nu}{p}) \left\{ (\beta \epsilon^{-\nu})^{\frac{p}{p-\nu}} \left(3C\left(\frac{\nu}{p}\right) C\left(\frac{p(1-\nu)}{2(p-\nu)}\right) \right)^{\frac{p(1-\nu)}{2(p-\nu)}} \right. \\
&\quad \times \alpha_n^{\frac{p(1+\nu)}{2(p-\nu)}} \left(\frac{\alpha_n D_q^{x_0}(x^\dagger, x_n^\delta)}{3C\left(\frac{\nu}{p}\right) C\left(\frac{p(1-\nu)}{2(p-\nu)}\right)} \right)^{\frac{p(1-\nu)}{2(p-\nu)}} \\
&\quad \left. + \left(\epsilon \frac{1}{(1 - C(c_1)K)^\nu} (\|F(x_n^\delta) - y^\delta\| + \delta) \right)^p \right\} \\
&\quad + \beta \alpha_n \left(\frac{C(c_1)K}{(1 - C(c_1)K)} \right)^\nu D_q^{x_0}(x^\dagger, x_n^\delta)^{\frac{1-\nu-c_1+\nu c_1+2\nu c_2}{2(1-c_1)}} \\
&\leq C\left(\frac{\nu}{p}\right) \left\{ C\left(\frac{p(1-\nu)}{2(p-\nu)}\right) \left[\left(\beta \epsilon^{-\nu} \left(3C\left(\frac{\nu}{p}\right) C\left(\frac{p(1-\nu)}{2(p-\nu)}\right) \right)^{\frac{1-\nu}{2}} \right)^{\frac{2p}{p(\nu+1)-2\nu}} \right. \right. \\
&\quad \left. \times \alpha_n^{\frac{p(1+\nu)}{p(\nu+1)-2\nu}} + \left(\frac{\alpha_n D_q^{x_0}(x^\dagger, x_n^\delta)}{3C\left(\frac{\nu}{p}\right) C\left(\frac{p(1-\nu)}{2(p-\nu)}\right)} \right) \right] \\
&\quad \left. + \left(\epsilon \frac{1}{(1 - C(c_1)K)^\nu} (\|F(x_n^\delta) - y^\delta\| + \delta) \right)^p \right\} \\
&\quad + \frac{1}{3} \alpha_n D_q^{x_0}(x^\dagger, x_n^\delta),
\end{aligned}
\tag{7.62}
$$

where we have used (7.53) twice and $\epsilon > 0$ will be chosen to be a sufficiently small number below. Moreover, by (7.42), the exponent

$$\frac{1-\nu-c_1+\nu c_1+2\nu c_2}{2(1-c_1)} = 1+\frac{1+\nu}{2(1-c_1)}(c_1+\frac{2\nu}{\nu+1}c_2-1)$$

is larger or equal to one and β is sufficiently small, so that

$$\beta(\frac{C(c_1)K}{(1-C(c_1)K)})^{\nu}\rho^{\frac{1-\nu-c_1+\nu c_1+2\nu c_2}{2(1-c_1)}-1} < \frac{1}{3}.$$

Finally, we have that

$$\begin{aligned}\langle J_q^X(x^\dagger-x_0)-J_q^X(x_n^\delta-x_0), x^\dagger-x_n^\delta\rangle_{X^*\times X} &= D_q^{x_0}(x^\dagger,x_n^\delta)+D_q^{x_0}(x_n^\delta,x^\dagger)\\ &\geq D_q^{x_0}(x^\dagger,x_n^\delta).\end{aligned} \tag{7.63}$$

Inserting estimates (7.57)-(7.63) with

$$\epsilon = 2^{p-1}\mu_n^{1/p}(\frac{1-3C(c_1)K}{3(1-C(c_1)K)})^{1/p}\frac{(1-C(c_1)K)^{\nu}}{C(\frac{\nu}{p})}$$

into (7.56) and using boundedness away from zero of μ_n and the abbreviations

$$\begin{aligned}
\mathrm{d}_n &= D_q^{x_0}(x^\dagger,x_n^\delta)^{1/2},\\
C_0 &= 6^{p-1}C\left(\frac{p-1}{p}\right)^p\frac{(2C(c_1)K)^p}{(1-C(c_1)K)},\\
C_1 &= 2^{q^*+q-2}\frac{C_{q^*}}{c_q},\\
C_2 &= C\left(\frac{\nu}{p}\right)C\left(\frac{p(1-\nu)}{2(p-\nu)}\right)\left(\beta\epsilon^{-\nu}\left(3C\left(\frac{\nu}{p}\right)C\left(\frac{p(1-\nu)}{2(p-\nu)}\right)^{\frac{1-\nu}{2}}\right)^{\frac{2p}{p(\nu+1)-2\nu}}\right.,\\
C_3 &= 2^{q^*+q-2}C_{q^*}\|x^\dagger-x_0\|^q,\\
C_4 &= 2^{p-1}C(\frac{\nu}{p})\overline{\epsilon}\frac{1}{(1-C(c_1)K)^{\nu}}+6^{p-1}C(\frac{p-1}{p})^p\frac{(1+C(c_1)K)^p}{1-C(c_1)K},\\
\underline{\epsilon} &= 2^{p-1}\underline{\mu}^{1/p}\left(\frac{1-3C(c_1)K}{3(1-C(c_1)K)}\right)^{1/p}\frac{(1-C(c_1)K)^{\nu}}{C(\frac{\nu}{p})},\\
\overline{\epsilon} &= 2^{p-1}\overline{\mu}^{1/p}\left(\frac{1-3C(c_1)K}{3(1-C(c_1)K)}\right)^{1/p}\frac{(1-C(c_1)K)^{\nu}}{C(\frac{\nu}{p})}
\end{aligned}$$

we obtain

$$\mathrm{d}_{n+1}^2 \le C_0 \mathrm{d}_n^{\frac{2c_2 p}{1-c_1}} + \left(1 - \frac{1}{3}\alpha_n + C_1 \alpha_n^{q^*}\right) \mathrm{d}_n^2 + C_2 \alpha_n^{\frac{p(1+\nu)}{p(\nu+1)-2\nu}} + C_3 \alpha_n^{q^*}$$
$$+ C_4 \delta^p - \Big(\mu_n \frac{1 - 3C(c_1)K}{3(1 - C(c_1)K)} \|F(x_n^\delta) - y^\delta\|^p$$
$$- 2^{q^*+q-2} C_{q^*} \mu_n^{q^*} \|A_n^* j_p^Y (F(x_n^\delta) - y^\delta)\|^{q^*}\Big) .$$

Here, the last term is non-positive, due to the choice (7.43) of μ_n, so that we arrive at

$$\mathrm{d}_{n+1}^2 \le C_0 \mathrm{d}_n^{\frac{2c_2 p}{1-c_1}} + \left(1 - \frac{1}{3}\alpha_n + C_1 \alpha_n^{q^*}\right) \mathrm{d}_n^2$$
$$+ \underbrace{(C_2 + C_3 + C_4 \tau^{-p})}_{=:C_5} \alpha_n^{\frac{p(1+\nu)}{p(\nu+1)-2\nu}} , \tag{7.64}$$

where we have used (7.48) and the stopping rule (7.47). Denoting

$$\gamma_n := \frac{\mathrm{d}_n^2}{\alpha_n^{\frac{2\nu}{p(\nu+1)-2\nu}}}$$

we get the following recursion

$$\gamma_{n+1} \tag{7.65}$$
$$\le C_0 \left(\frac{\alpha_n}{\alpha_{n+1}}\right)^\theta \alpha_n^{\theta\omega} \gamma_n^\omega + \left(\frac{\alpha_n}{\alpha_{n+1}}\right)^\theta \left(1 - \frac{1}{3}\alpha_n + C_1 \alpha_n^{q^*}\right) \gamma_n$$
$$+ C_5 \left(\frac{\alpha_n}{\alpha_{n+1}}\right)^\theta \alpha_n$$

with

$$\theta = \frac{2\nu}{p(\nu+1) - 2\nu} \qquad \omega = \frac{c_2 p}{1 - c_1} ,$$

where

$$\omega \ge 1$$

by (7.41) and

$$\theta\omega = \frac{p}{p - \frac{2\nu}{\nu+1}} \frac{c_2 \frac{2\nu}{\nu+1}}{1 - c_1} \ge 1,$$

due to assumption (7.42). Hence, as sufficient conditions for uniform boundedness of $\{\gamma_n\}_{n \le n_*}$ by $\overline{\gamma}$ and for $x_{n+1}^\delta \in \overline{B}_\rho^D(x^\dagger)$ we get

$$\overline{\gamma} \le \rho^2 \tag{7.66}$$

$$C_0 \alpha_n^{\theta\omega - 1} \overline{\gamma}^\omega - \left\{ \left(\frac{\alpha_{n+1}}{\alpha_n}\right)^\theta + \frac{1}{3}\alpha_n - 1 - C_1 \alpha_n^{q^*} \right\} \alpha_n^{-1} \overline{\gamma} + C_5 \le 0 , \tag{7.67}$$

where, by $q^* > 1$, (7.42), the factors $C_0\alpha_n^{\theta\omega-1}$, $C_1\alpha_n^{q^*-1}$ and C_5 can be made small, for small $\alpha_{\max}$, β, $\|x^\dagger - x_0\|$ and large τ. We use this fact to arrive at

$$C_0\alpha_n^{\theta\omega-1}\rho^{\omega-1} + C_1\alpha_n^{q^*-1} \le \tilde{c} < c,$$

with $\tilde{c}$ independent of n, which, together with (7.49), yields sufficiency of

$$\frac{C_5}{c-\tilde{c}} \le \overline{\gamma} \le \rho^2,$$

for (7.66), (7.67), which, for any (even small) prescribed ρ, is indeed enabled by possibly decreasing β, $\|x^\dagger - x_0\|$, τ^{-1}, and therewith C_5.

In case $c_1 = 1$, estimates (7.54), (7.55) simplify to

$$\left\|A(x_n^\delta - x^\dagger)\right\| \le \frac{1}{1-\rho^{2c_2}K}\left\|F(x_n^\delta) - F(x^\dagger)\right\| \tag{7.68}$$

and

$$\left\|F(x_n^\delta) - F(x^\dagger) - A_n(x_n^\delta - x^\dagger)\right\| \le \frac{2\rho^{2c_2}K}{1-\rho^{2c_2}K}\left\|F(x_n^\delta) - F(x^\dagger)\right\|. \tag{7.69}$$

With this, the terms containing $D_q^{x_0}(x^\dagger, x_n^\delta)^{\frac{c_2}{1-c_1}}$ are removed and $C(c_1)$ is replaced by ρ^{2c_2} in (7.57)–(7.63), so that we end up with a recursion of the form (7.65) as before, with C_0 replaced by zero. Hence, the remainder of the proof of uniform boundedness of γ_n can be done in the same way as for $c_1 < 1$.

In case $\delta = 0$, i.e., $n_* = \infty$, uniform boundedness of $\{\gamma_n\}_{n\in\mathbb{N}}$ implies (7.51). For $\delta > 0$, we get (7.50) by using (7.47) in

$$D_q^{x_0}(x^\dagger, x_{n_*}) = \gamma_{n_*}\alpha_{n_*}^{\frac{2\nu}{p(\nu+1)-2\nu}} \le \overline{\gamma}\alpha_{n_*}^{\frac{2\nu}{p(\nu+1)-2\nu}} \le \overline{\gamma}(\tau\delta)^{\frac{2\nu}{\nu+1}}. \qquad \square$$

Remark 7.6. Note that the rate exponent in (7.51)

$$\frac{2\nu}{p(\nu+1)-2\nu} = \frac{2\nu}{\nu+1}\left(p - \frac{2\nu}{\nu+1}\right)^{-1}$$

always lies in the interval $[0, \frac{1}{p-1}]$, since $\frac{2\nu}{\nu+1} \in [0,1]$.

Moreover, note that Theorem 7.5 provides a result on rates only, but produces no convergence result without variational inequality. This corresponds to the situation from [207] in a Hilbert space setting.

Remark 7.7. In view of estimate (7.64), an optimal choice of α_n would be one that minimizes the right hand side. At least in the special case where the same power of α_n appears in the last two terms, i.e., $\frac{p(1+\nu)}{p(\nu+1)-2\nu} = q^*$, elementary calculus yields

$$(\alpha_n^{opt})^{\frac{2\nu}{p(\nu+1)-2\nu}} = \frac{D_q^{x_0}(x^\dagger, x_n^\delta)}{3q^*(C_1 D_q^{x_0}(x^\dagger, x_n^\delta) + C_5)},$$

which shows that the obtained relation $D_q^{x_0}(x^\dagger, x_n^\delta) \sim \alpha_n^{\frac{2\nu}{p(\nu+1)-2\nu}}$ is indeed reasonable and probably even optimal.

Remark 7.8. A possible choice of $\{\alpha_n\}_{n\in\mathbb{N}}$, satisfying (7.49) and smallness of $\alpha_{\max}$, is given by

$$\alpha_n = \frac{\alpha_0}{(n+1)^x}$$

with $x \in (0,1]$ such that $3x\theta < \alpha_0$ sufficiently small, since in this case, with $c := \frac{1}{3} - \frac{x\theta}{\alpha_0} > 0$, using the abbreviation $\theta = \frac{2\nu}{p(\nu+1)-2\nu} \in [0, \frac{1}{p-1}]$ we get by the Mean Value Theorem

$$\begin{aligned}
&\left(\frac{\alpha_{n+1}}{\alpha_n}\right)^\theta + \left(\frac{1}{3} - c\right)\alpha_n - 1 \\
&= \frac{\alpha_n}{\alpha_0}\left\{\alpha_0\left(\frac{1}{3} - c\right) - \frac{(n+2)^{x\theta} - (n+1)^{x\theta}}{(n+2)^{x\theta}}(n+1)^x\right\} \\
&= \frac{\alpha_n}{\alpha_0}\left\{\alpha_0\left(\frac{1}{3} - c\right) - \frac{x\theta(n+1+t)^{x\theta-1}}{(n+2)^{x\theta}}(n+1)^x\right\} \\
&\geq \frac{\alpha_n}{\alpha_0}\left\{\alpha_0\left(\frac{1}{3} - c\right) - x\theta\frac{(n+1)^x}{n+1+t}\right\} \geq 0,
\end{aligned}$$

for some $t \in [0,1]$.

7.3 The iteratively regularized Gauss–Newton method

The iteratively regularized Gauss–Newton method (IRGNM) can be generalized to a Banach space setting by computing iterates $x_{n+1}^\delta = x_{n+1}^\delta(\alpha_n)$ in a variational form, as a minimizer $x_{n+1}^\delta(\alpha)$ of

$$\begin{aligned}
&\left\|A_n(x - x_n^\delta) + F(x_n^\delta) - y^\delta\right\|^p + \alpha\,\|x - x_0\|^q \to \min, \quad \text{subject to } x \in \mathcal{D}(F), \\
&\quad n = 0, 1, \ldots, \qquad (7.70)
\end{aligned}$$

where $q, p \in (1, \infty)$, $\{\alpha_n\}_n$ is a sequence of regularization parameters, x_0 is some a priori guess, and we abbreviate

$$A_n = F'(x_n^\delta).$$

Thus, each Newton step is defined as a convex minimization problem (7.70) of the type already extensively studied in Chapter 5 of this book. Well-definedness of $x_{n+1}^\delta(\alpha)$ for fixed $\alpha > 0$ is implied by Proposition 4.1, where the case of a linear operator $F = A_n$ was considered. We provide an extension to the following result, which will also be useful in the convergence analysis below.

Lemma 7.9.

(a) *Let X be a reflexive Banach space, let $D \subseteq X$ be nonempty, and assume that either*

(i) $A : D \subseteq X \to Y$ *is weakly closed and* Y *reflexive*

or

(ii) $A : D \subseteq X \to Y$ *is weak-to-weak continuous and* D *weakly closed.*

Then, for $\alpha > 0$, the functional Φ_α, defined on D by $\Phi_\alpha(x) := \|Ax\|_Y^p + \alpha \|x - x_0\|_X^q$, has a minimizer over D, which is unique, if X is additionally strictly convex.

(b) *Let the assumptions of* (a) *and strict convexity of X hold and $x(\alpha)$, for $\alpha > 0$, denote the minimizer of*

$$\Phi_\alpha(x) \to \min, \quad \textit{subject to } x \in D\,.$$

Then, $\hat{\psi} : \alpha \mapsto \|Ax(\alpha)\|_Y^p$ is a continuous function on $(0, \infty)$. Furthermore, the mapping $\alpha \mapsto x(\alpha)$ is continuous, if X is uniformly convex.

Proof. By standard arguments, existence of $x(\alpha)$ follows from (i) or (ii) and its uniqueness follows from strict convexity of X.

At first, we prove monotonicity of the mappings $\hat{\psi}$ and $\alpha \mapsto \|x(\alpha) - x_0\|_X$, in the sense that

$$\alpha_1 \le \alpha_2 \quad \Rightarrow \quad \left\{ \begin{array}{rcl} \|x(\alpha_1) - x_0\|_X & \ge & \|x(\alpha_2) - x_0\|_X \\ \hat{\psi}(\alpha_1) & \le & \hat{\psi}(\alpha_2) \end{array} \right. . \tag{7.71}$$

Monotonicity of $\alpha \mapsto \|x(\alpha) - x_0\|_X$ follows from

$$\begin{aligned} \Phi_{\alpha_1}(x(\alpha_1)) &\le \Phi_{\alpha_1}(x(\alpha_2)) \\ &= \Phi_{\alpha_2}(x(\alpha_2)) + (\alpha_1 - \alpha_2)\, \|x(\alpha_2) - x_0\|_X^q \\ &\le \Phi_{\alpha_2}(x(\alpha_1)) + (\alpha_1 - \alpha_2)\, \|x(\alpha_2) - x_0\|_X^q \end{aligned} \tag{7.72}$$

which implies $(\alpha_1 - \alpha_2)(\|x(\alpha_2) - x_0\|_X^q - \|x(\alpha_1) - x_0\|_X^q) \ge 0$. Monotonicity of $\hat{\psi}$ follows from (7.72) and the monotonicity of $\alpha \mapsto \|x(\alpha - x_0)\|_X$.

To show continuity, let $\alpha > 0$, $\alpha_n \to \alpha$, which implies $\underline{\alpha} \le \alpha_n \le \overline{\alpha}$ for some $\underline{\alpha}, \overline{\alpha} > 0$. For all $n \in \mathbb{N}$ we have, by minimality of $x(\alpha_n)$,

$$\Phi_{\alpha_n}(x(\alpha_n)) \le \Phi_{\alpha_n}(x(\alpha)) \le C\,.$$

Hence, $\|x(\alpha_n)\|_X$, $\|Ax(\alpha_n)\|_Y$ are uniformly bounded by C, $C/\underline{\alpha}$, respectively, and there exists a subsequence α_{n_k}, such that $x(\alpha_{n_k})$ converges weakly to some $\bar{x} \in D$.

In case (i), by reflexivity of Y, a subsequence of $Ax(\alpha_{n_k})$, denoted again by $Ax(\alpha_{n_k})$, converges weakly to some $\bar{y} \in Y$ and, by weak closedness $\bar{x} \in D$, $A\bar{x} = \bar{y}$ holds.

In case (ii), by weak closedness of D, we have $\bar{x} \in D$ and, by weak continuity of A, $Ax(\alpha_{n_k})$, converges weakly to $A\bar{x}$.

By the weak lower semicontinuity of the norms, we get

$$\begin{aligned}\Phi_\alpha(\bar{x}) &\le \liminf_{k\to\infty} \Phi_{\alpha_{n_k}}(x(\alpha_{n_k})) \le \limsup_{k\to\infty} \Phi_{\alpha_{n_k}}(x(\alpha_{n_k})) \\ &\le \limsup_{k\to\infty} \Phi_{\alpha_{n_k}}(x(\alpha)) = \Phi_\alpha(x(\alpha)),\end{aligned}$$

where we used minimality of the $x(\alpha_{n_k})$ in the third inequality. Since, in a strictly convex X, the minimizer of Φ_α is unique, we must have $\bar{x} = x(\alpha)$ and thus it also follows that

$$\lim_{k\to\infty} \Phi_{\alpha_{n_k}}(x(\alpha_{n_k})) = \Phi_\alpha(x(\alpha)). \tag{7.73}$$

In case $\alpha_{n_k} \ge \alpha$ for all k, we get, by weak convergence of $x(\alpha_{n_k})$ to $\bar{x} = x(\alpha)$, weak lower semicontinuity of the norm and (7.71)

$$\begin{aligned}\|x(\alpha) - x_0\|_X^q &\le \liminf_{k\to\infty} \left\|x(\alpha_{n_k}) - x_0\right\|_X^q \\ &\le \limsup_{k\to\infty} \left\|x(\alpha_{n_k}) - x_0\right\|_X^q \le \|x(\alpha) - x_0\|_X^q .\end{aligned}$$

Hence, $\lim_{k\to\infty} \left\|x(\alpha_{n_k}) - x_0\right\|_X^q = \|x(\alpha) - x_0\|_X^q$ and from (7.73) we further deduce $\lim_{k\to\infty} \hat{\psi}(\alpha_{n_k}) = \hat{\psi}(\alpha)$.

In case $\alpha_{n_k} \le \alpha$ for all k, we similarly conclude at first, by the monotonicity of $\hat{\psi}$ (7.71), that $\lim_{k\to\infty} \hat{\psi}(\alpha_{n_k}) = \hat{\psi}(\alpha)$ and then, again with (7.73), that

$$\lim_{k\to\infty} \left\|x(\alpha_{n_k}) - x_0\right\|_X^q = \|x(\alpha) - x_0\|_X^q .$$

Finally, subsequence arguments yield continuity of $\alpha \mapsto \hat{\psi}(\alpha)$ and $\alpha \mapsto \|x(\alpha) - x_0\|_X$. The latter, together with the weak convergence of $x(\alpha_n)$ to $x(\alpha)$, implies strong convergence in a uniformly convex X. □

Throughout the remainder of this section, we will assume, in view of Lemma 7.9, that

$$F'(x) : X \to Y \quad \text{weakly closed for all } x \in \mathcal{D}(F) \text{ and } Y \text{ reflexive} \tag{7.74}$$

or

$$\mathcal{D}(F) \quad \text{weakly closed} \tag{7.75}$$

holds.

In the following Subsection 7.3.1, we will consider the IRGNM (7.70), with a priori choice of $\{\alpha_n\}_{n\in\mathbb{N}}$ and n_*. Combination of (7.70) with an a posteriori choice of $\{\alpha_n\}_{n\in\mathbb{N}}$ and n_*, according to the discrepancy principle, will be studied in Subsection 7.3.2. In both cases, we will prove optimal convergence rates under Hölder type source conditions

$$\begin{aligned}&\exists \beta > 0 \,\forall x \in \overline{B} :\\ &|\langle j_q^X(x^\dagger - x_0), x - x^\dagger\rangle_{X^*\times X}| \le \beta D_q^{x_0}(x, x^\dagger)^{\frac{1-\nu}{2}} \|F'(x^\dagger)(x - x^\dagger)\|^\nu\end{aligned} \tag{7.76}$$

with $0 < \nu < 1$ and under general variational inequalities

$$\begin{aligned}&\forall x \in \overline{B},\ x \neq x^\dagger :\\ &|\langle j_q^X(x^\dagger - x_0), x - x^\dagger\rangle_{X^*\times X}|\\ &\le D_q^{x_0}(x, x^\dagger)^{1/2}\kappa\left(\frac{\|F'(x^\dagger)(x - x^\dagger)\|^2}{D_q^{x_0}(x, x^\dagger)}\right)\end{aligned} \tag{7.77}$$

with κ as in (7.9). The presentation partly follows [129] and [126].

7.3.1 Convergence with a priori parameter choice

The results of this subsection are based on an a priori choice of $\{\alpha_n\}_{n\in\mathbb{N}}$ and n_*, according to

$$\alpha_0 \le 1\,, \quad \alpha_n \to 0 \text{ as } n \to \infty\,, \quad 1 \le \frac{\alpha_n}{\alpha_{n+1}} \le \hat{C} \text{ for all } n \tag{7.78}$$

and

$$n_*(\delta) = \min\left\{n \in \mathbb{N} \,:\, \alpha_n^{\frac{\nu+1}{p(\nu+1)-2\nu}} \le \tau\delta\right\}, \quad \text{in case of (7.76)} \tag{7.79}$$

$$n_*(\delta) = \min\{n \in \mathbb{N} \,:\, \alpha_n \le \varphi_p(\tau\delta)\}, \quad \text{in case of (7.77)} \tag{7.80}$$

with

$$\varphi_p(t) = t^{p-2}\Theta^{-1}(t)\,, \quad \Theta(\lambda) := \kappa(\lambda)\sqrt{\lambda}. \tag{7.81}$$

In this subsection, like in Subsection 7.2.2, see Remark 7.6 there, we obtain only convergence rates, but no convergence without source conditions. Note that, in Subsection 7.3.2 below, with an a posteriori choice of α_n and n_*, we will be able to prove both convergence and convergence rates.

First of all, we will study rates under Hölder type source conditions (7.76). As for the iteratively regularized Landweber iteration with a priori parameter choice, the nonlinearity conditions may again depend on the smoothness index ν in (7.76), with weaker conditions for larger ν:

$$\begin{aligned}
&\left\|(F'(x^\dagger+\tilde{v})-F'(x^\dagger))v\right\| \\
&\le K\left\|F'(x^\dagger)v\right\|^{\tilde{c}_1} D_q^{x_0}(v+x^\dagger,x^\dagger)^{\tilde{c}_2}\left\|F'(x^\dagger)\tilde{v}\right\|^{\tilde{c}_3} D_q^{x_0}(\tilde{v}+x^\dagger,x^\dagger)^{\tilde{c}_4} \\
&v,\tilde{v}\in X,\ x^\dagger+v\in\overline{B}\ x^\dagger+\tilde{v}\in\overline{B}
\end{aligned} \tag{7.82}$$

with

$$\tilde{c}_1+\tilde{c}_2\frac{2\nu}{\nu+1}\ge\frac{1}{2},\quad \tilde{c}_3+\tilde{c}_4\frac{2\nu}{\nu+1}\ge\frac{1}{2} \tag{7.83}$$

as well as

$$\begin{aligned}
&\left(\tilde{c}_1=\frac{1}{2}\quad\text{and}\quad\tilde{c}_3=\frac{1}{2}\right)\ \text{or} \\
&\left(\tilde{c}_1+\tilde{c}_2 p>\frac{1}{2}\quad\text{and}\quad\tilde{c}_3+\tilde{c}_4 p>\frac{1}{2}\right)
\end{aligned} \tag{7.84}$$

or

$$\left(\tilde{c}_1+\tilde{c}_2 p\ge\frac{1}{2}\quad\text{and}\quad\tilde{c}_3+\tilde{c}_4 p\ge\frac{1}{2}\quad\text{and}\quad K\quad\text{sufficiently small}\right).$$

The necessity of using a slightly stronger condition here, compared to Subsection 7.2.2, results from the need to estimate the difference between the derivatives of F in the proof of Theorem 7.10, see (7.91) below.

Theorem 7.10. *Assume that X is reflexive, that x_0 is sufficiently close to $x^\dagger$, i.e., $x_0\in\overline{B}_\rho^D(x^\dagger)$, and that F satisfies* (7.82)*, with* (7.83) *and* (7.84) *for some $\nu\in[0,1]$,* (7.11)*, and* (7.74) *or* (7.75)*. Moreover, let $q,p\in(1,\infty)$, and τ be chosen sufficiently large.*

Let a variational inequality (7.76)*, with $\nu\in[0,1]$ and β sufficiently small, hold.*

Then, with the a priori choice (7.47)*, the iterates remain in $\overline{B}_\rho^D(x^\dagger)$ and we obtain optimal convergence rates*

$$D_q^{x_0}(x_{n_*},x^\dagger)=O(\delta^{\frac{2\nu}{\nu+1}}),\quad \textit{as}\ \delta\to 0 \tag{7.85}$$

as well as in the noise-free case $\delta=0$

$$\begin{aligned}
&\left\|T(x_n-x^\dagger)\right\|=O\left(\alpha_n^{\frac{\nu+1}{p(\nu+1)-2\nu}}\right), \\
&D_q^{x_0}(x_n,x^\dagger)=O\left(\alpha_n^{\frac{2\nu}{p(\nu+1)-2\nu}}\right)\quad \textit{as}\ n\to\infty.
\end{aligned} \tag{7.86}$$

Proof. Observe that, under the assumption (7.76), we get, with the notation $T = F'(x^\dagger)$

$$
\begin{aligned}
&\left\|x_{n+1}^\delta - x_0\right\|^q - \left\|x^\dagger - x_0\right\|^q \\
&= pD_{j_q}(x_{n+1}^\delta - x_0, x^\dagger - x_0) + p\langle j_q^X(x^\dagger - x_0), x_{n+1}^\delta - x^\dagger\rangle_{X^*\times X} \\
&\geq pD_q^{x_0}(x_{n+1}^\delta, x^\dagger) - q\beta D_q^{x_0}(x_{n+1}^\delta, x^\dagger)^{(1-\nu)/2}\left\|T(x_{n+1}^\delta - x^\dagger)\right\|^\nu \\
&\geq pD_q^{x_0}(x_{n+1}^\delta, x^\dagger) \\
&\quad - q\beta\left(\epsilon D_q^{x_0}(x_{n+1}^\delta, x^\dagger) + C\left(\epsilon, \frac{\nu+1}{2}\right)\left\|T(x_{n+1}^\delta - x^\dagger)\right\|^{2\nu/(\nu+1)}\right)
\end{aligned} \tag{7.87}
$$

with $\epsilon > 0$ to be chosen sufficiently small later on,

$$
C(\epsilon, 1) = 1\,,
$$

and

$$
\begin{aligned}
C(\epsilon, \mu) &= \max\left\{1, \phi\left(\left(\frac{\epsilon}{1-\mu}\right)^{1/\mu}\right)\right\} \\
&= \max\left\{1, \frac{\mu}{(1-\mu)^{(\mu+1)/\mu}}\epsilon^{-(1-\mu)/\mu}\right\}
\end{aligned}
$$

for $\mu \in (0,1)$, where $\phi(\lambda) = \frac{\lambda^\mu - \epsilon}{\lambda}$, so that

$$
\lambda^\mu \leq \epsilon + C(\epsilon, \mu)\lambda \quad \text{for all } \lambda > 0\,. \tag{7.88}
$$

Due to minimality in (7.70), we have, for any solution $x^\dagger \in \mathcal{D}(F)$ of (7.1), which needs not necessarily be an x_0-minimum norm solution,

$$
\begin{aligned}
&\left\|A_n(x_{n+1}^\delta - x_n^\delta) + F(x_n^\delta) - y^\delta\right\|^p + \alpha_n\left\|x_{n+1}^\delta - x_0\right\|^q \\
&\leq \left\|A_n(x^\dagger - x_n^\delta) + F(x_n^\delta) - y^\delta\right\|^p + \alpha_n\left\|x^\dagger - x_0\right\|^q\,.
\end{aligned} \tag{7.89}
$$

Combining (7.87) and (7.89), we get, by the simple inequality $(a-b)^p + b^p \geq \frac{1}{2^{p-1}}a^p$,

$$
\begin{aligned}
&\frac{1}{2^{p-1}}\left\|T(x_{n+1}^\delta - x^\dagger)\right\|^p + \alpha_n p(1-\beta\epsilon)D_q^{x_0}(x_{n+1}^\delta, x^\dagger) \\
&\leq \left\|A_n(x^\dagger - x_n^\delta) + F(x_n^\delta) - y^\delta\right\|^p \\
&\quad + \left(\left\|(A_n - T)(x_{n+1}^\delta - x^\dagger)\right\| + \left\|A_n(x^\dagger - x_n^\delta) + F(x_n^\delta) - y^\delta\right\|\right)^p \\
&\quad + \alpha_n p\beta C\left(\epsilon, \frac{\nu+1}{2}\right)\left\|T(x_{n+1}^\delta - x^\dagger)\right\|^{2\nu/(\nu+1)}\,.
\end{aligned}
$$

The terms on the right hand side can be estimated using (7.82),

$$\begin{aligned}
&\left\| A_n(x^\dagger - x_n^\delta) + F(x_n^\delta) - y^\delta \right\| \\
&\le \left\| (A_n - T)(x_n^\delta - x^\dagger) \right\| + \left\| F(x_n^\delta) - F(x^\dagger) - T(x_n^\delta - x^\dagger) \right\| + \delta \\
&\le 2K \left\| T(x_n^\delta - x^\dagger) \right\|^{\tilde{c}_1+\tilde{c}_3} D_q^{x_0}(x_n^\delta, x^\dagger)^{\tilde{c}_2+\tilde{c}_4} + \delta \qquad (7.90)
\end{aligned}$$

$$\begin{aligned}
&\left\| (A_n - T)(x_{n+1}^\delta - x^\dagger) \right\| \qquad (7.91) \\
&\le K \left\| T(x_{n+1}^\delta - x^\dagger) \right\|^{\tilde{c}_1} \left\| T(x_n^\delta - x^\dagger) \right\|^{\tilde{c}_3} D_q^{x_0}(x_{n+1}^\delta, x^\dagger)^{\tilde{c}_2} D_q^{x_0}(x_n^\delta, x^\dagger)^{\tilde{c}_4}
\end{aligned}$$

which, together with the simple inequality $(a+b)^p \le 2^{p-1}(a^p + b^p)$, yields

$$\begin{aligned}
&\frac{1}{2^{p-1}} \left\| T(x_{n+1}^\delta - x^\dagger) \right\|^p + \alpha_n q(1 - \beta\epsilon) D_q^{x_0}(x_{n+1}^\delta, x^\dagger) \\
&\le (1 + 2^{p-1}) \left(2K \left\| T(x_n^\delta - x^\dagger) \right\|^{\tilde{c}_1+\tilde{c}_3} D_q^{x_0}(x_n^\delta, x^\dagger)^{\tilde{c}_2+\tilde{c}_4} + \delta \right)^p \\
&\quad + 2^{p-1} \Big(K \left\| T(x_{n+1}^\delta - x^\dagger) \right\|^{\tilde{c}_1} D_q^{x_0}(x_{n+1}^\delta, x^\dagger)^{\tilde{c}_2} \\
&\quad \times \left\| T(x_n^\delta - x^\dagger) \right\|^{\tilde{c}_3} D_q^{x_0}(x_n^\delta, x^\dagger)^{\tilde{c}_4} \Big)^p \\
&\quad + \alpha_n q \beta C\left(\epsilon, \frac{\nu+1}{2}\right) \left\| T(x_{n+1}^\delta - x^\dagger) \right\|^{2\nu/(\nu+1)} .
\end{aligned}$$

Applying the estimate

$$a^\zeta b \le \tilde{\epsilon} a + C(\tilde{\epsilon}, 1 - \zeta) b^{1/(1-\zeta)} \qquad (7.92)$$

for $\zeta \in (0, 1]$, which follows from (7.88), with $\lambda := \frac{b^{1/(1-\zeta)}}{a}$ and $\mu = 1 - \zeta$ in the last term, and $ab \le \frac{1}{2}a^2 + \frac{1}{2}b^2$ in the second term on the right hand side, we get

$$\begin{aligned}
&\left(\frac{1}{2^{p-1}} - \tilde{\epsilon} \right) \left\| T(x_{n+1}^\delta - x^\dagger) \right\|^p + \alpha_n q(1 - \beta\epsilon) D_q^{x_0}(x_{n+1}^\delta, x^\dagger) \qquad (7.93) \\
&\le (1 + 2^{p-1}) \left(2K \left\| T(x_n^\delta - x^\dagger) \right\|^{\tilde{c}_1+\tilde{c}_3} D_q^{x_0}(x_n^\delta, x^\dagger)^{\tilde{c}_2+\tilde{c}_4} + \delta \right)^p \qquad (7.94) \\
&\quad + \frac{2^{p-1} K^p}{2} \left\| T(x_{n+1}^\delta - x^\dagger) \right\|^{2p\tilde{c}_1} D_q^{x_0}(x_{n+1}^\delta, x^\dagger)^{2p\tilde{c}_2} \qquad (7.95) \\
&\quad + \frac{2^{p-1} K^p}{2} \left\| T(x_n^\delta - x^\dagger) \right\|^{2p\tilde{c}_3} D_q^{x_0}(x_n^\delta, x^\dagger)^{2p\tilde{c}_4} \qquad (7.96) \\
&\quad + C\left(\tilde{\epsilon}, \frac{p(\nu+1) - 2\nu}{p(\nu+1)} \right) \left(\alpha_n q \beta C\left(\epsilon, \frac{\nu+1}{2}\right) \right)^{\frac{p(\nu+1)}{p(\nu+1)-2\nu}} , \qquad (7.97)
\end{aligned}$$

where we choose $\tilde{\epsilon} < \frac{1}{2^{p-1}}$. Considering (7.93) and (7.97) and neglecting the rest for a moment, which is just an estimate of the nonlinearity error, we expect that (7.86) can be obtained, which we prove as follows: Dividing (7.93)–(7.97) by $\alpha_{n+1}^{\frac{p(\nu+1)}{p(\nu+1)-2\nu}}$, using (7.78), (7.83), (7.47), and defining

$$\gamma_n := \max\left\{ \frac{\left\| T(x_n^\delta - x^\dagger) \right\|^p}{\alpha_n^{\frac{p(\nu+1)}{p(\nu+1)-2\nu}}}, \frac{D_q^{x_0}(x_n^\delta, x^\dagger)}{\alpha_n^{\frac{2\nu}{p(\nu+1)-2\nu}}} \right\},$$

we get the following estimate

$$\begin{aligned}
&\min\left\{\frac{1}{2^{p-1}} - \tilde{\epsilon}, q(1-\beta\epsilon)\right\} \gamma_{n+1} \\
&\le (1+2^{p-1})2^{p-1}(2K)^p \hat{C}^{\frac{p(\nu+1)}{p(\nu+1)-2\nu}} \gamma_n^{\tilde{c}_1+\tilde{c}_3+(\tilde{c}_2+\tilde{c}_4)p} \\
&\quad + \frac{2^{p-1}K^p}{2} \gamma_{n+1}^{2(\tilde{c}_1+p\tilde{c}_2)} \\
&\quad + \frac{2^{p-1}K^p}{2} \hat{C}^{\frac{p(\nu+1)}{p(\nu+1)-2\nu}} \gamma_n^{2(\tilde{c}_3+p\tilde{c}_4)} \\
&\quad + C\left(\tilde{\epsilon}, \frac{p(\nu+1)-2\nu}{p(\nu+1)}\right) \left(q\beta C\left(\epsilon, \frac{\nu+1}{2}\right)\right)^{\frac{p(\nu+1)}{p(\nu+1)-2\nu}} \hat{C}^{\frac{p(\nu+1)}{p(\nu+1)-2\nu}} \\
&\quad + \frac{(1+2^{p-1})2^{p-1}}{\tau^p}.
\end{aligned}$$

With this, we get a recursive estimate of the form

$$\begin{aligned}
\left(1 - E\gamma_{n+1}^{2(\tilde{c}_1+p\tilde{c}_2)-1}\right) \gamma_{n+1} \le B(&\gamma_n^{\tilde{c}_1+\tilde{c}_3+(\tilde{c}_2+\tilde{c}_4)p-1} \\
&+ \gamma_n^{2(\tilde{c}_3+p\tilde{c}_4)-1})\gamma_n + c,
\end{aligned} \tag{7.98}$$

where c can be made small by making β small and τ large.

From this, we can now derive an induction step of the form

$$\gamma_n \le \bar{\gamma} \Rightarrow \gamma_{n+1} \le \bar{\gamma} \tag{7.99}$$

as follows: Using (7.84) and the fact that E and B will be small if K is small, we can conclude first of all that, for $\bar{\gamma}, \bar{\zeta}$ sufficiently small, the function

$$\begin{aligned}
h(\gamma):\quad (0,\bar{\gamma}) &\to (0,\bar{\zeta}) \\
\gamma &\mapsto \left(1 - E\gamma^{2(\tilde{c}_1+p\tilde{c}_2)-1}\right)\gamma
\end{aligned}$$

is strictly monotonically increasing and invertible, with

$$h^{-1}(\zeta) \le 2\zeta.$$

By using the induction hypothesis $\gamma_n \le \bar{\gamma}$, with a possibly reduced value of $\bar{\gamma}$, we can cause the right hand side in (7.98) to be smaller than $\bar{\zeta}$, so that, by applying h^{-1} to both sides of (7.98), we can conclude

$$\begin{aligned}\gamma_{n+1} &\le 2B(\gamma_n^{(\tilde{c}_1+\tilde{c}_3)+(\tilde{c}_2+\tilde{c}_4)p-1} + \gamma_n^{2(\tilde{c}_3+\tilde{c}_4 p)-1})\,\gamma_n + 2c \\ &\le 2B(\bar{\gamma}^{(\tilde{c}_1+\tilde{c}_3)+(\tilde{c}_2+\tilde{c}_4)p-1} + \bar{\gamma}^{2(\tilde{c}_3+\tilde{c}_4 p)-1})\,\bar{\gamma} + \frac{1}{2}\bar{\gamma}\,, \qquad (7.100)\end{aligned}$$

where we use the fact that we can make β small and τ large, so that $c < \frac{\bar{\gamma}}{4}$. Now, we use (7.84) again to arrive at

$$2B(\bar{\gamma}^{(\tilde{c}_1+\tilde{c}_3)+(\tilde{c}_2+\tilde{c}_4)p-1} + \bar{\gamma}^{2(\tilde{c}_3+p\tilde{c}_4)-1}) \le \frac{1}{2},$$

by possibly decreasing $\bar{\gamma}$. Inserting this into (7.100), this yields $\gamma_{n+1} \le \bar{\gamma}$.

Applying (7.99) as an induction step, we can conclude that

$$\gamma_n \le \bar{\gamma} \text{ for all } n \le n_*$$

and therewith, by possibly decreasing $\bar{\gamma}$ to below ρ^2,

$$D_q^{x_0}(x_n^\delta, x^\dagger) \le \gamma_n \alpha_n^{\frac{2\nu}{p(\nu+1)-2\nu}} \le \bar{\gamma} \le \rho^2 \quad \text{for all } n \le n_*,$$

provided γ_0 and $D_q^{x_0}(x_0, x^\dagger)$ are sufficiently small. By the assumption (7.11), this yields well-definedness of the iterates. Moreover

$$D_q^{x_0}(x_{n_*}^\delta, x^\dagger) \le \bar{\gamma}\alpha_{n_*}^{\frac{2\nu}{p(\nu+1)-2\nu}} \le \bar{\gamma}(\tau\delta)^{\frac{2\nu}{\nu+1}}\,. \qquad \square$$

In the case of the general variational inequality (7.77), we have to apply somewhat different techniques, compared to the special case (7.76). Moreover, in (7.82), we have to assume the strongest case $\tilde{c}_1 = \tilde{c}_3 = \frac{1}{2}$, $\tilde{c}_2 = \tilde{c}_4 = 0$, as well as K sufficiently small:

$$\begin{aligned}&\left\|(F'(x^\dagger + \tilde{v}) - F'(x^\dagger))v\right\| \le K \left\|F'(x^\dagger)v\right\|^{1/2} \left\|F'(x^\dagger)\tilde{v}\right\|^{1/2} \\ &v, \tilde{v} \in X,\ x^\dagger + v \in \overline{B}\ x^\dagger + \tilde{v} \in \overline{B}\,, \qquad (7.101)\end{aligned}$$

which makes sense, in light of the fact that we may thus take a logarithmic function κ, with slower decay at zero than any function of the form $t \mapsto t^\nu$, $\nu > 0$.

Theorem 7.11. *Assume that X is reflexive, that x_0 is sufficiently close to $x^\dagger$, i.e., $x_0 \in \overline{B}_\rho^D(x^\dagger)$, and that F satisfies* (7.101) (7.11), *and* (7.74) *or* (7.75). *Moreover, let $q, p \in (1, \infty)$, and let τ be chosen sufficiently large.*

Let a variational inequality (7.77)*, with*

$$t \mapsto \frac{\kappa(t)}{\sqrt{t}} \quad \textit{monotonically decreasing} \tag{7.102}$$

hold and assume that there exists a $\bar{t} > 0$ *sufficiently large, such that, for all* $C > 0$*, there exists* $C_\kappa > 0$*, satisfying*

$$\forall 0 < t \le \bar{t}: \quad \kappa(C C_\kappa^{2-p} t) \le C_\kappa \kappa(t) . \tag{7.103}$$

Then, with the a priori choice (7.80)*, the iterates remain in* $\overline{B}_\rho^D(x^\dagger)$ *and we obtain optimal convergence rates*

$$D_q^{x_0}(x_{n_*}, x^\dagger) = O\left(\kappa^2(\Theta^{-1}(\delta))\right) = O\left(\frac{\delta^2}{\Theta^{-1}(\delta)}\right) \quad \textit{as } \delta \to 0 \tag{7.104}$$

with Θ *as in* (7.81)*, as well as in the noise-free case* $\delta = 0$

$$\begin{aligned} \left\| T(x_n - x^\dagger) \right\| &= O\left(\varphi_p^{-1}(\alpha_n)\right) , \\ D_q^{x_0}(x_n, x^\dagger) &= O\left(\kappa(\Theta^{-1}(\varphi_p^{-1}(\alpha_n)))^2\right) \quad \textit{as } n \to \infty . \end{aligned} \tag{7.105}$$

Remark 7.12. It can be seen from the proof that it suffices to assume the rescaling condition (7.103) for specific constants C, namely

$$\forall 0 < t \le \hat{t}: \quad \kappa((\hat{C}\hat{C}_\kappa^{2-p})^{2/p} t) \le \hat{C}_\kappa \kappa(t) \tag{7.106}$$

$$\forall 0 < t \le \tilde{t}: \quad \kappa((\tilde{C}\tilde{C}_\kappa^{2-p})^{2/p} t) \le \tilde{C}_\kappa \kappa(t) \tag{7.107}$$

with

$$\begin{aligned} &1 \le \hat{C}_\varphi := (\hat{C}\hat{C}_\kappa^2)^{1/p} < C_0^{-\frac{1}{p}}, \quad 1 \le \tilde{C}_\varphi := (\tilde{C}\tilde{C}_\kappa^2)^{1/p} , \\ &\tilde{C} = (2C_2)^{2-p}\hat{C}, \quad \hat{C}, C_0, C_2 \text{ as in (7.78), (7.112), (7.113)} \\ &\hat{t} = \Theta^{-1}(\hat{C}_\varphi \varphi_p^{-1}(\alpha_0))(\hat{C}\hat{C}_\kappa^{2-p})^{-2/p}, \\ &\tilde{t} = \Theta^{-1}(\tilde{C}_\varphi \varphi_p^{-1}((2C_2)^{p-2}\alpha_0))(\tilde{C}\tilde{C}_\kappa^{2-p})^{-2/p}, \end{aligned} \tag{7.108}$$

where C_0 can be made small by making K small.

Condition (7.102) implies the inequality

$$\kappa(\Theta^{-1}(Ct)) \le \max\{\sqrt{C}, 1\}\, \kappa(\Theta^{-1}(t)) \quad \forall t \ge 0\, \forall C > 0 , \tag{7.109}$$

a fact which can be seen as follows. Because of the monotonicity of the index functions κ and Θ^{-1}, we have $\kappa(\Theta^{-1}(Ct)) \le \kappa(\Theta^{-1}(t))$ for $0 < C \le 1$. On the other hand, by substituting $u := \Theta(t)$, we have $\frac{\kappa(\Theta^{-1}(\tau))}{\sqrt{\tau}} = \frac{\kappa(u)}{\sqrt{\Theta(u)}} = \sqrt{\frac{\kappa(u)}{\sqrt{u}}}$ showing that these quotient functions, with positive arguments τ and u, respectively, are both monotonically increasing, in light of (7.102). Consequently, we have $\frac{\kappa(\Theta^{-1}(Ct))}{\sqrt{Ct}} \le \frac{\kappa(\Theta^{-1}(t))}{\sqrt{t}}$ for $C > 1$. Both facts together imply (7.109).

Moreover, condition (7.102) means that the variational inequality condition, determined by the index function f, is not too strong, i.e., the decay rate of $\kappa(t) \to 0$ as $t \to 0$ is not faster than the corresponding decay rate of $\sqrt{t}$. A sufficient condition for this is the concavity of κ^2, which is equivalent to condition (7.9).

Proof. In place of (7.87), we get the estimate

$$\begin{aligned}&\left\|x_{n+1}^\delta - x_0\right\|^q - \left\|x^\dagger - x_0\right\|^q \\ &= pD_{j_q}(x_{n+1}^\delta - x_0, x^\dagger - x_0) + q\langle j_q^X(x^\dagger - x_0), x_{n+1}^\delta - x^\dagger\rangle_{X^*\times X} \\ &\ge qD_q^{x_0}(x_{n+1}^\delta, x^\dagger) - qD_q^{x_0}(x_{n+1}^\delta, x^\dagger)^{1/2}\kappa\left(\frac{\left\|F'(x^\dagger)(x_{n+1}^\delta - x^\dagger)\right\|^2}{D_q^{x_0}(x_{n+1}^\delta, x^\dagger)}\right),\end{aligned} \tag{7.110}$$

which, together with (7.89), (7.90) and (7.91), implies

$$\begin{aligned}&\frac{1}{2^{p-1}}\left\|T(x_{n+1}^\delta - x^\dagger)\right\|^p + \alpha_n q D_q^{x_0}(x_{n+1}^\delta, x^\dagger) \\ &\le (1 + 2^{p-1})\left(2K\left\|T(x_n^\delta - x^\dagger)\right\| + \delta\right)^p \\ &\quad + \frac{2^{p-1}K^p}{2}\left\|T(x_{n+1}^\delta - x^\dagger)\right\|^p \\ &\quad + \frac{2^{p-1}K^p}{2}\left\|T(x_n^\delta - x^\dagger)\right\|^p \\ &\quad + \alpha_n q D_q^{x_0}(x_{n+1}^\delta, x^\dagger)^{1/2}\kappa\left(\frac{\left\|T(x_{n+1}^\delta - x^\dagger)\right\|^2}{D_q^{x_0}(x_{n+1}^\delta, x^\dagger)}\right)\end{aligned}$$

in place of (7.93)–(7.97), which, by moving the second term on the right-hand side to the left-hand side, using $K^p < \frac{2}{2^{p-1}}$ and (7.80), yields an inequality of the form

$$\mathrm{t}_{n+1}^p + \alpha_n \mathrm{d}_{n+1}^2 \le C_0 \mathrm{t}_n^p + m(\varphi_p^{-1}(\alpha_n))^p + M\alpha_n \mathrm{d}_{n+1}\kappa\left(\frac{\mathrm{t}_{n+1}^2}{\mathrm{d}_{n+1}^2}\right) \tag{7.111}$$

for all $n \leq n_* - 1$, where we use the abbreviations

$$\begin{aligned}
\mathrm{d}_n &= D_q^{x_0}(x_n^\delta, x^\dagger)^{1/2}\,,\\
\mathrm{t}_n &= \left\| T(x_n^\delta - x^\dagger) \right\|\,,\\
C_0 &= \frac{2^p(1+2^{p-1})2^{p-1} + 2^{p-1}/2}{\tilde{c}}\, K^p\,, && (7.112)\\
C_1 &= \frac{(1+2^{p-1})2^{p-1}}{\tau^p \tilde{c}}\,,\\
C_2 &= \frac{q}{\tilde{c}}\,, && (7.113)\\
\tilde{c} &= \min\left\{ \frac{1}{2^{p-1}} - \frac{2^{p-1}K^p}{2}, q \right\}\,,
\end{aligned}$$

where C_0, C_1 can be made small by making K small and τ large.

Now we prove by induction that, for all $n \leq n_*$, respectively, if $\delta = 0$, for all $k \in \mathbb{N}$,

$$\mathrm{d}_n \leq C_3 \kappa(\Theta^{-1}(\varphi_p^{-1}(\alpha_n)))\,, \tag{7.114}$$

$$\mathrm{t}_n \leq C_4 \varphi_p^{-1}(\alpha_n)\,, \tag{7.115}$$

where C_4 is sufficiently large, so that

$$\hat{C}_\varphi \leq (2(C_0 + m/C_4^p))^{-1/p}\,, \quad \tilde{C}_\varphi \leq \frac{C_4}{2C_2}\,, \tag{7.116}$$

cf. (7.108), and $C_3 := \sqrt{\frac{C_4^p}{\min\{1,\hat{C}\}}}$ so that

$$C_3^2 \hat{C} \geq C_4^p\,, \quad C_3^2 \geq C_4^p\,. \tag{7.117}$$

For this purpose, observe that (7.111), together with the induction hypothesis, implies

$$\mathrm{t}_{n+1}^p + \alpha_n \mathrm{d}_{n+1}^2 \leq (C_0 C_4^p + C_1)(\varphi_p^{-1}(\alpha_n))^p + C_2 \alpha_n \mathrm{d}_{n+1} \kappa\left(\frac{\mathrm{t}_{n+1}^2}{\mathrm{d}_{n+1}^2}\right). \tag{7.118}$$

We distinguish between two cases:

If $(C_0 C_4^p + C_1)(\varphi_p^{-1}(\alpha_n))^p \leq C_2 \alpha_n \mathrm{d}_{n+1} \kappa\left(\frac{\mathrm{t}_{n+1}^2}{\mathrm{d}_{n+1}^2}\right)$, then we get from (7.118)

$$\mathrm{t}_{n+1}^p + \alpha_n \mathrm{d}_{n+1}^2 \leq 2C_2 \alpha_n \mathrm{d}_{n+1} \kappa\left(\frac{\mathrm{t}_{n+1}^2}{\mathrm{d}_{n+1}^2}\right). \tag{7.119}$$

Since, for $\mathrm{d}_{n+1} = 0$ and thus $\mathrm{t}_{n+1} = 0$, and for $\mathrm{t}_{n+1} = 0$ and thus $\mathrm{d}_{n+1} = 0$, by $\mathrm{d}_{n+1}^2 \le 2C_2\mathrm{d}_{n+1}\kappa(\frac{\mathrm{t}_{n+1}^2}{\mathrm{d}_{n+1}^2})$, the assertions (7.114), (7.115) hold trivially, for k replaced with $k+1$, we may assume, without loss of generality, that $\mathrm{d}_{n+1} \neq 0$ and $\mathrm{t}_{n+1} \neq 0$. Multiplying (7.119) by t_{n+1} and dividing by d_{n+1}^2 we get

$$\frac{\mathrm{t}_{n+1}^2}{\mathrm{d}_{n+1}^2}\mathrm{t}_{n+1}^{p-1} + \alpha_n \mathrm{t}_{n+1} \le 2C_2\alpha_n\Theta\left(\frac{\mathrm{t}_{n+1}^2}{\mathrm{d}_{n+1}^2}\right),$$

which implies

$$\Phi\left(\frac{\mathrm{t}_{n+1}^2}{\mathrm{d}_{n+1}^2}\right)\mathrm{t}_{n+1}^{p-1} \le 2C_2\alpha_n,$$

with the monotonically increasing function (cf. (7.102))

$$\Phi : u \mapsto \frac{\sqrt{u}}{\kappa(u)} = \frac{u}{\Theta(u)}$$

and

$$\mathrm{t}_{n+1} \le 2C_2\Theta\left(\frac{\mathrm{t}_{n+1}^2}{\mathrm{d}_{n+1}^2}\right), \quad \text{i.e.} \quad \Theta^{-1}(\frac{\mathrm{t}_{n+1}}{2C_2}) \le \frac{\mathrm{t}_{n+1}^2}{\mathrm{d}_{n+1}^2}, \tag{7.120}$$

leading to

$$\Phi\left(\Theta^{-1}(\frac{\mathrm{t}_{n+1}}{2C_2})\right)\mathrm{t}_{n+1}^{p-1} \le 2C_2\alpha_n .$$

Since $\Phi(\Theta^{-1}(\frac{t}{C}))t^{p-1} = C\Theta^{-1}(\frac{t}{C})t^{p-2} = \varphi_p(\frac{t}{C})C^{p-1}$, this implies

$$\mathrm{t}_{n+1} \le 2C_2\varphi_p^{-1}((2C_2)^{2-p}\alpha_n) \tag{7.121}$$

from which, by (7.120), we get

$$\begin{aligned} \mathrm{d}_{n+1}^2 &\le \frac{\mathrm{t}_{n+1}^2}{\Theta^{-1}(\frac{\mathrm{t}_{n+1}}{2C_2})} = (2C_2)^2\left(\kappa(\Theta^{-1}(\frac{\mathrm{t}_{n+1}}{2C_2})\right)^2 \\ &\le (2C_2)^2\left(\kappa(\Theta^{-1}(\varphi_p^{-1}((2C_2)^{2-p}\alpha_n))\right)^2 . \end{aligned} \tag{7.122}$$

Otherwise, if $(C_0C_4^p + C_1)(\varphi_p^{-1}(\alpha_n))^p \ge C_2\alpha_n\mathrm{d}_{n+1}\kappa\left(\frac{\mathrm{t}_{n+1}^2}{\mathrm{d}_{n+1}^2}\right)$, we get, from (7.118),

$$\mathrm{t}_{n+1}^p + \alpha_n\mathrm{d}_{n+1}^2 \le 2(C_0C_4^p + C_1)(\varphi_p^{-1}(\alpha_n))^p . \tag{7.123}$$

From (7.121), (7.122) and (7.123), using the identity

$$\begin{aligned} \kappa(\Theta^{-1}(\underbrace{\varphi_p^{-1}(\alpha)}_{=:z}) &= \frac{z}{\sqrt{\Theta^{-1}(z)}} = z^{p/2}\frac{1}{\sqrt{z^{p-2}\Theta^{-1}(z)}} \\ &= \frac{1}{\sqrt{\varphi_p(z)}}z^{p/2} = \frac{1}{\sqrt{\alpha}}\left(\varphi_p^{-1}(\alpha)\right)^{p/2} \end{aligned}$$

and (7.78), we see that, in order to complete the induction proof of (7.114) and (7.115), it suffices to show

$$\varphi_p^{-1}(\alpha) \le \hat{C}_\varphi \varphi_p^{-1}(\alpha/\hat{C}) \quad \forall 0 < \alpha \le \alpha_0\,, \tag{7.124}$$

$$\varphi_p^{-1}(\alpha) \le \tilde{C}_\varphi \varphi_p^{-1}(\alpha/\tilde{C}) \quad \forall 0 < \alpha \le (2C_2)^{p-2}\alpha_0\,, \tag{7.125}$$

and use (7.116), (7.117). By the definition of φ_p, (7.124) can be concluded from (7.106) as follows: With $\hat{C}_\varphi = \sqrt{\hat{C}_p \hat{C}_\kappa}$, $\hat{C}_p = \hat{C}\hat{C}_\varphi^{2-p}$, cf. (7.108), $\lambda = \hat{C}_\varphi \varphi_p^{-1}(\alpha/\hat{C})$, $t = \Theta^{-1}(\lambda)/\hat{C}_p$, we have for any $\alpha \in (0, \alpha_0]$:

$$\begin{aligned}
&\kappa(\hat{C}_p t) \le \hat{C}_\kappa \kappa(t)\\
&\Leftrightarrow \underbrace{\Theta(\hat{C}_p t)}_{\lambda} \le \underbrace{\sqrt{\hat{C}_p \hat{C}_\kappa}}_{=\hat{C}_\varphi}\, \Theta(t)\\
&\Leftrightarrow \Theta^{-1}(\lambda/\hat{C}_\varphi) \le t = \frac{1}{\hat{C}_p}\Theta^{-1}(\lambda)\\
&\Leftrightarrow (\lambda/\hat{C}_\varphi)^{p-2}\Theta^{-1}(\lambda/\hat{C}_\varphi) \le \frac{1}{\hat{C}_p \hat{C}_\varphi^{p-2}} \lambda^{p-2}\Theta^{-1}(\lambda)\\
&\Leftrightarrow \underbrace{\hat{C}_p \hat{C}_\varphi^{p-2}}_{=\hat{C}} \underbrace{\varphi_p(\lambda/\hat{C}_\varphi)}_{=\alpha/\hat{C}} \le \varphi_p(\lambda)\\
&\Leftrightarrow \varphi_p^{-1}(\alpha) \le \lambda = \hat{C}_\varphi \varphi_p^{-1}(\alpha/\hat{C})\,,
\end{aligned}$$

where we have used the fact that the functions φ_p, Θ, as well as their inverses, are strictly monotonically increasing. Analogously, (7.125) follows from (7.107). With this, the induction proof of (7.114), (7.115) is finished.

The estimates (7.114), (7.115) immediately yield (7.105).

Inserting (7.80) into (7.114) for $n = n_*$, with (7.109), directly yields

$$\begin{aligned}
\mathrm{d}_{n_*} &\le C_3 \kappa(\Theta^{-1}(\varphi_p^{-1}(\alpha_{n_*})) \le C_3 \kappa(\Theta^{-1}(\tau\delta)) \le C_3 \max\{\sqrt{\tau}, 1\}\, \kappa(\Theta^{-1}(\delta))\\
&= C_3 \max\{\sqrt{\tau}, 1\} \frac{\Theta(\Theta^{-1}(\delta))}{\sqrt{\Theta^{-1}(\delta)}} = C_3 \max\{\sqrt{\tau}, 1\} \frac{\delta}{\sqrt{\Theta^{-1}(\delta)}}\,.
\end{aligned}$$

This provides us with the convergence rate assertion (7.104) and completes the proof of 7.11. □

7.3.2 Convergence with a posteriori parameter choice

Under the η-condition

$$\left\| F(x) - F(\bar{x}) - F'(x)(x - \bar{x}) \right\| \le \eta \, \|F(x) - F(\bar{x})\| \quad \forall x, \bar{x} \in \overline{B} \tag{7.126}$$

for some $0 < \eta < 1$, cf. (3.42), we can prove convergence and convergence rates, with a posteriori choices of the regularization parameters α_n

$$\begin{aligned}\underline{\sigma}\left\|F(x_n^\delta)-y^\delta\right\| &\le \left\|A_n(x_{n+1}^\delta(\alpha_n)-x_n^\delta)+F(x_n^\delta)-y^\delta\right\| \\ &\le \overline{\sigma}\left\|F(x_n^\delta)-y^\delta\right\|,\end{aligned} \tag{7.127}$$

cf. [87], and of the stopping index n_*, by the discrepancy principle:

$$n_*(\delta) = \min\left\{n \in \mathbb{N} : \left\|F(x_n^\delta)-y^\delta\right\| \le \tau\delta\right\}. \tag{7.128}$$

A sufficient condition for α_n being well-defined by (7.127) is

$$\left\|A_n(x_0-x_n^\delta)+F(x_n^\delta)-y^\delta\right\| \ge \overline{\sigma}\left\|F(x_n^\delta)-y^\delta\right\|, \tag{7.129}$$

see the first part of the proof of the following theorem.

Theorem 7.13. *Assume that X is smooth and uniformly convex, and that F satisfies* (7.126)*, with η sufficiently small, as well as* (7.12) *or* (7.13)*, and* (7.74) *or* (7.75)*. Let*

$$\eta < \underline{\sigma} < \overline{\sigma} < 1,$$

and let τ be chosen sufficiently large, so that

$$\eta + \frac{1+\eta}{\tau} \le \underline{\sigma} \text{ and } \eta < \frac{1-\overline{\sigma}}{2}. \tag{7.130}$$

Moreover, assume that

$$\delta < \frac{\|F(x_0)-y^\delta\|}{\tau}.$$

Then, for all $n \le n_*(\delta)-1$, with $n_*(\delta)$ according to (7.29)*, the iterates*

$$x_{n+1}^\delta := \begin{cases} x_{n+1}^\delta = x_{n+1}^\delta(\alpha_n),\ \alpha_n \text{ as in (7.127)}, & \text{if (7.129) holds} \\ x_0 & \text{else}\end{cases}$$

are well-defined.

Moreover, there exists a weakly convergent subsequence of

$$\begin{cases} x_{n_*(\delta)}^\delta, & \text{if (7.12) holds} \\ J_q^X(x_{n_*(\delta)}^\delta - x_0), & \text{if (7.13) holds}\end{cases}$$

and along every such weakly convergent subsequence $x_{n_*(\delta)}^\delta$ converges strongly to a solution of (7.1) *as $\delta \to 0$. If the solution $x^\dagger$ to* (7.1) *is unique, then $x_{n_*(\delta)}^\delta$ converges strongly to $x^\dagger$ as $\delta \to 0$.*

Proof. Well-definedness of α_n for (7.129) can be seen as follows: By minimality of $x_{n+1}^\delta(\alpha)$ we have, for

$$\psi(\alpha) = \left\| A_n(x_{n+1}^\delta(\alpha) - x_n^\delta) + F(x_n^\delta) - y^\delta \right\| ,$$

that

$$\psi(\alpha)^p + \alpha \left\| x_{n+1}^\delta(\alpha) - x_0 \right\|^q \le \left\| A_n(x^\dagger - x_n^\delta) + F(x_n^\delta) - y^\delta \right\|^p + \alpha \left\| x^\dagger - x_0 \right\|^q ,$$

hence

$$\limsup_{\alpha\to 0} \psi(\alpha) \le \left\| A_n(x^\dagger - x_n^\delta) + F(x_n^\delta) - y^\delta \right\|^p \le (\eta + \frac{1+\eta}{\tau}) \left\| F(x_n^\delta) - y^\delta \right\| ,$$

by (7.2), (7.29), (7.126). On the other hand, minimality of $x_{n+1}^\delta(\alpha)$ again yields

$$\psi(\alpha)^p + \alpha \left\| x_{n+1}^\delta(\alpha) - x_0 \right\|^q \le \left\| A_n(x_0 - x_n^\delta) + F(x_n^\delta) - y^\delta \right\|^p ,$$

so that

$$\left\| x_{n+1}^\delta(\alpha) - x_0 \right\|^q \le \frac{1}{\alpha} \left\| A_n(x_0 - x_n^\delta) + F(x_n^\delta) - y^\delta \right\|^p \to 0 \text{ as } \alpha \to \infty$$

so, by continuity of A_n and the norms, there exists an $\bar{\alpha} > 0$ such that

$$\psi(\bar{\alpha}) > \frac{\overline{\sigma}}{\overline{\sigma}} \lim_{\alpha\to\infty} \psi(\alpha) = \frac{\overline{\sigma}}{\overline{\sigma}} \left\| A_n(x_0 - x_n^\delta) + F(x_n^\delta) - y^\delta \right\| \ge \underline{\sigma} \left\| F(x_n^\delta) - y^\delta \right\| .$$

To conclude existence of an α_n satisfying (7.127), it remains to show continuity of ψ, which we do by using the fact that the uniformly convex Banach space X is reflexive and strictly convex, and by setting $Ax = A_n(x - x_n^\delta) + F(x_n^\delta) - y^\delta$, $D = \mathcal{D}(F)$ in Lemma 7.9.

In case α_n can be chosen according to (7.127), we see that, by (7.70), (7.89) holds which, together with (7.2), (7.127), (7.29) and (7.126) yields

$$\begin{aligned} &\underline{\sigma}^p \left\| F(x_n^\delta) - y^\delta \right\|^p + \alpha_n \left\| x_{n+1}^\delta - x_0 \right\|^q \\ &\le (\eta + \frac{1+\eta}{\tau})^p \left\| F(x_n^\delta) - y^\delta \right\|^p + \alpha_n \left\| x^\dagger - x_0 \right\|^q \end{aligned} \tag{7.131}$$

for all $n \le n_*(\delta) - 1$ provided $x_n^\delta \in \overline{B}_\rho(x_0)$. By (7.130) this implies

$$\left\| x_{n+1}^\delta - x_0 \right\| \le \left\| x^\dagger - x_0 \right\| , \tag{7.132}$$

which holds trivially in the alternative case

$$\left\| A_n(x_0 - x_n^\delta) + F(x_n^\delta) - y^\delta \right\| < \overline{\sigma} \left\| F(x_n^\delta) - y^\delta \right\| ,$$

in which we set $x_{n+1}^\delta = x_0$.

Estimate (7.132) allows us to conclude, by an inductive argument, that $x_n^\delta \in \overline{B}_\rho(x_0)$, for all $n \le n_*(\delta)$.

According to (7.29), the stopping index $n_*(\delta)$ is finite since, on the one hand, the situation where $\|A_n(x_0 - x_n^\delta) + F(x_n^\delta) - y^\delta\| < \overline{\sigma}\|F(x_n^\delta) - y^\delta\|$, and hence $x_{n+1}^\delta := x_0$, can happen at most every second step:

$$x_{n+1}^\delta = x_0 \Rightarrow \|A_{n+1}(x_0 - x_{n+1}^\delta) + R_{n+1}\| = \|R_{n+1}\| \ge \overline{\sigma}\|R_{n+1}\|,$$

so α_{n+1} can be chosen as in (7.127), with n replaced by $n+1$. On the other hand, in steps where α_n is chosen as in (7.127), the residual norm decreases by a factor of $\frac{\overline{\sigma}+\eta}{1-\eta}$, which is smaller than 1 by (7.130):

$$\begin{aligned}
&\|R_{n+1}\| \\
&= \left\| A_n(x_{n+1}^\delta - x_n^\delta) + F(x_n^\delta) - y^\delta + F(x_{n+1}^\delta) - F(x_n^\delta) - A_n(x_{n+1}^\delta - x_n^\delta)\right\| \\
&\le \overline{\sigma} \left\| F(x_n^\delta) - y^\delta \right\| + \eta \left\| F(x_{n+1}^\delta) - F(x_n^\delta)\right\| \\
&\le (\overline{\sigma} + \eta) \left\| F(x_n^\delta) - y^\delta \right\| + \eta \, \|R_{n+1}\|
\end{aligned}$$

Hence,

$$\left\| F(x_n^\delta) - y^\delta \right\| \le \left(\frac{\overline{\sigma}+\eta}{1-\eta}\right)^{[n/2]} \le \tau\delta$$

for n sufficiently large.

Setting $n = n_*(\delta) - 1$ in (7.132), we arrive at

$$\left\| x_{n_*(\delta)}^\delta - x_0 \right\| \le \left\| x^\dagger - x_0 \right\| . \tag{7.133}$$

Hence, there exist weakly convergent subsequences $\{x^l\}_{l\in\mathbb{N}} := \{x_{n_*(\delta_l)}^{\delta_l}\}_{l\in\mathbb{N}}$ and $\{J_q^X(x^l - x_0)\}_{l\in\mathbb{N}} := \{J_q^X(x_{n_*(\delta_l)}^{\delta_l} - x_0)\}_{l\in\mathbb{N}}$. The weak limit $\bar{x}$ of any weakly convergent subsequence $\{x^l\}_{l\in\mathbb{N}}$ (or $\bar{x} := J_{q^*}^{X^*}(\bar{x}^*) + x_0$ with the weak limit $\bar{x}^*$ of $\{J_q^X(x^l - x_0)\}_{l\in\mathbb{N}}$), by

$$\left\| F(x^l) - y \right\| \le (\tau + 1)\delta_l \to 0 \text{ as } l \to \infty$$

and the (weak) sequential closedness of F (7.12) (or (7.13)), defines a solution $\bar{x}$ of (7.1). Hence, we can insert $\bar{x}$ in place of $x^\dagger$ in (7.133) to obtain, in case of (7.12),

$$\|\bar{x} - x_0\| \le \liminf_{l\to\infty} \left\| x^l - x_0 \right\| \le \limsup_{l\to\infty} \left\| x^l - x_0 \right\| \le \|\bar{x} - x_0\|$$

due to the weak lower semicontinuity of the norm, i.e., convergence of $\|x^l - x_0\|$ to $\|\bar{x} - x_0\|$. Since X is uniformly convex and x_l weakly converges to $\bar{x}$, this yields norm convergence of x_l to $\bar{x}$. In case of $\{J_q^X(x^l - x_0)\}_{l\in\mathbb{N}}$ weakly converging to $\bar{x}^*$ and (7.13), convergence in the Bregman distance can be established by the argument

$$0 \le D_{j_q}(\bar{x} - x_0, x^l - x_0)$$
$$= \frac{1}{q}\Big(\underbrace{\left\|x^l - x_0\right\|^q}_{\le \|\bar{x}-x_0\|^q} - \|\bar{x} - x_0\|^q\Big) + \underbrace{\langle x^* - J_q^X(x^l - x_0), \bar{x} - x_0\rangle_{X^*\times X}}_{\to 0 \text{ as } l\to\infty},$$

which, by Theorem 2.60 (e), implies strong convergence.

In case of uniqueness of $x^\dagger$, a subsequence-subsequence argument yields overall convergence. □

The proof of convergence rates under Hölder type and general variational inequalities is much shorter and less technical than with a priori parameter choice, so we provide both rates results within one theorem:

Theorem 7.14. *Let the assumptions of Theorem* 7.13 *be satisfied.*

(a) *Under a variational inequality* (7.76), *we obtain optimal convergence rates*

$$D_q^{x_0}(x_{n_*}, x^\dagger) = O(\delta^{\frac{2\nu}{\nu+1}}), \quad \textit{as } \delta \to 0. \tag{7.134}$$

(b) *Under a variational inequality* (7.77), *we obtain optimal convergence rates*

$$D_q^{x_0}(x_{n_*}, x^\dagger) = O\left(\kappa^2(\Theta^{-1}(\delta))\right) = O\left(\frac{\delta^2}{\Theta^{-1}(\delta)}\right) \quad \textit{as } \delta \to 0 \tag{7.135}$$

with Θ *as in* (7.81).

Proof. In the proof of Theorem 7.13, we have seen that estimates (7.87), (7.89) together with (7.126), (7.2), (7.127) and (7.29), yield (7.131), for all $n \le n_*(\delta) - 1$, provided $x_n^\delta \in \overline{B}_\rho(x_0)$.

Inserting (7.87) into (7.131) and taking into account (7.130) and (7.126), we get

$$\begin{aligned}&(1 - \beta\epsilon) D_q^{x_0}(x_{n+1}^\delta, x^\dagger)\\ &\le \beta C\left(\epsilon, \frac{\nu+1}{2}\right)\left((1+\eta)\left\|F(x_{n+1}^\delta) - F(x^\dagger)\right\|\right)^{2\nu/(\nu+1)}\end{aligned} \tag{7.136}$$

in case α_n is chosen according to (7.127). Hence, with $\epsilon < \beta^{-1}$, for $n = n_* - 1$, the discrepancy principle (7.29) yields the optimal rate

$$D_q^{x_0}(x_{n_*}^\delta, x^\dagger) \le \frac{\beta C(\epsilon, \frac{\nu+1}{2})}{1 - \beta\epsilon}\Big((1+\eta)(1+\tau)\Big)^{2\nu/(\nu+1)} \delta^{2\nu/(\nu+1)},$$

since, due to the signal to noise ratio assumption $\delta < \|F(x_0) - y^\delta\|/\tau$, we can exclude the case $x_{n_*}^\delta = x_0$, i.e., the case where α_{n_*-1} is not chosen according to (7.127).

In the general case (7.77) we get, in place of (7.87) and (7.136), the estimates (7.110) and

$$D_q^{x_0}(x_{n+1}^\delta, x^\dagger)^{1/2} \le \kappa \left(\frac{(1+\eta)^2 \left\| F(x_{n+1}^\delta) - F(x^\dagger) \right\|^2}{D_q^{x_0}(x_{n+1}^\delta, x^\dagger)} \right),$$

respectively. Hence, with $n = n_* - 1$, using (7.126) and (7.29), we get

$$\begin{aligned} C\delta &= \frac{C\delta}{D_q^{x_0}(x_{n_*}^\delta, x^\dagger)^{1/2}} D_q^{x_0}(x_{n_*}^\delta, x^\dagger)^{1/2} \\ &\le \frac{C\delta}{D_q^{x_0}(x_{n_*}^\delta, x^\dagger)^{1/2}} \kappa \left(\frac{C^2\delta^2}{D_q^{x_0}(x_{n_*}^\delta, x^\dagger)} \right) = \Theta\left(\frac{(C\delta)^2}{D_q^{x_0}(x_{n_*}^\delta, x^\dagger)} \right) \end{aligned}$$

with $C := (1+\eta)(1+\tau)$ so, taking the inverse of Θ on both sides, we get

$$D_q^{x_0}(x_{n_*}^\delta, x^\dagger) \le \frac{C^2\delta^2}{\Theta^{-1}(C\delta)} \le C^2 \frac{\delta^2}{\Theta^{-1}(\delta)},$$

since $C > 1$ and Θ^{-1} is strictly monotonically increasing. □

7.3.3 Numerical illustration

To illustrate performance of the iteratively regularized Gauss–Newton method in a Banach space setting, we return to the parameter identification example from Section 1.3, where we recovered c from measurements of u, with

$$-\Delta u + c\,u = 0 \text{ in } \Omega\,.$$

As a first test case we consider $\Omega = (0,1)^2 \subseteq \mathbb{R}^2$ and a smooth exact parameter

$$c(x,y) = 10 \exp\left(-\frac{(x-0.3)^2 + (y-0.3)^2}{0.04} \right), \tag{7.137}$$

see Figure 7.1 for the exact parameter c and state u.

Computations were carried out on a 30×30 grid, using finite differences, whereas synthetic data was generated on a different grid, in order to avoid an inverse crime. Figure 7.2 shows the reconstructions from noisy data with 1 percent L^∞-noise in a Hilbert space setting, as well as in a Banach space setting. The latter takes into account the fact that the noise bound is given, not only with respect to the L^2-norm but – as often relevant in practice – with respect to the L^∞-norm, by using an L^P-norm with large P ($P = 10$) as data space. While the reconstruction with $Y = L^2$ is overdamped, in spite of an optimal choice of the stopping index, the height of the parameter values is better resolved in the reconstruction with $Y = L^{10}$. An additional

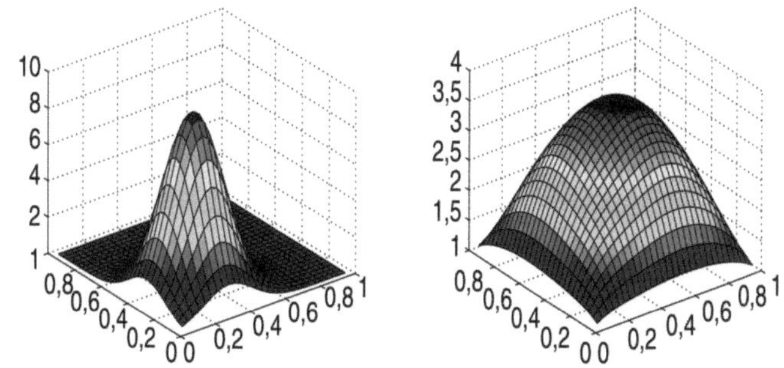

Figure 7.1. Exact parameter c (left) and state u (right) for the test example (7.137).

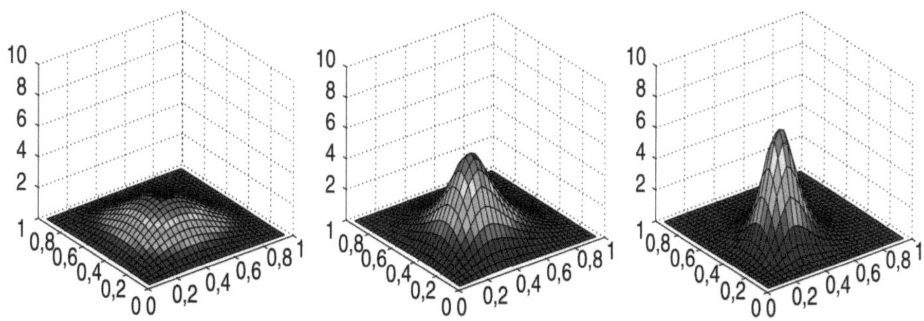

Figure 7.2. Reconstruction for example (7.137) from data with 1 percent L^∞-noise with $X = Y = L^2$ (left), $X = L^2$, $Y = L^{10}$ (middle), $X = 1.5$, $Y = L^{10}$ (right).

improvement appears, due to the decrease of the norm in preimage space to $L^{1.5}$. Recall that a combination $X = L^{1.5}$, $Y = L^{10}$ yields a reduction of the ill-posedness of the parameter identification problem, compared to the Hilbert space combination $X = Y = L^2$. These effects are also visible for higher noise levels, see Figure 7.3.

In our second test case, we consider again $\Omega = (0,1)^2 \subseteq \mathbb{R}^2$, but this time refer to a "sparse" exact parameter

$$c(x,y) = 40\chi_{[0.19,0.24]^2}(x,y)\,, \tag{7.138}$$

see Figure 7.4 for the exact parameter c and state u. Note that the similarity of the states in Figures 7.1 and 7.4 also illustrates the ill-posedness of the inverse problem of identifying c. Also for this test example, the use of an L^P-norm with large P in data space obviously yields an improvement, compared to the L^2 setting. As an additional effect, we see the considerably better reconstruction of the sparse solution by using an L^Q-norm, with Q close to one in preimage space, instead of L^2, see Figure 7.5.

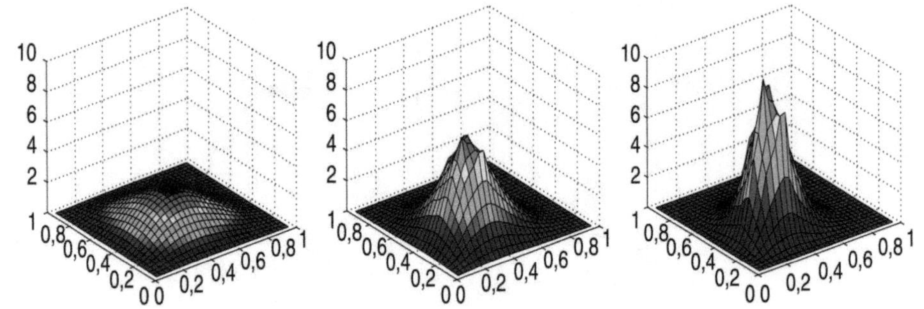

Figure 7.3. Reconstruction for example (7.137) from data with 10 percent L^∞-noise with $X = Y = L^2$ (left), $X = L^2$, $Y = L^{10}$ (middle), $X = 1.5$, $Y = L^{10}$ (right).

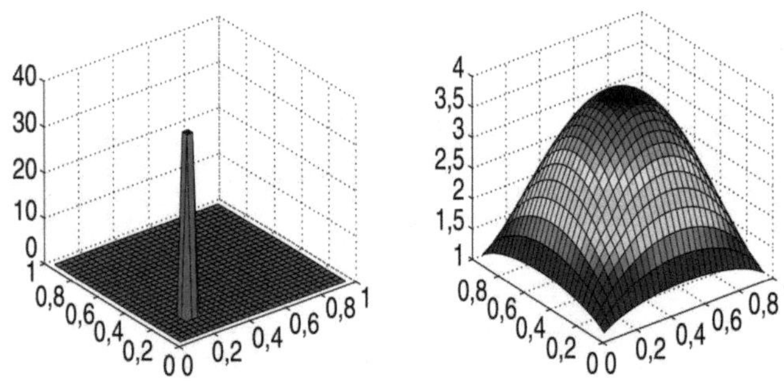

Figure 7.4. Exact parameter c (left) and state u (right) for the test example (7.138).

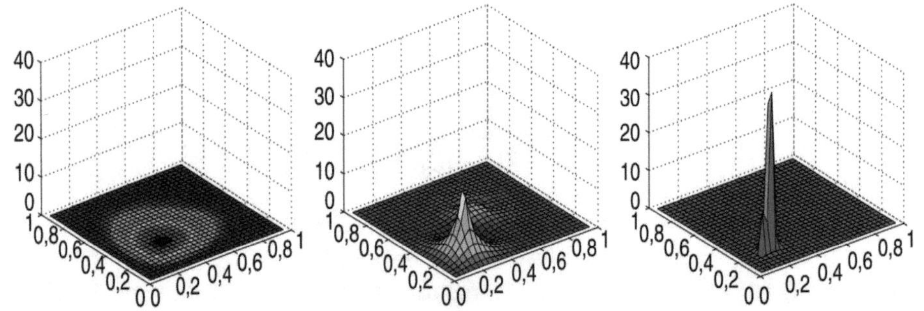

Figure 7.5. Reconstruction for example (7.138) from data with 1 percent L^∞-noise with $X = Y = L^2$ (left), $X = L^2$, $Y = L^{10}$ (middle), $X = L^{1.1}$, $Y = L^{10}$ (right).

Part V

The method of approximate inverse

This part is devoted to the solution of linear operator equations of the first kind

$$Af = g\,, \qquad g \in \mathcal{R}(A)$$

by the *method of approximate inverse* which, in contrast to the methods considered in the previous part, is non-iterative. In Chapters 8 and 9, we consider linear, bounded mappings A, acting on Banach spaces that are function spaces. This is why we prefer to use the notations $g \in Y$ for the given data and $f \in X$ for the searched solution, in order to avoid confusion with the points x and y the functions are evaluated at.

The method of approximate inverse is a mollification method for stably solving inverse problems. In its original form, it has been developed to solve operator equations in L^2-spaces and general Hilbert spaces. We describe its extension to linear, ill-posed problems in Banach spaces. The method itself consists of evaluations of dual pairings of the given data with *reconstruction kernels*, which are associated with so-called *mollifiers* and the adjoint of the operator. This technique offers several features, compared to other regularization methods: The reconstruction kernel can be precomputed before the measurement starts and the computation is done independently of any noise, since the mollifer is given analytically. Furthermore, invariance properties of the underlying operator A might be used, in order to improve the numerical efficiency. At last, evaluation of dual pairings is very stable with respect to noisy data g^δ. Chapter 8 provides all necessary definitions and notations to state the method. The method is then analyzed in Chapter 9, which follows along the lines of [223]. There, we investigate two settings more exactly: the case of L^p-spaces and the case of the Banach space of continuous functions on a compact set. For both settings, we present criteria that turn the method of approximate inverse into a regularization method and prove convergence with rates. It will be shown that the absolute error has the order $\mathcal{O}(\delta^{1-\nu})$, for any $0 < \nu < 1$, provided that f satisfies specific smoothness assumptions. As an application, we refer to X-ray diffractometry, as it was introduced in Section 1.1. Since we know that the stress tensor is smooth, X-ray diffractometry can be appropriately modeled by a Banach space setting, using continuous functions. Actually, X-ray diffractometry is described by a semi-discrete operator equation, since the stress tensor is to be computed from finitely many observations. Consequently, Chapter 10 gives an outlook on how the method of approximate inverse can be used in general to tackle semi-discrete operator equations, that means operator equations where the range of the forward operator has a finite dimension. Such equations are especially important in practical applications, since there only a finite number of data is available; a fact we will take into account when applying a so-called *observation operator* Ψ_n to A. This observation operator may be seen as the mathematical model of the measurement process, which intrinsically boils the infinite dimensional setting down to a finite dimensional one. We give criteria to get strong convergence in general Banach spaces.

Chapter 8

Setting of the method

The method of approximate inverse was originally developed by Louis [151], in order to solve ill-posed operator equations of first kind. It consists of the evaluation of the given and maybe noisy data with precomputed reconstruction kernels. These reconstruction kernels are solutions of a dual equation, associated with a so-called mollifier that implies the regularization property. The method is well established in Hilbert space settings and was successfully applied to different problems, such as computerized tomography [155], inverse scattering [1], Doppler tomography [217, 218, 224], X-ray diffractometry [221], thermoacoustic tomography [86], sonar [188, 222] and even image processing [153]. A recent publication [154] shows how an appropriate choice of the mollifier can be used to extract features from the reconstruction. In [152], an interesting unification concept is presented, for general regularization methods, by means of the approximate inverse. A concise monograph on the method is [220].

Let X, Y be Banach spaces with norms $\|\cdot\|_X$, $\|\cdot\|_Y$ and $A : X \to Y$ a linear, bounded and injective operator. Our aim is to solve the equation

$$Af = g^\delta \tag{8.1}$$

with given noisy data g^δ, satisfying

$$\|g - g^\delta\|_Y < \delta\,, \quad g \in \mathcal{R}(A)\,.$$

In this and the following chapter, we only consider Banach spaces X, Y consisting of functions on a domain $\Omega \subset \mathbb{R}^q$. As typical examples, we refer to L^p-spaces with $1 \le p \le \infty$ or the Banach space of functions continuous on a compact set. As a first step towards constructing the method of approximate inverse, we introduce the concept of mollifiers, which is essential for the method.

Definition 8.1 (mollifier). Let X be a Banach space consisting of functions with domain $\Omega \subset \mathbb{R}^q$. We call a family $\{e_\gamma\}_{\gamma>0}$ of mappings $e_\gamma : \Omega \to X^*$ a *mollifier* for X (or X-mollifier), if the two following conditions hold:

(a) For each function $f \in X$, the family $\{f_\gamma\}_{\gamma>0}$ of mappings

$$f_\gamma(x) := \langle e_\gamma(x), f\rangle_{X^*\times X} \tag{8.2}$$

belongs to X.

(b) The functions f_γ converge to f in X, that is

$$\lim_{\gamma\to 0} \|f_\gamma - f\|_X = 0\,, \quad f \in X\,. \tag{8.3}$$

We will show specific criteria for such a family $\{e_\gamma\}$ to be $L^p(\Omega)$- and $\mathcal{C}(K)$-mollifiers. Once we have chosen a mollifier $\{e_\gamma\}$, we associate a *reconstruction kernel* with it. A reconstruction kernel is a family $\{v_\gamma\}_{\gamma>0}$ of mappings $v_\gamma : \Omega \to Y^*$ solving the dual equation

$$A^*[v_\gamma(x)] = e_\gamma(x)\,, \quad x \in \Omega\,. \tag{8.4}$$

In the following, we assume that $e_\gamma(x)$ is in the range of A^*. If we have the reconstruction kernel v_γ at hand, we are able to formulate the *method of approximate inverse* in Banach spaces.

Definition 8.2 (approximate inverse). Let $A \in \mathcal{L}(X,Y)$, $\{e_\gamma : \Omega \to X^*\}$ be a mollifier and $\{v_\gamma : \Omega \to Y^*\}$ the corresponding reconstruction kernel, according to (8.4). The family of mappings $\{\mathcal{A}_\gamma\}$, defined by

$$[\mathcal{A}_\gamma g](x) := \langle v_\gamma(x), g\rangle_{Y^*\times Y}\,, \quad g \in Y\,, x \in \Omega$$

is called the *approximate inverse* of A associated with the mollifier $\{e_\gamma\}$.

In shortened form, we say that $\mathcal{A}_\gamma$ is the approximate inverse of A, if it is clear which mollifier we have chosen. Obviously, $\mathcal{A}_\gamma$ is a linear mapping from Y into the vector space of real-valued functions on Ω and weak convergence of a sequence $\{g_n\} \subset Y$ implies pointwise convergence of $\{\mathcal{A}_\gamma g_n\}_n$. Moreover, if $Af = g$, we may deduce from (8.4) that

$$[\mathcal{A}_\gamma g](x) = \langle v_\gamma(x), g\rangle_{Y^*\times Y} = \langle e_\gamma(x), f\rangle_{X^*\times X}$$

and (8.3) tells us that

$$\lim_{\gamma\to 0} \mathcal{A}_\gamma g = f \tag{8.5}$$

with respect to the strong (norm-) topology in X. In this sense, $\mathcal{A}_\gamma$ is, in fact, an approximate inverse to A. Under specific assumptions, it turns out to be a regularization method.

Theorem 8.3. *Let the assumptions of Definiton* 8.2 *hold. Assume that A is injective, that the mappings $\mathcal{A}_\gamma$ map Y into X and are bounded with estimates*

$$\|\mathcal{A}_\gamma\|_{Y\to X} \le l_\gamma \quad \textit{for all } \gamma > 0\,. \tag{8.6}$$

Let $f \in X$ be the solution of $Af = g$ for $g \in \mathcal{R}(A)$ and $g^\delta \in Y$ be noisy data, with $\|g^\delta - g\|_Y < \delta$.

If the a priori parameter choice rule $\gamma = \gamma(\delta)$ *is chosen such that* $\gamma(\delta) \to 0$ *for* $\delta \to 0$ *and*

$$l_{\gamma(\delta)} = o(\delta^{-1}) \quad \text{as } \delta \to 0 \tag{8.7}$$

then

$$\lim_{\delta \to 0} \|f^{\delta}_{\gamma(\delta)} - f\|_X = 0\,,$$

where $f^{\delta}_{\gamma(\delta)} = \mathcal{A}_{\gamma(\delta)} g^{\delta}$. *This means, that the approximate inverse* $\{\mathcal{A}_\gamma\}$ *represents a regularization method.*

Proof. To prove the regularization property, we estimate

$$\begin{aligned} \|f^{\delta}_{\gamma} - f\|_X &\le \|\mathcal{A}_\gamma g - f\|_X + \|\mathcal{A}_\gamma (g^{\delta} - g)\|_X \\ &\le \|\mathcal{A}_\gamma g - f\|_X + l_\gamma \delta\,. \end{aligned}$$

The first summand tends to 0 as $\gamma \to 0$, due to (8.5), the second summand tends to zero, because of (8.7). □

From the proof of Theorem 8.3, we can easily deduce convergence rates.

Corollary 8.4. *Let the assumptions of Theorem* 8.3 *hold and let* $f_\gamma = \mathcal{A}_\gamma g$ *be the approximate inverse, applied to exact data. If the parameter choice rule* $\gamma = \gamma(\delta)$ *is chosen according to*

$$\lim_{\delta \to 0} \gamma(\delta) = 0\,, \quad l_{\gamma(\delta)} = \mathcal{O}(\delta^{-\nu})\,, \quad 0 < \nu < 1$$

and

$$\|f_{\gamma(\delta)} - f\|_X \le c\delta^{1-\nu} \quad \text{as } \delta \to 0$$

for a constant $c > 0$, *it follows that*

$$\|f^{\delta}_{\gamma(\delta)} - f\|_X = \mathcal{O}(\delta^{1-\nu}) \quad \text{as } \delta \to 0\,.$$

As in Hilbert spaces, see [151], operator invariances can help to decrease the computational costs.

Lemma 8.5. *If to each* $x \in \Omega$ *there exist mappings* $T_1^x \in \mathcal{L}(X^*)$, $T_2^x \in \mathcal{L}(Y^*)$ *and* $x_0 \in \Omega$ *such that*

$$e_\gamma(x) = T_1^x e_\gamma(x_0) \quad \text{and} \quad T_1^x A^* = A^* T_2^x \quad x \in \Omega\,, \tag{8.8}$$

then

$$v_\gamma(x) = T_2^x v_\gamma(x_0)\,. \tag{8.9}$$

Proof. Assertion (8.9) becomes obvious from (8.4) and (8.8). □

Hence, it suffices to compute one single reconstruction kernel $v_\gamma(x_0)$, in order to generate all reconstruction kernels, by applying T_2^x, if only (8.8) is satisfied.

Chapter 9

Convergence analysis in $L^p(\Omega)$ and $\mathcal{C}(K)$

In the following, we consider two specific choices for X in detail, namely $X = L^p(\Omega)$, for $1 \le p < \infty$ and $X = \mathcal{C}(K)$, for a compact set $K \subset \mathbb{R}^q$, and investigate under which conditions the assertions of Theorem 8.3 and Corollary 8.4 are valid.

9.1 The case $X = L^p(\Omega)$

We set $X = L^p(\Omega)$, with $1 \le p < \infty$ and an open subset $\Omega \subset \mathbb{R}^q$. We recall that we denote by p^* the dual exponent to p, meaning $1 \le p^* \le \infty$, with

$$\frac{1}{p} + \frac{1}{p^*} = 1 \,.$$

For $p = 1$ we set $p^* = \infty$. Then, $X^* = L^{p^*}(\Omega)$ and the dual pairing is given by

$$\langle f^*, f \rangle_{L^{p^*} \times L^p} = \int_\Omega f^*(x) f(x) \, \mathrm{d}x \,.$$

We want to present a recipe for constructing mollifiers in this situation. Furthermore, we prove that, under certain conditions,

$$\|f_\gamma - f\|_{L^p(\Omega)} \le C\gamma^m \|f\|_{\mathcal{C}^m(\overline{\Omega})} \quad \text{for all } \gamma > 0 \,,$$

where Ω is bounded. This estimate is sufficient to achieve the convergence rate $\mathcal{O}(\delta^{1-\nu})$, see Corollary 8.4.

We first present a general recipe to generate mollifiers in bounded domains Ω.

Theorem 9.1. *Let $\Omega \subset \mathbb{R}^q$ be a bounded domain, $\bar{e} \in L^{p^*}(\mathbb{R}^q)$ be a function with ess supp$(\bar{e}) = B_\rho(0) \subset \Omega$, for some $\rho > 0$, and*

$$\int_\Omega \bar{e}(x) \, \mathrm{d}x = \int_{B_\rho(0)} \bar{e}(x) \, \mathrm{d}x = 1 \,. \tag{9.1}$$

Define

$$e_\gamma(x, y) := \gamma^{-q} \bar{e}\left(\frac{x - y}{\gamma}\right) \quad a.e.. \tag{9.2}$$

Then, $\{e_\gamma\}$ is a mollifier for $L^p(\Omega)$.

Proof. Since Ω is bounded, we may estimate for $k_\gamma(x) := \|e_\gamma(x,\cdot)\|_{L^{p^*}}$

$$\begin{aligned}
\|k_\gamma\|_{L^p}^p &= \int_\Omega |k_\gamma(x)|^p \,\mathrm{d}x = \int_\Omega \|e_\gamma(x,\cdot)\|_{L^{p^*}}^p \,\mathrm{d}x \\
&\le \int_\Omega \left(\int_{\mathbb{R}^q} \left| \gamma^{-q} \bar{e}\left(\frac{x-y}{\gamma}\right)\right|^{p^*} \mathrm{d}y \right)^{p/p^*} \mathrm{d}x \qquad (9.3)\\
&= \gamma^{-q} \int_\Omega \left(\int_{\mathbb{R}^q} |\bar{e}(z)|^{p^*} \,\mathrm{d}z \right)^{p/p^*} \mathrm{d}x = \gamma^{-q} |\Omega| \|\bar{e}\|_{L^{p^*}(\mathbb{R}^q)}^p ,
\end{aligned}$$

where we applied the substitution $z \leftarrow (x-y)/\gamma$. Hence, $k_\gamma \in L^p$ implying $f_\gamma \in L^p$.

It remains to show the convergence $f_\gamma \to f$ in L^p. To this end, we first assume that $f \in \mathcal{C}_0(\Omega)$ is a continuous function, with compact support in Ω, and then use a density argument. Hence, let $f \in \mathcal{C}_0(\Omega)$. We have

$$\|f_\gamma - f\|_{L^p}^p = \int_\Omega \left| f(x) - \int_\Omega \gamma^{-q} f(y) \bar{e}\left(\frac{x-y}{\gamma}\right) \mathrm{d}y \right|^p \mathrm{d}x .$$

We first investigate the integrand and estimate

$$\begin{aligned}
I_\gamma(x) &:= \left| f(x) - \int_\Omega \gamma^{-q} f(y) \bar{e}\left(\frac{x-y}{\gamma}\right) \mathrm{d}y \right| \\
&= \left| f(x) - \int_{B_{\gamma\rho}(x)} \gamma^{-q} f(y) \bar{e}\left(\frac{x-y}{\gamma}\right) \mathrm{d}y \right| \\
&= \left| f(x) - \int_{B_\rho(0)} f(x-\gamma z) \bar{e}(z) \,\mathrm{d}z \right| \\
&= \left| \int_{B_\rho(0)} (f(x) - f(x-\gamma z))\, \bar{e}(z) \,\mathrm{d}z \right| \\
&\le \left(\int_{B_\rho(0)} |f(x) - f(x-\gamma z)|^p \,\mathrm{d}z \right)^{1/p} \left(\int_{B_\rho(0)} |\bar{e}(z)|^{p^*} \,\mathrm{d}z \right)^{1/p^*} \\
&= \|\bar{e}\|_{L^{p^*}} \left(\int_{B_\rho(0)} |f(x) - f(x-\gamma z)|^p \,\mathrm{d}z \right)^{1/p}
\end{aligned}$$

using Hölders inequality and the substitution $z \leftarrow (x-y)/\gamma$. Since $|a+b|^p \le 2^p(|a|^p + |b|^p)$ for real values a, b, we have that

$$|f(x) - f(x-\gamma z)|^p \le 2^p \left(|f(x)|^p + |f(x-\gamma z)|^p \right).$$

Furthermore, the function

$$\tilde{f}(x) := \sup_{z \in B_\rho(0)} |f(x-\gamma z)|$$

is a continuous function with compact support, satisfying

$$\tilde{f}(x) \geq |f(x - \gamma z)| .$$

As a summary of our results, we obtain

(i) $I_\gamma(x)^p \to 0$ pointwise as $\gamma \to 0$

(ii) $I_\gamma(x)^p \leq \|\bar{e}\|^p_{L^{p*}} |B_\rho(0)| 2^p (|f(x)|^p + \tilde{f}(x)^p)$ and the right-hand side of that estimate is in $L^1(\Omega)$.

Finally, Lebesgue's theorem of dominated convergence implies

$$\lim_{\gamma \to 0} \int_\Omega I_\gamma(x)^p \, dx = 0 .$$

A standard argument, using the density of $\mathcal{C}_0(\Omega)$ in $L^p(\Omega)$, completes the proof. For a detailed outline, we refer the interested reader to [223]. □

Another criterion for f_γ to be a L^p-function is the boundedness of k_γ.

Lemma 9.2. *Let $\{e_\gamma : \Omega \to L^{p^*}\}$ be a family of mappings such that the function*

$$k_\gamma(x) := \|e_\gamma(x)\|_{L^{p*}} \in L^p(\Omega) . \tag{9.4}$$

Then,

$$f_\gamma(x) = \langle e_\gamma(x), f \rangle_{L^{p*} \times L^p} \quad \in L^p(\Omega) . \tag{9.5}$$

Condition (9.4) is satisfied, if $k_\gamma \in L^\infty(\Omega)$, where, if Ω is unbounded, we have to postulate in addition

$$k_\gamma(x) \leq c|x|^s \quad \text{for } x \in \{x \in \mathbb{R}^q : |x| > R\} \cap \Omega \tag{9.6}$$

for constants $c, R > 0$ and $s < -q/p$.

Proof. Let $f \in L^p$. The estimate

$$\begin{aligned} \int_\Omega |f_\gamma(x)|^p \, dx &= \int_\Omega \left|\langle e_\gamma(x), f \rangle_{L^{p*} \times L^p}\right|^p dx \\ &\leq \|f\|^p_{L^p} \int_\Omega \|e_\gamma(x)\|^p_{L^{p*}} \, dx = \|f\|^p_{L^p} \|k_\gamma\|^p_{L^p} \end{aligned}$$

and (9.4) imply $f_\gamma \in L^p$.

If Ω is bounded and $k_\gamma \in L^\infty(\Omega)$, then obviously (9.4) holds. Now, let Ω be unbounded. We have

$$\begin{aligned}\int_\Omega |k_\gamma(x)|^p \,\mathrm{d}x &= \int_{\Omega\cap B_R(0)} |k_\gamma(x)|^p \,\mathrm{d}x + \int_{\Omega\cap\{x\in\mathbb{R}^q:|x|>R\}} |k_\gamma(x)|^p \,\mathrm{d}x \\ &\le \int_{\Omega\cap B_R(0)} |k_\gamma(x)|^p \,\mathrm{d}x + c^p \int_{\Omega\cap\{x\in\mathbb{R}^q:|x|>R\}} |x|^{sp} \,\mathrm{d}x\,.\end{aligned}$$

The first integral is bounded, since $k_\gamma \in L^\infty(\Omega)$. The second integral can be estimated by using spherical coordinates $x = r\omega$, $r > 0$, $\omega \in S^{q-1}$ and

$$\begin{aligned}\int_{\Omega\cap\{x\in\mathbb{R}^q:|x|>R\}} |x|^{sp} \,\mathrm{d}x &\le \int_{\{x\in\mathbb{R}^q:|x|>R\}} |x|^{sp} \,\mathrm{d}x \\ &= |S^{q-1}| \int_R^\infty r^{q-1} r^{sp} \,\mathrm{d}r < \infty\,,\end{aligned}$$

since $sp + q - 1 < -1 \Leftrightarrow s < -q/p$. Here, $|S^{q-1}|$ denotes the surface measure of the unit sphere in $\mathbb{R}^q$. □

We finish this section by proving rates for the convergence $f_\gamma \to f$, when Ω is bounded and the mollifier $\{e_\gamma\}$ is generated by (9.2). These rates are important to get the stability rates in Corollary 8.4. To obtain these convergence rates, a momentum condition on $\bar{e}$ is required.

Definition 9.3 (order of mollifier, L^p-case). Let $\bar{e} \in L^{p^*}(\mathbb{R}^q) \cap L^1(\mathbb{R}^q)$. We say that $\bar{e}$ has *order* m, if the following three conditions are true

(a) $\displaystyle\int_{\mathbb{R}^q} \bar{e}(z)\,\mathrm{d}z = 1\,,$

(b) $\displaystyle\int_{\mathbb{R}^q} \bar{e}(z) z^\alpha \,\mathrm{d}z = 0$

for all multiindices $\alpha \in \mathbb{N}_0^q$ with $1 \le |\alpha| < m$,

(c) $\displaystyle\mu_\alpha := \int_{\mathbb{R}^q} \bar{e}(z) z^\alpha \,\mathrm{d}z/\alpha! \ne 0$

for all multiindices $\alpha \in \mathbb{N}_0^q$ with $|\alpha| = m$.

Now, we have all ingredients together to prove convergence rates for f_γ, if we apply mollifiers constructed in accordance with Theorem 9.1.

Theorem 9.4. *Adopt the assumptions of Theorem* 9.1 *where, additionally,* $\bar{e}$ *has order* $m \in \mathbb{N}$ *and ess supp*$(\bar{e}) \subseteq B_\rho(0) \subset \Omega$. *Furthermore, let* $f \in \mathcal{C}_0^m(\Omega)$. *Then, there exists a constant* $C > 0$, *such that*

$$\|f_\gamma - f\|_{L^p} \le C\gamma^m \|f\|_{\mathcal{C}^m(\overline{\Omega})} \quad \textit{for all } \gamma > 0\,. \tag{9.7}$$

Proof. We show (9.7) for $q = 1$ only, since this assertion does not depend on the dimension. Then, $B_\rho(0) = (-\rho, \rho)$ and we have

$$\|f_\gamma - f\|_{L^p}^p = \int_\Omega I_\gamma(x)^p \, dx$$

with $I_\gamma(x)$ as in the proof of Theorem 9.1. Using the momentum conditions for $\bar{e}$ and the Taylor approximation

$$f(x - \gamma z) = \sum_{j=0}^{m-1} \frac{f^{(j)}(x)}{j!}(-\gamma z)^j + \frac{f^{(m)}(x_{\gamma,z})}{m!}(-\gamma z)^m ,$$

where $x_{\gamma,z}$ denotes an intermediate value between x and $x - \gamma z$, we compute

$$\begin{aligned} I_\gamma(x) &= \left| \int_{-\rho}^{\rho} \left[f(x) - f(x - \gamma z)\right]\bar{e}(z) \, dz \right| \\ &= \left| \int_{-\rho}^{\rho} \frac{(-\gamma z)^m}{m!} f^{(m)}(x_{\gamma,z})\bar{e}(z) \, dz \right| \\ &\le \gamma^m K_m \|f^{(m)}\|_\infty , \end{aligned}$$

where

$$K_m := \int_{-\rho}^{\rho} |\bar{e}(z) z^m| \, dz .$$

From this, we get

$$\|f_\gamma - f\|_{L^p}^p = \int_\Omega I_\gamma(x)^p \, dx \le \gamma^{mp} K_m^p |\Omega| \|f\|_{\mathcal{C}^m(\overline{\Omega})}^p$$

for all $\gamma > 0$. Hence, (9.7) is satisfied with $C := |\Omega|^{1/p} K_m$. □

With the help of Theorem 9.4, we are able to formulate conditions for $\bar{e}$ and f, necessary to obtain the convergence rate $\mathcal{O}(\delta^{1-\nu})$ of the approximate inverse $f_\gamma^\delta = \mathcal{A}_\gamma g^\delta$ in L^p-spaces.

Theorem 9.5. *Let the assumptions of Theorems* 8.3 *and* 9.4 *hold and* ν *be given with* $0 < \nu < 1$. *Furthermore, assume that*

$$l_\gamma \sim \gamma^{-k} \quad \text{as } \gamma \to 0, \quad k > 0 .$$

If the parameter choice rule $\gamma = \gamma(\delta)$ *is chosen such that*

$$\gamma(\delta) = \mathcal{O}(\delta^{\nu/k}) \quad \text{as } \delta \to 0, \tag{9.8}$$

$f \in \mathcal{C}_0^m(\Omega)$ *and* $\bar{e}$ *has order* m, *where* $m \in \mathbb{N}$ *satisfies*

$$m \ge k \frac{1-\nu}{\nu} , \tag{9.9}$$

then

$$\|f_{\gamma(\delta)}^{\delta} - f\|_{L^p} \le c\delta^{1-\nu}\|f\|_{\mathcal{C}^m(\overline{\Omega})} \quad \text{as } \delta \to 0\,.$$

Proof. Because of (9.8), we have that $l_{\gamma(\delta)} = \mathcal{O}(\delta^{-\nu})$. From Corollary 8.4 and (9.7), we deduce that

$$\begin{aligned}\|f_{\gamma(\delta)}^{\delta} - f\|_{L^p} &\le \|f_{\gamma(\delta)} - f\|_{L^p} + l_{\gamma(\delta)}\delta \\ &\le C\left(\delta^{\frac{m\nu}{k}}\|f\|_{\mathcal{C}^m(\overline{\Omega})} + \delta^{1-\nu}\right).\end{aligned}$$

Balancing the terms, we arrive at

$$\|f_{\gamma(\delta)}^{\delta} - f\|_{L^p} \le c\delta^{1-\nu}\|f\|_{\mathcal{C}^m(\overline{\Omega})} \quad \text{as } \delta \to 0\,,$$

if $m\nu/k \ge 1-\nu$, which gives (9.9). □

Remark 9.6. Condition (9.9) shows that high convergence rates correspond to strong assumptions on $\bar{e}$ and f. It is remarkable that the smoothness condition on f coincides with the required order of $\bar{e}$.

9.2 The case $X = \mathcal{C}(K)$

Let $K \subset \mathbb{R}^q$ be a nonempty, compact set, with $\operatorname{int}(K) \neq \emptyset$ and let $\mathcal{C}(K)$ be the Banach space of continuous functions, defined on K, equipped with the sup-Norm

$$\|f\|_\infty = \|f\|_{\infty,K} = \sup_{x\in K}|f(x)|\,.$$

We want to describe mollifiers and derive convergence rates for this situation. As an analogue to (9.7), we prove the estimate

$$\|f_\gamma - f\|_\infty \le C\gamma^m\|f\|_{\mathcal{C}^m(K)}\,,$$

which is again sufficient to show convergence with rates in case of noisy data g^δ.

According to the Riesz Representation Theorem, see e.g. [206, Theorem 2.14], the dual space $\mathcal{C}(K)^*$ of $\mathcal{C}(K)$ can be identified with the space of all regular and countable additive Borel measures on K

$$\operatorname{rca}(K) := \{\lambda : \lambda \text{ is a regular and countable additive Borel measure on } \mathcal{B}(K)\}$$

equipped with the total variation

$$\|\lambda\|_{\mathcal{B}} = \sup\left\{\sum_{i=1}^{k}|\lambda(K_i)| : k \in \mathbb{N}\,,\ K_i \in \mathcal{B}(K)\,,\ K_i \text{ pairwise disjoint}\right\}$$

as norm, where $\mathcal{B}(K)$ denotes the σ-algebra of all Borel sets of K, that is, the σ-algebra generated by the closed subsets of K. We have

$$\lambda(E) = \int_E \mathrm{d}\lambda(x) \quad \text{for all } E \in \mathcal{B}(K)\,.$$

There exists an isometric isomorphism $\mathcal{C}(K)^* \cong \mathrm{rca}(K)$ in the sense that any linear, bounded functional $\varphi \in \mathcal{C}(K)^*$ is represented by a unique measure $\lambda_\varphi \in \mathcal{B}(K)$ such that

$$\lambda_\varphi(f) = \int_K f(x)\,\mathrm{d}\lambda(x) \quad \text{for all } f \in \mathcal{C}(K)$$

and $\sup\{|\varphi(f)| : \|f\|_\infty = 1\} = \|\lambda_\varphi\|_{\mathcal{B}}$. With respect to this isomorphism, the dual pairing on X is given by

$$\langle \lambda, f\rangle_{\mathrm{rca}(K)\times\mathcal{C}(K)} = \int_K f(x)\,\mathrm{d}\lambda(x)\,.$$

Note that any function $f^* \in L^1(K)$ generates a measure $\lambda_{f^*} \in \mathrm{rca}(K)$ through

$$\lambda_{f^*}(E) := \int_E f^*(x)\,\mathrm{d}x\,, \qquad E \in \mathcal{B}(K)\,.$$

We present a concept to generate $\mathcal{C}(K)$-mollifiers, similar to (9.2).

Theorem 9.7. *Let $\bar\lambda \in \mathrm{rca}(K)$ be non-negative and suppose that*

$$\bar\lambda(K) = \int_K \mathrm{d}\bar\lambda(x) = 1\,. \tag{9.10}$$

We define a family of mappings $\{\lambda_\gamma : K \to \mathrm{rca}(K)\}$ by

$$\int_K f(x)\,\mathrm{d}[\lambda_\gamma(y)](x) := \int_K f(\gamma x + y)\,\mathrm{d}\bar\lambda(x) \quad f \in \mathcal{C}(K)\,. \tag{9.11}$$

Then, the family $\{\lambda_\gamma\}$ represents a mollifier for $\mathcal{C}(K)$.

Note that, in (9.11), we silently set $f(z) := 0$ for all $z \in \mathbb{R}^q \backslash K$. We assume this throughout this section, without always mentioning it explicitly. Definition (9.11) is then well-defined, since the translated function $f(\gamma x + y)$ is $\bar\lambda$-measurable for all $y \in K$.

Proof. At first we show that $\lambda_\gamma(y) \in \mathrm{rca}(K)$, for each $y \in K$. Let $f \in \mathcal{C}(K)$. For $\gamma > 0$ and $y \in K$ fixed we have

$$\sup_{x\in K} |f(\gamma x + y)| \le \|f\|_\infty\,.$$

Defining $Tf := \int_K f(x)\,\mathrm{d}[\lambda_\gamma(y)](x)$, we see that $T : \mathcal{C}(K) \to \mathbb{R}$ is linear. Since

$$|Tf| = \left|\int_K f(x)\,\mathrm{d}[\lambda_\gamma(y)](x)\right| = \left|\int_K f(\gamma x + y)\,\mathrm{d}\bar{\lambda}(x)\right| \le \|\bar{\lambda}\|_{\mathcal{B}} \|f\|_\infty$$

we see that T is also bounded. Hence, $T \in \mathcal{C}(K)^*$ and thus $\lambda_\gamma(y) \in \mathrm{rca}(K)$.

Next, we show that $f_\gamma(y) = \langle \lambda_\gamma(y), f\rangle_{\mathrm{rca}(K)\times\mathcal{C}(K)}$ is in $\mathcal{C}(K)$ for $f \in \mathcal{C}(K)$. Let $\varepsilon > 0$ be given. Since K is compact, f is even uniformly continuous. Hence, there exists a $\eta = \eta(\varepsilon) > 0$ such that $|f(y) - f(y')| < \varepsilon$, whenever $\|y - y'\| < \eta$ for $y, y' \in K$. This implies

$$\begin{aligned} |f_\gamma(y) - f_\gamma(y')| &\le \int_K \left|f(\gamma x + y) - f(\gamma x + y')\right| \mathrm{d}\bar{\lambda}(x) \\ &< \int_K \varepsilon\, \mathrm{d}\bar{\lambda}(x) = \varepsilon \quad \text{for all } y, y' \in K \end{aligned}$$

proving the (uniform) continuity of f_γ. In the latter estimate, we used the fact that $\bar{\lambda}$ is non-negative.

It remains for us to show the convergence $f_\gamma \to f$ in $\mathcal{C}(K)$, as $\gamma \to 0$. Again, let $\varepsilon > 0$ and $\eta = \eta(\varepsilon)$ as above. We set $M := \sup_{x\in K} |x|$ and choose a $\gamma_0 > 0$ such that $\gamma_0 M < \eta$. Such a γ_0 exists, since $M < \infty$ because of the compactness of K. Taking into account that $\bar{\lambda}(K) = 1$, we may estimate for $\gamma < \gamma_0$

$$\begin{aligned} \|f_\gamma - f\|_\infty &= \sup_{y\in K} \left|\int_K \big(f(\gamma x + y) - f(y)\big)\, \mathrm{d}\bar{\lambda}(x)\right| \\ &\le \sup_{y\in K} \int_K \left|f(\gamma x + y) - f(y)\right| \mathrm{d}\bar{\lambda}(x) \\ &< \sup_{y\in K} \int_K \varepsilon\, \mathrm{d}\bar{\lambda}(x) = \varepsilon\,, \end{aligned}$$

since $|\gamma x + y - y| = \gamma\|x\| \le \gamma M < \eta$ for $\gamma < \gamma_0$. This proves that $f_\gamma \to f$ in $\mathcal{C}(K)$, as $\gamma \to 0$. □

Remark 9.8. As mentioned before, any function $f^* \in L^1(K)$ generates a measure $\bar{\lambda}_{f^*} \in \mathrm{rca}(K)$ through

$$\bar{\lambda}_{f^*}(E) = \int_E f^*(x)\,\mathrm{d}x\,, \quad E \in \mathcal{B}(K)\,.$$

In this case, (9.10) means that $\int_K f^*(x)\,\mathrm{d}x = 1$ and (9.11) reads as

$$f_\gamma^*(x, y) = \gamma^{-q} f^*\left(\frac{x - y}{\gamma}\right),$$

which is the same construction as in Theorem 9.1.

A further criterion for general measures $\{\lambda_\gamma\} \subset \mathrm{rca}(K)$, leading to continuous functions f_γ, is the continuity of the mappings $y \mapsto \lambda_\gamma(y)$.

Lemma 9.9. *If the mappings $\lambda_\gamma : K \to \mathrm{rca}(K)$ are continuous as functions from K to $\mathrm{rca}(K)$, then $f_\gamma \in \mathcal{C}(K)$ for each $f \in \mathcal{C}(K)$. The functions f_γ are then even uniformly continuous.*

Proof. The assertions become clear from

$$\begin{aligned} |f_\gamma(x) - f_\gamma(y)| &= |\langle \lambda_\gamma(x) - \lambda_\gamma(y), f\rangle_{\mathrm{rca}(K)\times\mathcal{C}(K)}| \\ &\le \|\lambda_\gamma(x) - \lambda_\gamma(y)\|_{\mathcal{B}(K)} \|f\|_\infty . \end{aligned}$$

Because K is compact, the functions f_γ are even uniformly continuous. □

To show convergence with rates for $f_\gamma \to f$ we again have to define what we mean by the *order* of a measure $\lambda \in \mathrm{rca}(K)$.

Definition 9.10 (order of mollifier, $\mathcal{C}(K)$-case). Let $\bar\lambda \in \mathrm{rca}(K)$, $K \subset \mathbb{R}^q$ compact and $m > 0$ an integer. We say that $\bar\lambda$ has *order* m, if the three following conditions hold true:

(a) $\bar\lambda(K) = 1$,

(b) $\displaystyle\int_K x^\alpha \,\mathrm{d}\bar\lambda(x) = 0$

for all multiindices $\alpha \in \mathbb{N}_0^q$ with $1 \le |\alpha| < m$,

(c) $\displaystyle\mu_\alpha := \int_K x^\alpha \mathrm{d}\bar\lambda(x)/\alpha! \neq 0$

for all multiindices $\alpha \in \mathbb{N}_0^q$ with $|\alpha| = m$.

Note that this definition coincides with Definition 9.3, if $\bar\lambda$ is represented by an integrable function $f^* \in L^1(K)$.

Theorem 9.11. *Adopt the assumptions of Theorem* 9.7 *where additionally $\bar\lambda$ has order $m \in \mathbb{N}$. Furthermore assume that $K = \overline{\Omega}$ for an open and bounded subset $\Omega \subset \mathbb{R}^q$ and $f \in \mathcal{C}_0^m(\Omega)$. Then there exists a constant $C > 0$ satisfying*

$$\|f_\gamma - f\|_\infty \le C\gamma^m \|f\|_{\mathcal{C}^m(K)} \quad \textit{for all } \gamma > 0 . \tag{9.12}$$

Proof. As in the proof of Theorem 9.4 we restrict the verification of (9.12) to $q = 1$ and may assume that $K = [a, b]$ is a closed interval. Again we use a Taylor series expansion up to the order m to prove the estimate

$$\begin{aligned}\|f_\gamma - f\|_\infty &= \sup_{y\in K}\left|\int_a^b \big(f(\gamma x + y) - f(y)\big)\,\mathrm{d}\bar{\lambda}(x)\right| \\ &= \sup_{y\in K}\left|\int_a^b \sum_{j=1}^{m-1} \frac{f^{(j)}(y)}{j!}(-\gamma x)^j + \frac{f^{(m)}(y_{\gamma,x})}{m!}(-\gamma x)^m\,\mathrm{d}\bar{\lambda}(x)\right| \\ &\le \gamma^m K_m \|f\|_{\mathcal{C}^m(K)}\,,\end{aligned}$$

where

$$K_m := \int_a^b |x^m|\,\mathrm{d}\bar{\lambda}(x)\,,$$

$y_{\gamma,x} \in [a,b]$ lies between y and $\gamma x + y$ and we used the fact that $\bar{\lambda}$ has order m. Hence we showed (9.12) with $C := K_m$. □

The following lemma stating convergence with rates in case of noisy data g^δ is the $\mathcal{C}(K)$-analogue of Theorem 9.5.

Theorem 9.12. *Let the assumptions of Theorems* 8.3 *and* 9.11 *hold true and* ν *be given with* $0 < \nu < 1$. *Furthermore assume that*

$$l_\gamma \sim \gamma^{-k} \quad \text{as } \gamma \to 0, \quad k > 0\,.$$

If the parameter choice rule $\gamma = \gamma(\delta)$ *is chosen such that*

$$\gamma(\delta) = \mathcal{O}(\delta^{\nu/k}) \quad \text{as } \delta \to 0\,,$$

$f \in \mathcal{C}^m(K)$ *and* $\bar{\lambda}$ *has order* m, *where* $m \in \mathbb{N}$ *satisfies*

$$m \ge k\frac{1-\nu}{\nu}\,, \tag{9.13}$$

then

$$\|f^\delta_{\gamma(\delta)} - f\|_\infty \le c\,\delta^{1-\nu}\|f\|_{\mathcal{C}^m(K)} \quad \text{as } \delta \to 0\,. \tag{9.14}$$

Proof. The proof is done as for Theorem 9.5 only by changing the norms accordingly. □

9.3 An application to X-ray diffractometry

We describe briefly the application of the concepts developed in the previous section to the problem of X-ray diffractometry as it was introduced in Section 1.1.

As we have seen, X-ray diffractometry means inverting the Laplace transform

$$Lf(\tau) = \check{f}(\tau) = \int_0^\infty f(z)\mathrm{e}^{-\tau z}\,\mathrm{d}z$$

for a finite number of penetration depths $\tau = \tau_j$, $j = 1,\dots,m$. We assume the stresses σ_{ij} to be at least continuous and close to surface, and hence consider L as a mapping between the Banach spaces $\mathcal{C}([\omega_1,\omega_2])$ and $\mathcal{C}([\tau_{\min},\tau_{\max}])$. Because $\exp(-\tau z) \in \mathcal{C}([\omega_1,\omega_2] \times [\tau_{\min},\tau_{\max}])$ we have that

$$L : \mathcal{C}([\omega_1,\omega_2]) \to \mathcal{C}([\tau_{\min},\tau_{\max}])$$

is linear and bounded and that we thus are in the situation of Section 9.2. Moreover we may assume that $\omega_1 > 0$ since the translation property of the Laplace transform

$$Lf(\tau) = \mathrm{e}^{r\tau} L\{f(\cdot - r)\}(\tau)$$

allows us to shift the support of f to any closed interval $[\omega_1 + r, \omega_2 + r]$, $r > 0$.

In [219] the author defines

$$\bar{e}(z) := \sum_{j=1}^{m} w_j v_j \mathrm{e}^{-\tau_j z} \tag{9.15}$$

as a mollifier, where

$$w_j = \begin{cases} h_1/2, & j = 1 \\ (h_{j-1} + h_j)/2, & 1 < j < m \\ h_{m-1}/2, & j = m \end{cases}$$

with $h_j = \tau_{j+1} - \tau_j$ and $v = (v_1,\dots,v_m)^t \in \mathbb{R}^m$ is chosen such that

$$\int_{\omega_1}^{\omega_2} \bar{e}(z)\,\mathrm{d}z = 1\,.$$

One possibility to define v_j is

$$v_j = \frac{1}{m}\left(w_j \int_{\omega_1}^{\omega_2} \mathrm{e}^{-\tau_j z}\,\mathrm{d}z\right)^{-1}.$$

The weights w_j are defined such that $\sum_{j=1}^{m} w_j v_j \exp(-\tau_j z)$ is the trapezoidal sum corresponding to the nodes $\{\tau_j\}$ applied to the integral

$$L^* v(z) = \int_{\tau_{\min}}^{\tau_{\max}} v(\tau)\mathrm{e}^{-z\tau}\,\mathrm{d}\tau$$

which is the adjoint of the Laplace transform applied to a continuous function $v \in \mathcal{C}([\tau_{\min}, \tau_{\max}])$ and $v_j = v(\tau_j)$. Via

$$\bar{\lambda}_{\bar{e}}(E) = \int_E \bar{e}(z)\,\mathrm{d}z\,, \quad E \in \mathcal{B}([\omega_1, \omega_2])$$

the mollifier $\bar{e}$ induces a regular and countable additive measure on the Borel sets of $[\omega_1, \omega_2]$.

In the following we pursue a semi-discrete setting for the solution of $Lf = g$ that takes into account that we measure the data Lf only at the scanning points $\tau = \tau_j$ although we want to recover f in the infinitely dimensional Banach space $\mathcal{C}([\omega_1, \omega_2])$. The connection between the discrete data $Lf(\tau_j)$, $j = 1, \dots, m$ is then established by appropriate interpolation operators mapping $\mathbb{R}^m$ to the Banach space of continuous functions.

We deduce that $\bar{e}$ has (at least) order 1 and by Theorem 9.7 generates a mollifier $\{e_\gamma : [\omega_1, \omega_2] \to \mathrm{rca}([\omega_1, \omega_2])\}$ for $\mathcal{C}([\omega_1, \omega_2])$. The associated reconstruction kernels $\{v_\gamma : [\omega_1, \omega_2] \to \mathrm{rca}([\tau_{\min}, \tau_{\max}])\}$ are computed in [219] with the help of a collocation method postulating that

$$L^*[v_\gamma(y)](z) = [e_\gamma(y)](z)$$

is satisfied only for finitely many points $z = z_k$, $k = 1, \dots, m$. Together with a numerical integration this leads to the system of linear equations

$$\sum_{j=1}^{m} w_j [\tilde{v}_\gamma(y)]_j \mathrm{e}^{-z_k \tau_j} = [e_\gamma(y)](z_k)\,, \quad 1 \le k, j \le m\,. \tag{9.16}$$

Article [219] presents a condition under which system (9.16) is solvable. Having the solution $\tilde{v}_\gamma(y) \in \mathbb{R}^m$ at hand we set

$$v_\gamma(y) := I_m \tilde{v}_\gamma(y) \in \mathcal{C}(\tau_{\min}, \tau_{\max}) \tag{9.17}$$

which denotes the piecewise linear interpolating function and hence is continuous. The following lemma is an immediate consequence of Lemma 4.1 from [219].

Lemma 9.13. *Let the reconstruction kernel $\{v_\gamma\}$ for the Laplace transform be determined by* (9.16) *and* (9.17). *Then there exists a constant $C_m > 0$ depending on m such that*

$$\|v_\gamma(y)\|_\infty \le C_m \gamma^{-1} y^{-1} \quad \textit{for all } y \in [\omega_1, \omega_2]\,. \tag{9.18}$$

Proof. Lemma 4.1 from [219] states that

$$\|\tilde{v}_\gamma(y)\|_2 \le \tilde{C}_m \gamma^{-1} y^{-1}$$

for a certain constant $\tilde{C}_m > 0$. Since the norms on $\mathbb{R}^m$ are equivalent this implies

$$\|\tilde{v}_\gamma(y)\|_\infty \le C_m \gamma^{-1} y^{-1}$$

with another constant $C_m > 0$. Together with

$$\|v_\gamma(y)\|_\infty = \|I_m \tilde{v}_\gamma(y)\|_\infty \le \|\tilde{v}_\gamma(y)\|_\infty$$

this proves (9.18). □

Of course as a continuous function $v_\gamma(y)$ induces a measure in $\mathrm{rca}([\tau_{\min}, \tau_{\max}])$ whose total variation can be estimated by the sup-norm.

Lemma 9.14. *We have that*

$$\|v_\gamma(y)\|_{\mathcal{B}} \le (\tau_{\max} - \tau_{\min}) C_m \gamma^{-1} y^{-1} \quad \textit{for all } y \in [\omega_1, \omega_2]. \tag{9.19}$$

Proof. Estimate (9.19) follows immediately from (9.18) and

$$\begin{aligned} \|v_\gamma(y)\|_{\mathcal{B}} &= \sup_{\|f\|_\infty \le 1} \left| \int_{\tau_{\min}}^{\tau_{\max}} f(x)[v_\gamma(y)](x)\,\mathrm{d}x \right| \\ &\le (\tau_{\max} - \tau_{\min}) \|v_\gamma(y)\|_\infty \end{aligned}$$ □

Finally we are able to estimate l_γ, the sup-norm of $\|v_\gamma(y)\|_\infty$.

Lemma 9.15. *From* (9.19) *we deduce*

$$l_\gamma \le c_m \gamma^{-1} \quad \textit{for all } \gamma > 0, \tag{9.20}$$

that is $l_\gamma = \mathcal{O}(\gamma^{-1})$.

Proof. The proof is a simple consequence of (9.19) and $y^{-1} \le \omega_1^{-1}$. We have that $c_m = (\tau_{\max} - \tau_{\min}) C_m$. □

We are now able to prove convergence rates for the method of approximate inverse applied to the Laplace transform. To this end we denote the approximate inverse of Lf by

$$f_\gamma^\delta(y) := \langle v_\gamma(y), g^\delta \rangle_{\mathrm{rca}([\tau_{\min},\tau_{\max}]) \times \mathcal{C}([\tau_{\min},\tau_{\max}])},$$

where $g^\delta \in \mathcal{C}([\tau_{\min}, \tau_{\max}])$ represents noise-contaminated measure data. Indeed applying piecewise linear interpolation to the noise-contaminated, discrete data g_m^δ results in a continuous function g^δ. We note that in this case the mapping $y \to v_\gamma(y)$ is continuous what follows from (9.16) and the continuity of $y \mapsto e_\gamma(y)$ and I_m. From Lemma 9.9 we deduce then that $f_\gamma^\delta \in \mathcal{C}([\omega_1, \omega_2])$ and the approximate inverse is a bounded operator, since

$$\sup_{y \in [\omega_1, \omega_2]} \|v_\gamma(y)\|_{\mathcal{B}} < \infty.$$

Theorem 9.16. *Let* $g = Lf$ *for* $f \in \mathcal{C}([\omega_1, \omega_2])$ *the exact measure data, which are given only at* $m \in \mathbb{N}$ *points*

$$g(\tau_j) = (g_m)_j\,, \quad j = 1, \ldots, m$$

and assume that only noise-contaminated data $g_m^\delta \in \mathbb{R}^m$ *are available satisfying* $\|g - I_m g_m^\delta\|_\infty < \delta$, *where* $I_m : \mathbb{R}^m \to \mathcal{C}([\tau_{\min}, \tau_{\max}])$ *denotes the piecewise linear interpolation operator. Furthermore assume that the mollifier* $\bar{e}$ *is given by* (9.15) *and that the associated reconstruction kernel is calculated via* (9.16), (9.17).

If the parameter choice rule $\gamma = \gamma(\delta)$ *is chosen such that*

$$\gamma(\delta) \sim \delta^{1/2} \quad \text{as } \delta \to 0\,,$$

then there exists a $c > 0$ *with*

$$\|f_{\gamma(\delta)}^\delta - f\|_\infty \le c\delta^{\frac{1}{2}} \|f\|_{\mathcal{C}^1([\omega_1,\omega_2])} \quad \text{as } \delta \to 0 \tag{9.21}$$

provided that $f \in \mathcal{C}^1(([\omega_1, \omega_2]))$.

Proof. We apply Theorem 9.12 to prove the convergence statements. From (9.20) we see that $k = 1$ in Theorem 9.12. For given ν with $0 < \nu < 1$ we have that due to (9.13) and $m = 1$

$$1 \ge \frac{1 - \nu}{\nu}\,.$$

Hence, the highest possible convergence order is achieved when $1 = (1 - \nu)/\nu \Leftrightarrow \nu = 1/2$. But for $\nu = 1/2$ we immediately get (9.21) from (9.14). □

Remark 9.17. Provided that in (9.15) we could $v = (v_1, \ldots, v_m)^t$ define such that $\bar{e}$ has the largest possible order m, then following the lines in the proof of Theorem 9.16 we would get a convergence rate $\mathcal{O}(\delta^{\frac{m}{m+1}})$. In that case the number of scanning points would determine the optimal convergence rate in X-ray diffractometry, but also the smoothness requirements to the exact solution f. That fits to the results of Plato and Vainikko [184], where they investigated the discretization of some regularization methods applied to general operator equations and proved that the optimal convergence rates are affected by the discetization step size.

Chapter 10

A glimpse of semi-discrete operator equations

This chapter addresses the situation where only a finite number of measurement data is available. This case is of large practical relevance since in any real world application only a finite set of data is acquired. For simplicity we assume A to be injective throughout this chapter. Further let X, Y be arbitrary Banach spaces if not indicated otherwise. With $\|\cdot\|_2$, $\langle\cdot,\cdot\rangle_2$ we denote the Euclidean norm and scalar product in $\mathbb{R}^n$.

The idea is to model the data acquisition by the so-called *observation operator* $\Psi_n : Y \to \mathbb{R}^n$. The map Ψ_n is assumed to be generated by n linear and continuous functionals $\psi_{n,k} \in Y^*$, that is

$$(\Psi_n v)_k = \langle \psi_{n,k}, v\rangle_{Y^*\times Y} = \psi_{n,k}(v)\,, \quad v \in Y\,, \quad k = 1,\dots,n$$

and thus is linear and continuous, too. If e.g. Y is a function space such functionals $\Psi_{n,k}$ are often given as evaluations at the data scanning points or as moments of order k. Concrete examples for $\psi_{n,k}$ are e.g. given in [201, 202]. Hence we have to investigate the semi-discrete equation

$$A_n f_n = g_n, \quad f_n \in X \tag{10.1}$$

with $A_n = \Psi_n A$, $g_n = \Psi_n g$ rather than $Af = g$. Equation (10.1) in general is not solvable for arbitrary $g_n \in \mathbb{R}^n$ and thus we rather consider the equation

$$A_n f_n = P_{\mathcal{R}(A_n)} g_n \tag{10.2}$$

which has a solution but is highly underdetermined. Let X be uniformly convex. Then we can search for the minimum norm solution $f_n^\dagger$ of (10.2) which exists and is unique due to Lemma 3.3. Note that equation (10.2) is equivalent to

$$A_n^* A_n f_n = A_n^* g_n$$

and solvable for any $g_n \in \mathbb{R}^n$ because of $\dim(\mathcal{R}(A_n)) < \infty$ and any solution of it minimizes the defect $\|A_n f_n - g_n\|_2$. Our aim is to extend the concept of approximate inverse to the given situation following the lines in Rieder, Schuster [201, 202] and thus to present a concept to approximate $f_n^\dagger$ in a stable way (see Theorem 10.12).

The key idea is to compute moments

$$\langle f_n^\dagger, e_i\rangle_{X\times X^*} = e_i(f_n^\dagger)\,, \quad i = 1,\dots,d \tag{10.3}$$

of $f_n^\dagger$ with mollifiers $e_i \in X^*$, $i = 1, \dots, d$, and then approximate $f_n^\dagger$ by

$$E_d f_n^\dagger := \sum_{i=1}^{d} \langle f_n^\dagger, e_i \rangle_{X \times X^*} \, b_i \, .$$

Here, $\{b_i\}_{i=1}^d \subset X$ is a family of elements in X which are associated with the mollifier $\{e_i\}_{i=1}^d$ and form a system in X which allows for an estimate as

$$\left\| \sum_{i=1}^{d} \alpha_i \, b_i \right\|_X^2 \leq \sigma(d) \sum_{i=1}^{d} |\alpha_i|^2 \, , \qquad \alpha \in \mathbb{R}^d \tag{10.4}$$

for a positive function $\sigma : \mathbb{N} \to \mathbb{R}_+$.

Example 10.1. For $X = (\mathcal{C}([0,1]), \| \cdot \|_\infty)$ a family $\{b_i\}_{i=0}^d$ is given by linear B-splines. Let

$$b(x) := \begin{cases} 1 - |x| \, , & |x| \leq 1 \\ 0 \, , & |x| > 1 \end{cases} .$$

Then we define $b_i(x) := b(dx - i)$ for $i = 1, \dots, d-1$ and

$$b_0(x) := \chi_{[0,1/d]}(x) b(dx) \, , \quad b_d(x) := \chi_{[1-1/d,1]}(x) b(dx - d) \, .$$

Obviously $b_i \in X$, $i = 0, \dots, d$, and we have for $\alpha \in \mathbb{R}^{d+1}$

$$\begin{aligned} \left\| \sum_{i=0}^{d} \alpha_i \, b_i \right\|_\infty^2 &\leq \sup_{x \in [0,1]} \left(\sum_{i=0}^{d} |\alpha_i| |b_i(x)| \right)^2 \leq \left(\sum_{i=0}^{d} |\alpha_i| \right)^2 \\ &\leq (d+1) \sum_{i=0}^{d} |\alpha_i|^2 \, . \end{aligned}$$

Thus (10.4) holds true with $\sigma(d) = d + 1$.

By now it is not clear what we understand by mollifiers e_i in a general Banach space X. The sequence $\{e_i\}_{i=1}^d$ and thus the associated sequence $\{b_i\}_{i=1}^d$ have to be chosen such that E_d satisfies the *mollifier property*

$$\lim_{d \to \infty} \| E_d w - w \|_X = 0 \, , \qquad w \in X \, , \tag{10.5}$$

which guarantees that $E_d w$ in fact approximates w for any $w \in X$. Thus, once we calculated $\langle f_n^\dagger, e_i \rangle_{X \times X^*}$ we can evaluate the approximation $E_d f_n^\dagger$. But $f_n^\dagger$ is not known and thus the moments (10.3) neither. Hence, we go one step farther and search for solutions of the dual equations

$$A_n^* v_i^n = e_i \, , \qquad i = 1, \dots, d \, , \tag{10.6}$$

where the adjoint operator $A_n^* : \mathbb{R}^n \to X^*$ is given by

$$A_n^* \alpha = \sum_{k=1}^{n} \alpha_k \, A^* \, \psi_{n,k} \,, \quad \alpha \in \mathbb{R}^n \,. \tag{10.7}$$

Assume for the moment that (10.6) has a solution. We deduce

$$\langle f_n^\dagger, e_i \rangle_{X \times X^*} = \langle A_n f_n^\dagger, v_i^n \rangle_2$$

and defining

$$\widetilde{A}_{n,d} : \mathbb{R}^n \to X \,, \quad \widetilde{A}_{n,d} \alpha := \sum_{i=1}^{d} \langle \alpha, v_i^n \rangle_2 \, b_i \tag{10.8}$$

we obtain

$$\lim_{d \to \infty} \widetilde{A}_{n,d} \, A_n f_n^\dagger = \lim_{d \to \infty} E_d \, f_n^\dagger = f_n^\dagger \,.$$

This motivates to call $\widetilde{A}_{n,d}$ the *(semi-discrete) approximate inverse* of A_n, a solution v_i^n of (10.6) is again called reconstruction kernel.

Remark 10.2. If X is a Hilbert space and thus uniformly convex, the minimum norm solution $f_n^\dagger$ of equation (10.2) exists and we have the interesting connection

$$\langle f_n^\dagger, e_i \rangle_{X \times X^*} = \langle g_n, v_i^n \rangle_2, \quad i = 1, \ldots, d,$$

if only $g_n \in \mathcal{R}(A_n)$ or $v_i^n \in \mathcal{N}(A_n^*)^\perp$. And this identity holds true even in the case where v_i^n only solves the normal equation $A_n A_n^* v_i^n = A_n e_i$, see [202, Lemma 2.1]. In Banach spaces this identity is valid only if $A_n^* v_i^n = e_i$ is solvable which can not be expected.

The image $\mathcal{R}(A_n^*)$ consists of the span of $\{A^* \psi_{n,k} : k = 1, \ldots, n\}$ and we can not expect e_i to be an element of it. We outline two different ways to calculate reconstruction kernels v_i^n. The first one is an iterative method where the iterates converge to a minimizer of $\| A_n^* v_i^n - e_i \|_{X^*}$; the second one uses an approximate solution of $A^* v_i = e_i$ to construct v_i^n. The latter one is the strategy that was also pursued in [201, 202] and for which we give criteria to obtain convergence and stability with respect to noisy data g_n^δ.

Let, for the moment, X be uniformly convex and smooth and thus reflexive. With $J^X := J_2^X$, $J^{X^*} := J_2^{X^*}$ we denote the single-valued (normalized) duality mappings on X, X^*, respectively. Any minimizer of $\| A_n^* v_i^n - e_i \|_{X^*}$ is then characterized by the optimality condition

$$A_n J^{X^*} (A_n^* v_i^n - e_i) = \partial \{ \| A_n^* v_i^n - e_i \|_{X^*}^2 / 2 \} = 0 \tag{10.9}$$

and this equation has a solution since $\mathcal{R}(A_n^*)$ is closed and hence

$$\mathcal{R}(A_n^*) + J^X (\mathcal{N}(A_n)) = X^* \,. \tag{10.10}$$

One possibility to approximately compute v_i^n is to adopt the Landweber method from Chapter 6.1 to this situation. We consider the following iteration scheme, where we drop the indices of v_i^n and e_i for a moment.

Algorithm 10.3.

(1) $v_0 = 0$,

(2) For $k = 0, 1, \dots$ iterate

$$v_{k+1} = v_k - \mu_k A_n J^{X^*}(A_n^* v_k - e) \tag{10.11}$$

with appropriately chosen μ_k.

Using the results from Chapter 6.1 we indeed can show convergence.

Theorem 10.4. *Let X be uniformly convex and smooth and $e \in X^*$. Then there is a choice of μ_k such that the iterates $\{v_k\}_k \subset \mathbb{R}^n$ from Algorithm* 10.3 *converge strongly to the unique minimizer $v \in \mathbb{R}^n$ of $\|A_n^* v - e\|_{X^*}$ having minimum $\|\cdot\|_2$-norm.*

Proof. Applying A_n^* to the iteration (10.11) and subtracting e yield

$$A_n^* v_{k+1} - e = A_n^* v_k - e - \mu_k A_n^* A_n J^{X^*}(A_n^* v_k - e)$$

which corresponds to the Landweber iteration

$$x_{k+1}^* = x_k^* - \mu_k A_n^* A_n x_k \tag{10.12a}$$

$$x_{k+1} = J^{X^*}(x_{k+1}^*), \quad k = 0, 1, \dots \tag{10.12b}$$

with the settings $x_0^* := -e$, $x_0 = -J^{X^*}(e)$, $x_k^* := A_n^* v_k - e$, $x_k := J^{X^*}(A_n^* v_k - e)$. Proposition 6.14 says that the step sizes μ_k can be chosen such that x_k tends strongly to $-\Pi_{\mathcal{N}(A)}^2 J^{X^*}(e)$ as $k \to \infty$. Note that here $Y = \mathbb{R}^n$ is finite-dimensional and that the Landweber method is included in the general framework of Algorithm 6.13. Since X is uniformly convex and smooth, the duality mapping J^X is norm-to-weak continuous, see Theorem 2.53, leading to

$$x_k^* = J^X(x_k) \rightharpoonup -J^X \Pi_{\mathcal{N}(A)} J^{X^*}(e) =: x^* \quad \text{as } k \to \infty$$

with respect to the weak topology. Hence $x_k^* + e \rightharpoonup x^* + e$ weakly, too. But $x_k^* + e \in \mathcal{R}(A_n^*)$ and since $\dim(\mathcal{R}(A_n^*)) < \infty$ this yields strong convergence of $x_k^* \to x^*$ as $k \to \infty$. Because of $x_k^* = A_n^* v_k - e$ we finally obtain

$$A_n^* v_k \to e - J^X \Pi_{\mathcal{N}(A)}^2 J^{X^*}(e) = P_{\mathcal{R}(A_n^*)}(e) \quad \text{as } k \to \infty. \tag{10.13}$$

Here we used the fact that

$$e = P_{\mathcal{R}(A_n^*)}(e) + J^X \Pi_{\mathcal{N}(A)}^2 J^{X^*}(e)$$

which is a consequence of (6.50) and (10.10). Furthermore by (10.11) all iterates v_k are in $\mathcal{R}(A_n) = \left(\mathcal{N}(A_n^*)\right)^\perp$. Since the restriction of A_n^* to $A_n^* : \left(\mathcal{N}(A_n^*)\right)^\perp \to \mathcal{R}(A_n^*)$ is bijective between finite-dimensional spaces, we have convergence of the sequence $\{v_k\}$ to some $v \in \left(\mathcal{N}(A_n^*)\right)^\perp$. But from (10.13) it is clear that

$$A_n^* v = P_{\mathcal{R}(A_n^*)}(e) .$$

Hence v minimizes $\|A_n^* v - e\|_{X^*}$. Finally we show that v has minimal $\|\cdot\|_2$-norm among all minimizers. Let $z \in \mathbb{R}^n$ be an arbitrary minimizer of $\|A_n^* z - e\|_{X^*}$. Since the metric projection $P_{\mathcal{R}(A_n^*)}(e)$ is unique, we must have $A_n^* z = A_n^* v$ and thus $z = v + u$ with some $u \in \mathcal{N}(A_n^*)$. Hence we have $\langle v, u\rangle_2 = 0$ and get

$$\|z\|_2^2 = \|v\|_2^2 + 2\,\langle v, u\rangle_2 + \|u\|_2^2 = \|v\|_2^2 + \|u\|_2^2 \geq \|v\|_2^2 ,$$

where the last inequality is strict for $u \neq 0$. □

Remark 10.5.

(a) The proof of Theorem 10.4 shows that iteration (10.11) approximates a reconstruction kernel v that minimizes $\|A_n^* v - e\|_{X^*}$ and that because of $\lim_{k\to\infty} A_n^* v_k = P_{\mathcal{R}(A_n^*)}(e)$ we easily get an approximation of the metric projection $P_{\mathcal{R}(A_n^*)}(e)$ by $A_n^* v_k$. Note that this proof strongly relies on the facts that X is uniformly convex and smooth and thus reflexive and that the range of A_n^* is of finite dimension.

(b) The iteration can be made more efficient by using the sequential subspace methods outlined in Section 6.2. Acceleration was there achieved by using more search directions $A_n^* w_l$ than just $A_n^* A_n x_k$ in iteration (10.12). The assertion of Theorem 10.4 still holds if one assures that $w_l \in \mathcal{R}(A_n)$, which is always fulfilled for the canonical search directions suggested in Section 6.2.

Although the calculation of the reconstruction kernels v_i^n can be done by Algorithm 10.3, this method has some drawbacks. Besides the conditions on X and A_n that have to be required in Theorem 10.4 to get convergence, the iteration (10.11) (numerically) converges very slowly and an approximate reconstruction kernel might cause heavy artifacts. Furthermore A_n^* in general does not fulfill invariance properties as in Lemma 10.9 and thus we had to perform the iteration for each mollifier e_i which is very time consuming.

To cure this dilemma we seek a replacement for the kernel v_i^n that relies on a (maybe even exact) kernel v_i for A. For the remainder of this chapter let X and Y be arbitrary Banach spaces. We recall that A is injective and thus $\mathcal{R}(A^*)$ is weak*-dense in X^*. This implies that for given numbers $\varepsilon_i > 0$ and mollifiers $e_i \in X$ we find some elements $v_i \in Y^*$ such that

$$|\langle A^* v_i - e_i, f\rangle_{X^*\times X}| = |(A^* v_i - e_i)(f)| < \varepsilon_i \,\|f\|_X , \quad i = 1, \ldots, d . \quad (10.14)$$

Here, $f \in X$ denotes the (unique) solution of $Af = g$. This solution exists since A is injective and we assumed that $g \in \mathcal{R}(A)$. Note, that (10.14) does not mean that $\|A^* v_i - e_i\|_{X^*} < \varepsilon_i$ since each v_i in (10.14) depends on f and we have no uniform boundedness.

Remark 10.6. Note that in case that X is reflexive $\mathcal{R}(A^*)$ is dense even with respect to the strong (norm-) topology of X and that there exist v_i satisfying

$$\|A^* v_i - e_i\|_{X^*} < \varepsilon_i$$

which is a stronger condition than (10.14).

Having elements v_i satisfying (10.14) at our disposal we define

$$v_i^n = G_n \, \Psi'_n \, v_i \, , \quad i = 1, \ldots, d \, , \tag{10.15}$$

where $\Psi'_n = (\psi'_{n,1}, \psi'_{n,2}, \ldots, \psi'_{n,n})^\top : Y^* \to \mathbb{R}^n$, $\psi'_{n,k} \in Y^{**}$, $k = 1, \ldots, n$, is linear and continuous and $G_n \in \mathbb{R}^{n \times n}$ is a matrix to be specified in the following. Note, that because of $Y \subset Y^{**}$, the $\psi'_{n,k}$ may even be elements of Y. Both, Ψ'_n as well as G_n will be defined such that we gain the convergence

$$\lim_{\substack{n\to\infty \\ d\to\infty}} \|\widetilde{A}_{n,d} \, A_n f - f\|_X = 0.$$

First we relate with Ψ_n a family $\{\varphi_k\}_{k=1}^n \subset Y$ and the operator $\mathcal{P}_n : Y \to Y$ given by

$$\mathcal{P}_n y := \sum_{k=1}^n \langle \psi_{n,k}, y \rangle_{Y^* \times Y} \, \varphi_k \, , \quad y \in Y \, .$$

The map $\mathcal{P}_n$ is supposed to fulfill the boundedness condition

$$\|\mathcal{P}_n y\|_Y \le C_b \, \|y\|_Y \quad \text{for } n \to \infty \, , \quad y \in Y \tag{10.16}$$

and the approximation condition

$$\|\mathcal{P}_n y - y\|_Y \le \rho_n \, \|y\|_Y \quad \text{for } n \to \infty \, , \quad y \in Y \, , \tag{10.17}$$

where $C_b > 0$ is a constant and ρ_n is a non-negative sequence converging to zero. Note that the conditions (10.16) and (10.17) implicitly are conditions on the space Y, too. In the same way we relate with Ψ'_n a family $\{\varphi'_k\} \subset Y^*$ and the operator $\mathcal{P}'_n : Y^* \to Y^*$ by

$$\mathcal{P}'_n y^* := \sum_{k=1}^n \langle \psi'_{n,k}, y^* \rangle_{Y^{**} \times Y^*} \, \varphi'_k \, , \quad y^* \in Y^* \, .$$

And analogously we postulate the existence of a constant $C_b' > 0$ and a sequence $\rho_n' \geq 0$ tending to zero such that

$$\|\mathcal{P}_n' y^*\|_{Y^*} \leq C_b' \, \|y^*\|_{Y^*} \quad \text{for } n \to \infty\,, \quad y^* \in Y^* \tag{10.18}$$

and

$$\|\mathcal{P}_n' y^* - y^*\|_{Y^*} \leq \rho_n' \, \|y^*\|_{Y^*} \quad \text{for } n \to \infty\,, \quad y^* \in Y^* \tag{10.19}$$

are fulfilled. Finally the matrix G_n is to be defined as

$$(G_n)_{j,k} := \langle \varphi_j', \varphi_k \rangle_{Y^* \times Y}\,, \quad j,k = 1,\ldots,n\,.$$

This yields the important relation

$$\langle \Psi_n w, G_n \, \Psi_n' v \rangle_2 = \langle \mathcal{P}_n w, \mathcal{P}_n' v \rangle_{Y \times Y^*}\,, \quad w \in Y\,, \quad v \in Y^*\,. \tag{10.20}$$

We have nothing said about Ψ_n' by now. The functionals $\psi_{n,k}'$ are to be chosen such that conditions (10.18) and (10.19) are fulfilled. We will present a particular choice in Remark 10.10.

We have now all ingredients together to formulate an estimate of the approximation error which comes from applying the approximate inverse $\widetilde{A}_{n,d}$.

Theorem 10.7 (Noise-free case). *Let A, E_d, Ψ_n, $\mathcal{P}_n$ and $\mathcal{P}_n'$ be as stated in this chapter. Further assume that the family $\{b_i\}_{i=1}^d \subset X$ satisfies* (10.4) *and that the triplets $\{(e_i, v_i, b_i)\}_{i=1}^d \subset X^* \times Y^* \times X$ fulfill the conditions* (10.5) *and* (10.14) *for $\varepsilon_i > 0$. Finally the discrete kernels in* (10.8) *are to be defined as in* (10.15).

Then there exists a $C > 0$ validating the estimate

$$\begin{aligned} &\|\widetilde{A}_{n,d}\, A_n f - f\|_X \\ &\leq \|(E_d - I) f\|_X + C \left(\sigma(d)\big((C_b' \rho_n)^2 + (\rho_n')^2\big) \sum_{i=1}^d \|v_i\|_{Y^*}^2 + \varepsilon_i^2 \right)^{1/2} \|f\|_X\,. \end{aligned}$$

Provided that

$$\sigma(d)((C_b' \rho_n)^2 + (\rho_n')^2) \sum_{i=1}^d \|v_i\|_{Y^*}^2 \to 0 \quad \text{as } n, d \to \infty$$

and

$$\sigma(d) \sum_{i=1}^d \varepsilon_i^2 \to 0 \quad \text{as } d \to \infty \tag{10.21}$$

we have the convergence

$$\lim_{\substack{n \to \infty \\ d \to \infty}} \|\widetilde{A}_{n,d}\, A_n f - f\|_X = 0.$$

Proof. An application of the triangle inequality gives

$$\|\widetilde{A}_{n,d}\, A_n f - f\|_X \le \|(E_d - I) f\|_X + \|\widetilde{A}_{n,d}\, A_n f - E_d f\|_X\,.$$

and thus we need only to estimate the second part. Property (10.4) of the system $\{b_i\}_{i=1}^d$ yields

$$\|\widetilde{A}_{n,d}\, A_n f - E_d f\|_X^2 \le \sigma(d) \sum_{i=1}^d \left|\langle A_n f, G_n \Psi_n' v_i\rangle_2 - \langle f, e_i\rangle_{X\times X^*}\right|^2.$$

Using (10.14) and the identity (10.20) we may estimate

$$\begin{aligned}
&|\langle A_n f, G_n \Psi_n' v_i\rangle_2 - \langle f, e_i\rangle_{X\times X^*}|\\
&\le |\langle f, A^* v_i - e_i\rangle_{X\times X^*}|\\
&\quad + |\langle \mathcal{P}_n Af, \mathcal{P}_n' v_i\rangle_{Y\times Y^*} - \langle Af, v_i\rangle_{Y\times Y^*}|\\
&\le \|\mathcal{P}_n Af - Af\|_Y \, \|\mathcal{P}_n' v_i\|_{Y^*} + \|\mathcal{P}_n' v_i - v_i\|_{Y^*} \, \|Af\|_Y + \varepsilon_i \, \|f\|_X\,.
\end{aligned}$$

We further apply the conditions (10.17), (10.18) and (10.19) and finally obtain

$$\begin{aligned}
&|\langle A_n f, G_n \Psi_n' v_i\rangle_2 - \langle f, e_i\rangle_{X\times X^*}|\\
&\le (\rho_n\, C_b' \,\|A\|_{X\to Y}\, \|v_i\|_{Y^*} + \rho_n' \,\|A\|_{X\to Y}\, \|v_i\|_{Y^*} + \varepsilon_i)\, \|f\|_X
\end{aligned}$$

completing the proof. □

Remark 10.8.

(a) The condition

$$\sigma(d)\big((C_b'\rho_n)^2 + (\rho_n')^2\big) \sum_{i=1}^d \|v_i\|_{Y^*}^2 \to 0 \quad \text{as } d, n \to \infty$$

implies a coupling of the regularization parameter d and the number of data n; a fact that is typical for an intertwining of regularization and discretization, compare e.g. with Plato and Vainikko [184]. We further remark that $\sigma(d)$ actually might be increasing, cf. Example 10.1. In the situation of Example 10.1 we can choose $\varepsilon_i = \exp(-d)$ for $i = 0, \dots, d$ to get

$$\sigma(d) \sum_{i=0}^d \varepsilon_i^2 = (d+1)^2 \exp(-2d) \to 0 \quad \text{as } d \to \infty$$

and (10.21) is fulfilled.

(b) The crux is to find a $v_i \in Y^*$ satisfying (10.14) since this condition depends on the solution f which is not known. If there exists a linear mapping $B : Y \to X$ such that $I_X = BA$, then obviously $v_i = B^* e_i$ solves $A^* v_i = e_i$. Note that this does in general not guarantee that $v_i \in Y^*$. This depends on the specific setting for A, B, X and Y. An alternative would be an operator $B : Y \to X$ with the property that $\Lambda = BA$ is a pseudodifferential operator on X and X a Banach space of functions. Then $v_i = B^* e_i$ is a reconstruction kernel with $A^* v_i = \Lambda^* e_i$ which can be used to compute the moments $\langle \Lambda f, e_i \rangle_{X \times X^*} = \langle y, v_i \rangle_{Y \times Y^*}$. This technique e.g. is applied in local tomography [200], sonar [188] or feature reconstruction [154]. Of course then e_i has to be substituted by $\Lambda^* e_i$ in (10.14). If this is not possible at least any a priori information such as e.g. $f \in B_r(f_*) = \{x \in X : \|x - f_*\|_X < r\}$ might be helpful.

Another problem is that in general we have to compute d kernels v_i what might be time consuming. But there is a remedy. Provided that A^* obeys a certain invariance property, then it is possible to solve (10.14) only once. Lemma 10.9 can be seen as a generalization of Lemma 2.3 in [202] to Banach spaces.

Lemma 10.9. *Assume that operators $T_i \in \mathcal{L}(X^*)$, $S_i \in \mathcal{L}(Y^*)$ are given with $T_i A^* = A^* S_i$, $i = 1, \dots, d$. Define $e_i = T_i e$, $i = 1, \dots, d$ for $e \in X^*$. Provided that*

$$|\langle A^* v - e, T_i^* f \rangle_{X^* \times X}| \le \varepsilon \, \|T_i^* f\|_X \,, \quad i = 1, \dots, d \tag{10.22}$$

for $v \in Y^*$, then

$$|\langle A^* v_i - e_i, f \rangle_{X^* \times X}| \le \varepsilon_i \, \|f\|_X \,, \quad i = 1, \dots, d \,, \tag{10.23}$$

where $v_i = S_i v$ and $\varepsilon_i = \varepsilon \, \|T_i\|_{X^* \to X^*}$.

Proof. First we remark, that the existence of a $v \in Y^*$ fulfilling (10.22) is guaranteed since the set $\{T_i^* f\}_{i=1}^d$ is finite and thus generates a neighborhood of zero in X^* with respect to the weak*-topology.

A simple calculation shows

$$|\langle A^* S_i v - T_i e, f \rangle_{X^* \times X}| = |\langle (A^* v - e), T_i^* f \rangle_{X^* \times X}| \le \varepsilon \, \|T_i^* f\|_X$$

which implies assertion (10.23) because of $\|T_i^*\|_{X \to X} = \|T_i\|_{X^* \to X^*}$, see, e.g., Rudin [205, Theorem 4.10]. □

Remark 10.10. A particular choice of $\mathcal{P}_n'$ is the adjoint $\mathcal{P}_n' = \mathcal{P}_n^* : Y^* \to Y^*$ of $\mathcal{P}_n$ which is given by

$$\mathcal{P}_n^* y^* = \sum_{k=1}^{n} \langle y^*, \varphi_k \rangle_{Y^* \times Y} \, \psi_{n,k} \,, \quad y^* \in Y^* \,.$$

We have then $\psi'_{n,k}(y^*) = \langle y^*, \varphi_k \rangle_{Y^* \times Y}$ and

$$\|\mathcal{P}_n^* - I\|_{Y^* \to Y^*} = \|\mathcal{P}_n - I\|_{Y \to Y} \le \rho_n$$

and hence $\rho'_n = \rho_n$. The matrix G_n has then the entries $(G_n)_{j,k} = \psi_{n,j}(\varphi_k)$.

Although this is the canonical way to generate Ψ'_n and $\mathcal{P}'_n$ another choice might be more convenient since the approximation power of $\mathcal{P}'_n$ has influence to the error estimate in Theorem 10.7.

Next we investigate the regularization property of the approximate inverse $\widetilde{A}_{n,d}$, that means the stability of the approximate solution with respect to noise in the given data. As in [201, 202] we interprete noise contaminated data as a perturbation of our observation operator Ψ_n. More explicitly, we define

$$(\Psi_n^\delta y)_k = (\Psi_n y)_k + \delta_k \, \|y\|_Y \,, \quad |\delta_k| \le \delta \,, \quad y \in Y \tag{10.24}$$

for a positive number δ representing the noise level. We can show that an appropriate coupling of the parameters n and d to the noise level δ gives convergence when δ goes to zero.

Theorem 10.11 (Regularization property). *Adopt the hypotheses of Theorem* 10.7. *Assume further that the triplets* $\{(e_i, v_i, b_i)\}_{i=1}^d \subset X^* \times Y^* \times X$ *imply the convergences*

$$\lim_{n\to\infty} \rho_n^2 \, \sigma(d_n) \sum_{i=1}^{d_n} \|v_i\|_{Y^*}^2 = 0 = \lim_{n\to\infty} (\rho'_n)^2 \, \sigma(d_n) \sum_{i=1}^{d_n} \|v_i\|_{Y^*}^2$$

and

$$\lim_{n\to\infty} \sigma(d_n) \sum_{i=1}^{d_n} \varepsilon_i^2 = 0$$

with $d = d_n$ *such that* $d_n \to \infty$ *whenever* $n \to \infty$.

If we couple $n = n_\delta$ *with the noise level* δ *such that* $n_\delta \to \infty$ *when* $\delta \to 0$, $\delta/\rho_{n_\delta} = \mathcal{O}(1) = \delta/\rho'_{n_\delta}$ *as well as*

$$\sigma(d_{n_\delta}) \, \delta^2 \, n_\delta \|G_{n_\delta} \Psi'_{n_\delta}\|_{Y^* \to \mathbb{R}^n} \sum_{i=1}^{d_{n_\delta}} \|v_i\|_2^2 \to 0 \quad \text{as } \delta \to 0 \,,$$

then

$$\lim_{\delta\to 0} \sup\{\|\widetilde{A}_{n_\delta, d_{n_\delta}} \Psi_{n_\delta}^\delta A f - f\|_X : \Psi_{n_\delta}^\delta \text{ fulfills } (10.24)\} = 0.$$

Proof. We denote by $g_n = A_n f$ the exact data and by $g_n^\delta = \Psi_n^\delta A f$ the noise contaminated data. Using again property (10.4) of the family $\{b_i\}$ we find that

$$\begin{aligned}\|\widetilde{A}_{n,d}(g_n - g_n^\delta)\|_X^2 &\le \sigma(d) \sum_{i=1}^{d} \left| \langle (\Psi_n - \Psi_n^\delta)\, A f, G_n\, \Psi_n' v_i \rangle_2 \right|^2 \\ &\le \sigma(d)\, \delta^2\, n\, \|A f\|_Y \sum_{i=1}^{d} \|G_n\, \Psi_n' v_i\|_2^2 \\ &\le \sigma(d)\, \delta^2\, n\, \|A\|_{X\to Y} \|f\|_X \|G_n \Psi_n'\|_{Y^*\to\mathbb{R}^n}^2 \sum_{i=1}^{d} \|v_i\|_{Y^*}^2\end{aligned}$$

yielding

$$\begin{aligned}\|\widetilde{A}_{n,d}\, g_n^\delta - f\|_X &\le \|E_d f - f\|_X + C\sigma(d) \\ &\times \left((\delta^2\, n\, \|G_n\, \Psi_n'\|_{Y^*\to\mathbb{R}^n}^2 + \rho_n^2 + (\rho_n')^2) \sum_{i=1}^{d} \|v_i\|_{Y^*}^2 + \varepsilon_i^2 \right)^{1/2} \|f\|_X,\end{aligned}$$

where $C > 0$ again denotes a properly chosen constant. Replacing n by n_δ and d by d_{n_δ} leads to the claimed convergence under the assumed coupling conditions. □

At the end of this chapter we come back to our promise that the approximate inverse $\widetilde{A}_{n,d}\, g_n$ approximates the minimum norm solution $f_n^\dagger$ of (10.2). At first we realize that the assertions of Theorem 10.7 and Theorem 10.11 remain valid even if we set

$$v_i^n := P_{\mathcal{R}(A_n)} G_n \Psi_n' v_i\,, \quad i = 1, \dots, d$$

instead of (10.15), since we always evaluate the kernels v_i^n with $A_n f$ which is in $\mathcal{R}(A_n)$. With this setting we are able to prove that the approximate inverse converges to $f_n^\dagger$ as $d \to \infty$ in case of a uniformly convex X.

Theorem 10.12. *Let X be uniformly convex and $f_n^\dagger \in X$ be the minimum norm solution of* (10.2). *Furthermore set*

$$v_i^n := P_{\mathcal{R}(A_n)} G_n \Psi_n' v_i\,, \quad i = 1, \dots, d \tag{10.25}$$

where the $v_i \in Y^*$ fulfill $\|A^* v_i - e_i\|_{X^*} < \varepsilon_i$ for given real numbers $\varepsilon_i > 0$, $i = 1, \dots, d$. Then

$$\lim_{d\to\infty} \|\widetilde{A}_{n,d}\, g_n - f_n^\dagger\|_X = 0 \tag{10.26}$$

for arbitrary data $g_n \in \mathbb{R}^n$ provided that $\sigma(d) \sum_{i=1}^{d} \varepsilon_i^2 \to 0$ as $d \to \infty$.

Proof. Using the triangle inequality yields

$$\|\widetilde{A}_{n,d} g_n - f_n^\dagger\|_X \le \|\widetilde{A}_{n,d} g_n - E_d f_n^\dagger\|_X + \|E_d f_n^\dagger - f_n^\dagger\|_X,$$

where the latter summand tends to zero as $d \to \infty$ due to the mollifier property (10.5). Since

$$g_n = P_{\mathcal{R}(A_n)} g_n + P_{\mathcal{R}(A_n)^\perp} g_n$$

and $A_n f_n^\dagger = P_{\mathcal{R}(A_n)} g_n$ we have

$$\langle v_i^n, g_n\rangle_2 = \langle A_n^* v_i^n, f_n^\dagger\rangle_{X^*\times X} + \langle v_i^n, P_{\mathcal{R}(A_n)^\perp} g_n\rangle_2 = \langle A_n^* v_i^n, f_n^\dagger\rangle_{X^*\times X}\,.$$

The last equation follows by (10.25). Taking this relation into account we may estimate

$$\begin{aligned}\|\widetilde{A}_{n,d} g_n - E_d f_n^\dagger\|_X &= \left\|\sum_{i=1}^{d} \langle A_n^* v_i^n - e_i, f_n^\dagger\rangle_{X^*\times X} b_i\right\|_X \\ &\le \sigma(d) \sum_{i=1}^{d} \|A_n^* v_i^n - e_i\|_{X^*}^2 \|f_n^\dagger\|_X^2 \\ &\le \sigma(d) \|f_n^\dagger\|_X^2 \sum_{i=1}^{d} \varepsilon_i^2\,,\end{aligned}$$

where we made use of property (10.4) and the particular setting (10.25). This finally proves (10.26). □

The orthogonal projection onto $\mathcal{R}(A_n)$ in (10.25) can be omitted if $g_n = A_n f$.

Remark 10.13. As the proofs in this section show things get easier if X is uniformly smooth. But throughout this section Y is arbitrary. The only condition to Y is that it allows for an approximation as (10.17). Condition (10.17) can be generalized to

$$\|\mathcal{P}_n y - y\|_Y \le \rho_n \|y\|_{Y_1} \quad \text{for } n \to \infty, \quad y \in Y_1\,,$$

where $Y_1 \subset Y$ is dense and continousely embedded in Y.

Bibliography

[1] H. Abdullah and A. K. Louis. The approximate inverse for solving an inverse scattering problem for acoustic waves in an inhomogeneous medium. *Inverse Problems*, 15:1213–1230, 1999.

[2] R. A. Adams and J. J. F. Fournier. *Sobolev Spaces*. Pure and applied mathematics. Elsevier, 2003.

[3] I. Y. Alber, A. N. Iusem, and M. V. Solodov. Minimization of nonsmooth convex functionals in Banach spaces. *Journal of Convex Analysis*, 4(2):235–255, 1997.

[4] Y. Alber and D. Butnariu. Convergence of Bregman projection methods for solving consistent convex feasibility problems in reflexive Banach rpaces. *Journal of Optimization Theory and Applications*, 92(1):33–61, 1997.

[5] Y. I. Alber and A. I. Notik. On some estimates for projection operators in Banach spaces. *Comm. Appl. Nonlinear Anal.*, 2(1):47–55, 1995.

[6] A. Ambrosetti and G. Prodi. *A Primer of Nonlinear Analysis*. Cambridge University Press, 1993.

[7] F. Andreu-Vaillo, V. Caselles, and J. M. Mazón. *Parabolic Quasilinear Equations Minimizing Linear Growth Functionals*, volume 223 of *Progress in Mathematics*. Birkhäuser Verlag, Basel, 2004.

[8] S. W. Anzengruber and R. Ramlau. Morozov's discrepancy principle for Tikhonov-type functionals with nonlinear operators. *Inverse Problems*, 26:025001, 17, 2010.

[9] S. W. Anzengruber and R. Ramlau. Convergence rates for Morozov's discrepancy principle using variational inequalities. *Inverse Problems*, 27:105007, 18, 2011.

[10] E. Asplund. Positivity of duality mappings. *Bulletin of the American Mathematical Society*, 73(2):200–203, 1967.

[11] M. Bachmayr and M. Burger. Iterative total variation schemes for nonlinear inverse problems. *Inverse Problems*, 25(10):105004, 26, 2009.

[12] A. Bakushinsky and A. Goncharsky. *Ill-Posed Problems: Theory and Applications*. Mathematics and Its Applications. Kluwer Academic Publishers, Dordrecht, 1994.

[13] A. B. Bakushinsky. Remarks on choosing a regularization parameter using the quasi-optimality and ratio criterion. *USSR Computational Mathematics and Mathematical Physics*, 24(4):181–182, 1984.

[14] A. B. Bakushinsky and M. Y. Kokurin. *Iterative Methods for Approximate Solution of Inverse Problems*. Springer, Dordrecht, 2004.

[15] V. Barbu and T. Precupanu. *Convexity and Optimization in Banach Spaces*, volume 10 of *Mathematics and its Applications (East European Series)*. D. Reidel Publishing Co., Dordrecht, second edition, 1986.

[16] H. H. Bauschke, J. M. Borwein, and P. L. Combettes. Essential smoothness, essential strict convexity, and Legendre functions in Banach spaces. *Communications in Contemporary Mathematics*, 3(4):615–647, 2001.

[17] M. Benning and M. Burger. Error estimates for general fidelities. *Electronic Transactions on Numerical Analysis*, 38:44–68, 2011.

[18] B. Blaschke, H. W. Engl, W. Grever, and M. V. Klibanov. An appication of Tikhonov regularization to phase retrieval. *Nonlinear World*, 3:771–786, 1996.

[19] B. Blaschke-Kaltenbacher and H. W. Engl. Regularization methods for nonlinear ill–posed problems with applications to phase reconstruction. In H. W. Engl, A. K. Louis, and W. Rundell, editors, *Inverse Problems in Medical Imaging and Nondestructive Testing*, pages 1–18, Wien-New York, 1997. Springer.

[20] R. I. Boţ and B. Hofmann. An extension of the variational inequality approach for obtaining convergence rates in regularization of nonlinear ill-posed problems. *Journal of Integral Equations and Applications*, 22(3):369–392, 2010.

[21] T. Bonesky. Morozov's discrepancy principle and Tikhonov-type functionals. *Inverse Problems*, 25(1):015015 (11pp), 2009.

[22] T. Bonesky, K. S. Kazimierski, P. Maaß, F. Schöpfer, and T. Schuster. Minimization of Tikhonov functionals in Banach spaces. *Abstract and Applied Analysis*, 2008:1–20, 2008.

[23] J. M. Borwein and Q. J. Zhu. *Techniques of Variational Analysis*. Springer-Verlag, New York, 2005.

[24] A. Böttcher, B. Hofmann, U. Tautenhahn, and M. Yamamoto. Convergence rates for Tikhonov regularization from different kinds of smoothness conditions. *Appl. Anal.*, 85(5):555–578, 2006.

[25] I. Bouchouev and V. Isakov. The inverse problem of option pricing. *Inverse Problems*, 13(5):L11–L17, 1997.

[26] K. Bredies and D. A. Lorenz. Iterated hard shrinkage for minimization problems with sparsity constraints. *SIAM Journal on Scientific Computing*, 30(2):657–683, 2008.

[27] K. Bredies and D. A. Lorenz. Regularization with non-convex separable constraints. *Inverse Problems*, 25(8):085011, 14, 2009.

[28] L. M. Bregman. The relaxation method for finding common points of convex sets and its application to the solution of problems in convex programming. *USSR Computational Mathematics and Mathematical Physics*, 7(3):200–217, 1967.

[29] J. Brodie, I. Daubechies, C. De Mol, D. Giannone, and I. Loris. Sparse and stable Markowitz portfolios. *PNAS*, 106:12267–12272, 2009.

[30] C. D. Brooks and P. K. Lamm. A generalized approach to local regularization of linear Volterra problems in L^p spaces. *Inverse Problems*, 27(5):055010, 26, 2011.

[31] C. D. Brooks, P. K. Lamm, and X. Luo. Local regularization of nonlinear Volterra equations of Hammerstein type. *J. Integral Equations Appl.*, 22(3):393–425, 2010.

[32] M. Burger and S. Osher. Convergence rates of convex variational regularization. *Inverse Problems*, 20(5):1411–1421, 2004.

[33] M. Burger and S. Osher. *A Guide to the TV Zoo*. University of Münster, Münster, 2012.

[34] M. Burger, E. Resmerita, and L. He. Error estimation for Bregman iterations and inverse scale space methods in image restoration. *Computing*, 81(2-3):109–135, 2007.

[35] D. Butnariu and E. Resmerita. Bregman distances, totally convex functions and a method for solving operator equations in Banach spaces. *Abstract and Applied Analysis*, 2006.

[36] R. H. Byrd and R. H. Tapia. An extension of Curry's theorem to steepest descent in normed linear spaces. *Math. Programming*, 9:247–254, 1975.

[37] C. Byrne. Iterative oblique projection onto convex sets and the split feasibility problem. *Inverse Problems*, 18:441–453, 2002.

[38] C. Byrne. A unified treatment of some iterative algorithms in signal processing and image reconstruction. *Inverse Problems*, 20:103–120, 2004.

[39] C. Canuto and K. Urban. Adaptive optimization of convex functionals in Banach spaces. *SIAM J. Numer. Anal.*, 42:2043–2075, 2005.

[40] Y. Censor, T. Elfving, N. Kopf, and T. Bortfeld. The multiple-sets split feasibility problem and its applications for inverse problems. *Inverse Problems*, 21:2071–2084, 2005.

[41] Y. Censor and G. T. Herman. Block-iterative algorithms with underrelaxed Bregman projections. *SIAM J. Optim.*, 13:283–297, 2002.

[42] Y. Censor, A. Motova, and A. Segal. Perturbed projections and subgradient projections for the multiple-sets split feasibility problem. *Journal of Mathematical Analysis and Applications*, 327(2):1244–1256, 2007.

[43] Y. Censor and A. Segal. Iterative projection methods in biomedical inverse problems. In M. Jiang Y. Censor and A. K. Louis, editors, *Mathematical Methods in Biomedical Imaging and Intensity-Modulated Radiation Therapy (IMRT)*. Edizioni della Normale, Pisa, Italy, 2008.

[44] K. Chadan and P. C. Sabatier. *Inverse Problems in Quantum Scattering Theory*. Springer, New York, 1989.

[45] G. Chavent and K. Kunisch. Convergence of Tikhonov regularization for constrained ill-posed inverse problems. *Inverse Problems*, 10(1):63–76, 1994.

[46] G. Chavent and K. Kunisch. Regularization of linear least squares problems by total bounded variation. *ESAIM Control Optim. Calc. Var.*, 2:359–376, 1997.

[47] J. Cheng and M. Yamamoto. One new strategy for a priori choice of regularizing parameters in Tikhonov's regularization. *Inverse Problems*, 16(4):L31–L38, 2000.

[48] I. Cioranescu. *Geometry of Banach spaces, duality mappings and nonlinear problems*. Kluwer, Dordrecht, 1990.

[49] C. Clason, B. Jin, and K. Kunisch. A semismooth Newton method for L^1 data fitting with automatic choice of regularization parameters and noise calibration. *SIAM J. Imaging Sci.*, 3:199–231, 2010.

[50] F. Colonius and K.Kunisch. Output least squares stability in elliptic systems. *Appl. Math. Optim.*, 19:33–63, 1989.

[51] P. L. Combettes. The foundations of set-theoretic estimation. In *Proceedings of the IEEE*, volume 81, pages 182–208, 1993.

[52] P. L. Combettes. The convex feasibility problem in image recovery. *Advances in Imaging and Electron Physics*, 95:155–270, 1996.

[53] S. Crépey. Calibration of the local volatility in a generalized Black-Scholes model using Tikhonov regularization. *SIAM J. Math. Anal.*, 34(5):1183–1206 (electronic), 2003.

[54] Z. Dai and P. K. Lamm. Local regularization for the nonlinear inverse autoconvolution problem. *SIAM J. Numer. Anal.*, 46(2):832–868, 2008.

[55] I. Daubechies, M. Defrise, and C. De Mol. An iterative thresholding algorithm for linear inverse problems with a sparsity constraint. *Communications in Pure and Applied Mathematics*, 57(11):1413–1457, 2004.

[56] A. De Cezaro, O. Scherzer, and Zubelli J. P. Convex regularization of local volatility models from option prices: Convergence analysis and rates. *Nonlinear Analysis*, 75:2398–2415, 2012.

[57] R. Deville, G. Godefroy, and V. Zizler. *Smoothness and Renormings in Banach Spaces*. Pitman Monographs and Surveys in Pure and Applied Mathematics. Longman Scientific & Technical, Harlow, UK, 1993.

[58] D. Dobson. Phase reconstruction via nonlinear least squares. *Inverse Problems*, 8:541–557, 1992.

[59] J. C. Dunn. Rates of convergence for conditional gradient algorithms near singular and nonsingular extremals. *SIAM Journal on Control and Optimization*, 17(2):187–211, 1979.

[60] J. C. Dunn. Global and asymptotic convergence rate estimates for a class of projected gradient processes. *SIAM Journal on Control and Optimization*, 19(3):368–400, 1981.

[61] D. Düvelmeyer, B. Hofmann, and M. Yamamoto. Range inclusions and approximate source conditions with general benchmark functions. *Numer. Funct. Anal. Optim.*, 28(11-12):1245–1261, 2007.

[62] H. Egger and H. W. Engl. Tikhonov regularization applied to the inverse problem of option pricing: convergence analysis and rates. *Inverse Problems*, 21(3):1027–1045, 2005.

[63] H. Egger, T. Hein, and B. Hofmann. On decoupling of volatility smile and term structure in inverse option pricing. *Inverse Problems*, 22(4):1247–1259, 2006.

[64] B. Eicke. Iteration methods for convexly constrained ill-posed problems in Hilbert space. *Numer. Funct. Anal. and Optimiz.*, 13:413–429, 1992.

[65] M. Elad, B. Matalon, and M. Zibulevsky. Coordinate and subspace optimization methods for linear least squares with non-quadratic regularization. *Appl. Comput. Harmon. Anal.*, 23:346–367, 2007.

[66] H. W. Engl. Discrepancy principles for Tikhonov regularization of ill-posed problems leading to optimal convergence rates. *Journal of Optimization Theory and Applications*, 52(2):209–215, 1987.

[67] H. W. Engl, M. Hanke, and A. Neubauer. *Regularization of Inverse Problems*. Mathematics and Its Applications. Kluwer Academic Publishers, Dordrecht, 1996.

[68] H.W. Engl, K. Kunisch, and A. Neubauer. Convergence rates for Tikhonov regularisation of non-linear ill-posed problems. *Inverse Problems*, 5(4):523–540, 1989.

[69] M. Fabian, P. Habala, P. Hájek, V. M. Santalucia, J. Pelant, and V. Zizler. *Functional Analysis and Infinite-dimensional Geometry*. CMS Books in Mathematics/Ouvrages de Mathématiques de la SMC, 8. Springer-Verlag, New York, 2001.

[70] G. Felcher and T. Russell. *Proceedings of the Workshop on Methods of Analysis and Interpretation of Neutron Reflectivity Data*. Phys. B., Vol. 173, 1991.

[71] T. Figiel. On the moduli of convexity and smoothness. *Stud. Math.*, 56:121–155, 1976.

[72] J. Flemming. Theory and examples of variational regularization with non-metric fitting functionals. *J. Inverse Ill-Posed Probl.*, 18(6):677–699, 2010.

[73] J. Flemming. *Generalized Tikhonov Regularization: Basic Theory and Comprehensive Results on Convergence Rates*. Technische Universität Chemnitz, Germany, 2011. PhD thesis (Dissertation),
`http://nbn-resolving.de/urn:nbn:de:bsz:ch1-qucosa-78152`.

[74] J. Flemming. *Generalized Tikhonov Regularization and Modern Convergence Rate Theory in Banach Spaces*. Shaker Verlag, Aachen, 2012.

[75] J. Flemming and B. Hofmann. A new approach to source conditions in regularization with general residual term. *Numer. Funct. Anal. Optim.*, 31(1-3):254–284, 2010.

[76] J. Flemming and B. Hofmann. Convergence rates in constrained Tikhonov regularization: equivalence of projected source conditions and variational inequalities. *Inverse Problems*, 27(8):085001, 11, 2011.

[77] J. Flemming, B. Hofmann, and P. Mathé. Sharp converse results for the regularization error using distance functions. *Inverse Problems*, 27(2):025006, 18, 2011.

[78] M. Grasmair. Generalized Bregman distances and convergence rates for non-convex regularization methods. *Inverse Problems*, 26(11):115014, 16, 2010.

[79] M. Grasmair. Linear convergence rates for Tikhonov regularization with positively homogeneous functionals. *Inverse Problems*, 27(7):075014, 16, 2011.

[80] M. Grasmair. Variational inequalities and improved convergence rates for Tikhonov regularization on Banach spaces. Technical report, University of Vienna, Preprint, 2011. `http://arxiv.org/abs/1107.2471`.

[81] M. Grasmair, M. Haltmeier, and O. Scherzer. Sparse regularization with l^q penalty term. *Inverse Problems*, 24(5):055020, 13, 2008.

[82] M. Grasmair, M. Haltmeier, and O. Scherzer. Necessary and sufficient conditions for linear convergence of ℓ^1-regularization. *Comm. Pure Appl. Math.*, 64(2):161–182, 2011.

[83] M. Grasmair, M. Haltmeier, and O. Scherzer. The residual method for regularizing ill-posed problems. *Applied Mathematics and Computation*, 218(5):2693–2710, 2011.

[84] C. W. Groetsch. *The Theory of Tikhonov Regularization for Fredholm Equations of the First Kind*. Pitman, Boston, MA, 1984.

[85] J. Hadamard. *Le problème de Cauchy et les équations aux dérivées partielles linéaires hyperboliques*. Hermann & Cie., Paris, 1932.

[86] M. Haltmeier, T. Schuster, and O. Scherzer. Filtered backprojection for thermoacoustic computed tomography in spherical geometry. *Math. Meth. Appl. Sci.*, 28:1919–1937, 2005.

[87] M. Hanke. A regularization Levenberg-Marquardt scheme, with applications to inverse groundwater filtration problems. *Inverse Problems*, 13:79–95, 1997.

[88] M. Hanke, A. Neubauer, and O. Scherzer. A convergence analysis of the Landweber iteration for nonlinear ill-posed problems. *Numer. Math.*, 72:21–37, 1995.

[89] P. C. Hansen. *Discrete Inverse Problems*, volume 7 of *Fundamentals of Algorithms*. Society for Industrial and Applied Mathematics (SIAM), Philadelphia, PA, 2010. Insight and algorithms.

[90] M. Hegland. An optimal order regularization method which does not use additional smoothness assumptions. *SIAM J. Numer. Anal.*, 29(5):1446–1461, 1992.

[91] M. Hegland. Variable Hilbert scales and their interpolation inequalities with applications to Tikhonov regularization. *Appl. Anal.*, 59(1-4):207–223, 1995.

[92] M. Hegland and B. Hofmann. Errors of regularisation under range inclusions using variable Hilbert scales. *Inverse Problems and Imaging*, 5(3):619–643, 2011.

[93] T. Hein. On Tikhonov regularization in Banach spaces – optimal convergence rates results. *Appl. Anal.*, 88(5):653–667, 2009.

[94] T. Hein. Tikhonov regularization in Banach spaces – improved convergence rates results. *Inverse Problems*, 25(3):035002 (18pp), 2009.

[95] T. Hein. *Regularization in Banach Spaces – Convergence Rates Theory*. Logos, Berlin, 2010.

[96] T. Hein and B. Hofmann. On the nature of ill-posedness of an inverse problem arising in option pricing. *Inverse Problems*, 19(6):1319–1338, 2003.

[97] T. Hein and B. Hofmann. Approximate source conditions for nonlinear ill-posed problems – chances and limitations. *Inverse Problems*, 25:035003 (16pp), 2009.

[98] T. Hein and K. S. Kazimierski. Accelerated Landweber iteration in Banach spaces. *Inverse Problems*, 26:055002 (17pp), 2010.

[99] Y. Heng, S. Lu, A. Mhamdi, and S. V. Pereverzev. Model functions in the modified L-curve method – case study: the heat flux reconstruction in pool boiling. *Inverse Problems*, 26(5):055006, 13, 2010.

[100] B. Hofmann. *Regularization for Applied Inverse and Ill-posed Problems*, volume 85 of *Teubner-Texte zur Mathematik [Teubner Texts in Mathematics]*. BSB B. G. Teubner Verlagsgesellschaft, Leipzig, 1986.

[101] B. Hofmann. Ill-posedness and local ill-posedness concepts in Hilbert spaces. *Optimization*, 48(2):219–238, 2000.

[102] B. Hofmann. Approximate source conditions in Tikhonov-Phillips regularization and consequences for inverse problems with multiplication operators. *Math. Methods Appl. Sci.*, 29(3):351–371, 2006.

[103] B. Hofmann, D. Düvelmeyer, and K. Krumbiegel. Approximate source conditions in Tikhonov regularization – new analytical results and some numerical studies. *Mathematical Modelling and Analysis*, 11(1):41–56, 2006.

[104] B. Hofmann, B. Kaltenbacher, C. Pöschl, and O. Scherzer. A convergence rates result for Tikhonov regularization in Banach spaces with non-smooth operators. *Inverse Problems*, 23(3):987–1010, 2007.

[105] B. Hofmann and R. Krämer. On maximum entropy regularization for a specific inverse problem of option pricing. *J. Inverse Ill-Posed Probl.*, 13(1):41–63, 2005.

[106] B. Hofmann and P. Mathé. Analysis of profile functions for general linear regularization methods. *SIAM J. Numer. Anal.*, 45(3):1122–1141, 2007.

[107] B. Hofmann, P. Mathé, and S. V. Pereverzev. Regularization by projection: approximation theoretic aspects and distance functions. *J. Inverse Ill-Posed Probl.*, 15(5):527–545, 2007.

[108] B. Hofmann, P. Mathé, and M. Schieck. Modulus of continuity for conditionally stable ill-posed problems in Hilbert space. *J. Inverse Ill-Posed Probl.*, 16(6):567–585, 2008.

[109] B. Hofmann, P. Mathé, and H. von Weizsäcker. Regularization in Hilbert space under unbounded operators and general source conditions. *Inverse Problems*, 25(11):115013, 15, 2009.

[110] B. Hofmann and O. Scherzer. Factors influencing the ill-posedness of nonlinear problems. *Inverse Problems*, 10(6):1277–1297, 1994.

[111] B. Hofmann and O. Scherzer. Local ill-posedness and source conditions of operator equations in Hilbert spaces. *Inverse Problems*, 14(5):1189–1206, 1998.

[112] B. Hofmann and M. Yamamoto. Convergence rates for Tikhonov regularization based on range inclusions. *Inverse Problems*, 21(3):805–820, 2005.

[113] B. Hofmann and M. Yamamoto. On the interplay of source conditions and variational inequalities for nonlinear ill-posed problems. *Appl. Anal.*, 89(11):1705–1727, 2010.

[114] T. Hohage. Logarithmic convergence rates of the iteratively regularized Gauß-Newton method for an inverse potential and an inverse scattering problem. *Inverse Problems*, 13:1279–1299, 1997.

[115] T. Hohage. Regularization of exponentially ill-posed problems. *Numer. Funct. Anal. Optim.*, 21(3-4):439–464, 2000.

[116] T. Hohage and F. Werner. Iteratively regularized Newton methods with general data misfit functionals and applications to Poisson data. *Preprint 2011–11*, 2011. Institut für Numerische und Angewandte Mathematik, University of Göttingen, `http://arxiv.org/abs/1105.2690v2`.

[117] N. E. Hurt. *Phase Retrieval and Zero Crossings*. Kluwer, Dordrecht, 1989.

[118] V. K. Ivanov, V. V. Vasin, and V. P. Tanana. *Theory of Linear Ill-posed Problems and its Applications*. Inverse and Ill-posed Problems Series. VSP, Utrecht, second edition, 2002. Translated from the 1978 Russian original ["Nauka", Moscow].

[119] J. Janno and U. Tautenhahn. On Lavrentiev regularization for ill-posed problems in Hilbert scales. *Numer. Funct. Anal. Optim.*, 24(5-6):531–555, 2003.

[120] B. Jin, T. Khan, and P. Maaß. A reconstruction algorithm for electrical impedance tomography based on sparsity regularization. *Internat. J. Numer. Methods Engrg.*, 89(3):337–353, 2012.

[121] B. Jin and J. Zou. Iterative parameter choice by discrepancy principle. Technical report, Dept. Math., The Chinese Univ. of Hong Kong, CUHK-2010-02 (369), Hong Kong, 2010.

[122] Q. Jin. Applications of the modified discrepancy principle to Tikhonov regularization of nonlinear ill-posed problems. *SIAM J. Numer. Anal.*, 36(2):475–490, 1998.

[123] Q. Jin. A convergence analysis for Tikhonov regularization of nonlinear ill-posed problems. *Inverse Problems*, 15(4):1087–1098, 1999.

[124] Q. Jin and Z.-Y. Hou. On the choice of the regularization parameter for ordinary and iterated Tikhonov regularization of nonlinear ill-posed problems. *Inverse Problems*, 13(3):815–827, 1997.

[125] M. Jung, E. Resmerita, and L. A. Vese. Dual norm based iterative methods for image restoration. *Journal of Math. Imaging and Vision*, 42:DOI: 10.1007/s10851–011–0318–7, 2012.

[126] B. Kaltenbacher and B. Hofmann. Convergence rates for the iteratively regularized Gauss-Newton method in Banach spaces. *Inverse Problems*, 26(3):035007, 21, 2010.

[127] B. Kaltenbacher and A. Neubauer. Convergence of projected iterative regularization methods for nonlinear problems with smooth solutions. *Inverse Problems*, 22:1105–1119, 2006.

[128] B. Kaltenbacher, A. Neubauer, and O. Scherzer. *Iterative Regularization Methods for Nonlinear Ill-posed Problems*. de Gruyter, 2007.

[129] B. Kaltenbacher, F. Schöpfer, and T. Schuster. Convergence of some iterative methods for the regularization of nonlinear ill-posed problems in Banach spaces. *Inverse Problems*, 25, 2009. 065003 doi: 10.1088/0266-5611/25/6/065003.

[130] K. S. Kazimierski. *Aspects of Regularization in Banach Spaces*. PhD thesis, Universität Bremen, Bremen, 2010.

[131] K. S. Kazimierski. Minimization of the Tikhonov functional in Banach spaces smooth and convex of power type by steepest descent in the dual. *Computational Optimization and Applications*, 48(2):309–324, 2011.

[132] B. T. Kien. The normalized duality mapping and two related characteristic properties of a uniformly convex Banach space. *Acta Mathematica Vietnamica*, 27(1):53–67, 2002.

[133] S. Kindermann and A. Neubauer. On the convergence of the quasioptimality criterion for (iterated) Tikhonov regularization. *Inverse Probl. Imaging*, 2(2):291–299, 2008.

[134] M. V. Klibanov and P. E. Sacks. Phaseless inverse scattering and the phase problem in optics. *J. Math. Phys.*, 33:3913–3821, 1992.

[135] M. V. Klibanov, P. E. Sacks, and A. V. Tikhonravov. The phase retrieval problem. *Inverse Problems*, 11:1–28, 1995.

[136] J. Köhler and U. Tautenhahn. Error bounds for regularized solutions of nonlinear ill-posed problems. *J. Inverse Ill-Posed Probl.*, 3(1):47–74, 1995.

[137] M. Y. Kokurin. Condition of sourcewise representability and estimates for convergence rate of methods for regularization of linear equations in Banach space I. *Russian Mathematics (Iz. VUZ)*, 45(8):49–57, 2001.

[138] R. Krämer and P. Mathé. Modulus of continuity of Nemytskiĭ operators with application to the problem of option pricing. *J. Inverse Ill-Posed Probl.*, 16:435–461, 2008.

[139] K. Kunisch and J. Zou. Iterative choices of regularization parameters in linear inverse problems. *Inverse Problems*, 14(5):1247–1264, 1998.

[140] R. Lagnado and S. Osher. A technique for calibrating derivative security pricing models: numerical solution of an inverse problem. *Journal of Computational Finance*, 1:13–25, 1997.

[141] P. K. Lamm. A survey of regularization methods for first-kind Volterra equations. In *Surveys on Solution Methods for Inverse Problems*, pages 53–82. Springer, Vienna, 2000.

[142] L. Landweber. An iteration formula for Fredholm integral equations of the first kind. *Amer. J. Math.*, 73:615–624, 1951.

[143] J. Lindenstrauß and L. Tzafriri. *Classical Banach Spaces, II.* Springer, Berlin, 1979.

[144] D. Lorenz. *Wavelet Shrinkage in Signal and Image Processing - An Investigation of Relations and Equivalences.* PhD thesis, Universität Bremen, Bremen, 2005.

[145] D. Lorenz, K. Bredies, T. Bonesky, and P. Maaß. A generalized conditional gradient method for non-linear operator equations with sparsity constraints. *Inverse Problems*, 23:2041–2058, 2007.

[146] D. Lorenz, S. Schiffler, and D. Trede. Beyond convergence rates: exact recovery with the Tikhonov regularization with sparsity constraints. *Inverse Problems*, 27(8):085009(17pp), 2011.

[147] D. A. Lorenz. Convergence rates and source conditions for Tikhonov regularization with sparsity constraints. *J. Inverse Ill-Posed Probl.*, 16(5):463–478, 2008.

[148] D. A. Lorenz and D. Trede. Optimal convergence rates for Tikhonov regularization in Besov scales. *Inverse Problems*, 24(5):055010, 14, 2008.

[149] I. Loris. On the performance of algorithms for the minimization of ℓ^1-penalized functionals. *Inverse Problems*, 25(3):035008, 2009.

[150] A. K. Louis. *Inverse und schlecht gestellte Probleme*. Teubner, Stuttgart, 1989.

[151] A. K. Louis. Approximate inverse for linear and some nonlinear problems. *Inverse Problems*, 12:175–190, 1996.

[152] A. K. Louis. A unified approach to regularization methods for linear ill-posed problems. *Inverse Problems*, 15:489–498, 1999.

[153] A. K. Louis. Combining image reconstruction and image analysis with an application to two-dimensional tomography. *SIAM J. Imaging Sciences*, 1(2):188–208, 2008.

[154] A. K. Louis. Feature reconstruction in inverse problems. *Inverse Problems*, 27:065010, 2011. DOI:10.1088/0266-5611/27/6/065010.

[155] A. K. Louis and T. Schuster. A novel filter design technique in 2D computerized tomography. *Inverse Problems*, 12:685–696, 1996.

[156] S. Lu, S. V. Pereverzev, and R. Ramlau. An analysis of Tikhonov regularization for nonlinear ill-posed problems under a general smoothness assumption. *Inverse Problems*, 23(1):217–230, 2007.

[157] S. Lu, S. V. Pereverzev, and U. Tautenhahn. A model function method in regularized total least squares. *Appl. Anal.*, 89(11):1693–1703, 2010.

[158] P. Mahale and M. T. Nair. Tikhonov regularization of nonlinear ill-posed equations under general source condition. *J. Inverse Ill-Posed Probl.*, 15(8):813–829, 2007.

[159] P. Mahale and M. T. Nair. Iterated Lavrentiev regularization for nonlinear ill-posed problems. *ANZIAM J.*, 51(2):191–217, 2009.

[160] P. Mathé and B. Hofmann. Direct and inverse results in variable Hilbert scales. *J. Approx. Theory*, 154(2):77–89, 2008.

[161] P. Mathé and B. Hofmann. How general are general source conditions? *Inverse Problems*, 24(1):015009, 5, 2008.

[162] P. Mathé and S. V. Pereverzev. Discretization strategy for linear ill-posed problems in variable Hilbert scales. *Inverse Problems*, 19(6):1263–1277, 2003.

[163] P. Mathé and S. V. Pereverzev. Geometry of linear ill-posed problems in variable Hilbert scales. *Inverse Problems*, 19(3):789–803, 2003.

[164] R. E. Megginson. *An Introduction to Banach Space Theory*. Springer, New York, 1998.

[165] V. A. Morozov. *Methods for Solving Incorrectly Posed Problems*. Springer-Verlag, New York, 1984. Translated from the Russian.

[166] U. Mosco. Approximation of the solutions of some variational inequalities. *Ann. Sc. Norm. Sup. Pisa*, 21:373–394, 1967.

[167] M. T. Nair. *Linear Operator Equations*. World Scientific Publishing Co. Pte. Ltd., Hackensack, NJ, 2009. Approximation and Regularization.

[168] M. T. Nair and U. Tautenhahn. Convergence rates for Lavrentiev-type regularization in Hilbert scales. *Comput. Methods Appl. Math.*, 8(3):279–293, 2008.

[169] G. Narkiss and M. Zibulevsky. Sequential subspace optimization method for large-scale unconstrained optimization. Technical report, Technion - The Israel Institute of Technology, Department of Electrical Engineering, 2005.

[170] M. Z. Nashed. Inner, outer, and generalized inverses in Banach and Hilbert spaces. *Numer. Funct. Anal. Optim.*, 9(3-4):261–325, 1987.

[171] M. Z. Nashed. A new approach to classification and regularization of ill-posed operator equations. In *Inverse and Ill-posed Problems (Sankt Wolfgang, 1986), volume 4 of Notes Rep. Math. Sci. Engrg.*, pages 53–75. Academic Press, Boston, MA, 1987.

[172] A. Neubauer. Tikhonov-regularization of ill-posed linear operator equations on closed convex sets. *J. Approx. Theory*, 53(3):304–320, 1988.

[173] A. Neubauer. On converse and saturation results for Tikhonov regularization of linear ill-posed problems. *SIAM J. Numer. Anal.*, 34(2):517–527, 1997.

[174] A. Neubauer. The convergence of a new heuristic parameter selection criterion for general regularization methods. *Inverse Problems*, 24(5):055005, 10, 2008.

[175] A. Neubauer. On enhanced convergence rates for Tikhonov regularization of nonlinear ill-posed problems in Banach spaces. *Inverse Problems*, 25(6):065009, 10, 2009.

[176] A. Neubauer. Modified Tikhonov regularization for nonlinear ill-posed problems in Banach spaces. *J. Integral Equations Appl.*, 22(2):341–351, 2010.

[177] A. Neubauer, T. Hein, B. Hofmann, S. Kindermann, and U. Tautenhahn. Improved and extended results for enhanced convergence rates of Tikhonov regularization in Banach spaces. *Appl. Anal.*, 89(11):1729–1743, 2010.

[178] S. Osher, M. Burger, D. Goldfarb, J. Xu, and W. Yin. An iterative regularization method for total variation based image restoration. *Multiscale Modelling and Simulation*, 4:460–489, 2005.

[179] S. Osher, A. Solé, and L. Vese. Image decomposition and restoration using total variation minimization and the H^{-1} norm. *Multiscale Model. Simul.*, 1(3):349–370, 2003.

[180] J. P. Penot. Continuity properties of projection operators. *Journal of Inequalities and Applications*, 5:509–521, 2003.

[181] J. P. Penot and R. Ratsimahalo. Characterizations of metric projections in Banach spaces and applications. *Abstr. Appl. Anal.*, 3(1-2):85–103, 1998.

[182] R. R. Phelps. Metric projections and the gradient projection method in Banach spaces. *SIAM J. Contr. Optim.*, 23, 1985.

[183] R. Plato. On the discrepancy principle for iterative and parametric methods to solve linear ill-posed equations. *Numerische Mathematik*, 75:99–120, 1996.

[184] R. Plato and G. Vainikko. On the regularization of projection methods for solving ill-posed problems. *Numer. Math.*, 57:63–79, 1990.

[185] C. Pöschl. *Tikhonov Regularization with General Residual Term.* PhD thesis (Dissertationsschrift). University of Innsbruck, Austria, 2008.

[186] C. Pöschl, E. Resmerita, and O. Scherzer. Discretization of variational regularization in Banach spaces. *Inverse Problems*, 26(10):105017, 18, 2010.

[187] B. Qu and N. Xiu. A note on the CQ algorithm for the split feasibility problem. *Inverse Problems*, 21:1655–1665, 2005.

[188] E. T. Quinto, A. Rieder, and T. Schuster. Local inversion of the sonar transform regularized by the approximate inverse. *Inverse Problems*, 27, 2011. DOI:10.1088/0266-5611/27/3/035006.

[189] R. Ramlau. Regularization properties of Tikhonov regularization with sparsity constraints. *Electron. Trans. Numer. Anal.*, 30:54–74, 2008.

[190] R. Ramlau and L. Justen. A general framework for soft-shrinkage with applications to blind deconvolution and wavelet denoising. *Appl. Comp. Harmon. Anal.*, 26(1):43–63, 2009.

[191] R. Ramlau and E. Resmerita. Convergence rates for regularization with sparsity constraints. *Electron. Trans. Numer. Anal.*, 37:87–104, 2010.

[192] R. Ramlau and G. Teschke. Sparse recovery in inverse problems. In *Theoretical Foundations and Numerical Methods for Sparse Recovery*, volume 9 of *Radon Ser. Comput. Appl. Math.*, pages 201–262. Walter de Gruyter, Berlin, 2010.

[193] S. Reich and A. J. Zaslavski. Generic convergence of descent methods in Banach spaces. *Mathematics of Operations Research*, 25:231–242, 2000.

[194] S. Reich and A. J. Zaslavski. The set of divergent descent methods in a Banach space is sigma-porous. *SIAM J. Optim.*, 11:1003–1018, 2001.

[195] E. Resmerita. On total convexity, Bregman projections and stability in Banach spaces. *Journal of Convex Analysis*, 11(1):1–16, 2004.

[196] E. Resmerita. Regularization of ill-posed problems in Banach spaces: convergence rates. *Inverse Problems*, 21(4):1303–1314, 2005.

[197] E. Resmerita and O. Scherzer. Error estimates for non-quadratic regularization and the relation to enhancement. *Inverse Problems*, 22(3):801–814, 2006.

[198] A. Rieder. On convergence rates of inexact Newton regularizations. *Numer. Math.*, 88:347–365, 2001.

[199] A. Rieder. *Keine Probleme mit inversen Problemen (No Problems with inverse Problems).* Vieweg, Wiesbaden, 2003.

[200] A. Rieder, R. Dietz, and T. Schuster. Approximate inverse meets local tomography. *Math. Meth. Appl. Sci.*, 23:1373–1387, 2000.

[201] A. Rieder and T. Schuster. The approximate inverse in action II: convergence and stability. *Mathematics of Computation*, 72:1399–1415, 2003.

[202] A. Rieder and T. Schuster. The approximate inverse in action III: 3D-Doppler tomography. *Numerische Mathematik*, 97:353–378, 2004.

[203] R. T. Rockafellar and R. J.-B. Wets. *Variational Analysis*. Springer, Berlin, 1998.

[204] L. I. Rudin, S. Osher, and E. Fatemi. Nonlinear total variation based noise removal algorithms. *Physica D*, 60(3):259–268, 1992.

[205] W. Rudin. *Functional Analysis*. Tata McGraw-Hill, New York, 1973.

[206] W. Rudin. *Real and Complex Analysis*. McGraw-Hill, New York, 1987.

[207] O. Scherzer. A modified Landweber iteration for solving parameter estimation problems. *Appl. Math. Optim.*, 38:45–68, 1998.

[208] O. Scherzer. A posteriori error estimates for the solution of nonlinear ill-posed operator equations. *Nonlinear Anal.*, 45, 2001.

[209] O. Scherzer, editor. *Handbook of Mathematical Methods in Imaging. In 3 Volumes.* Springer Science+Business Media LCC, New York, 2011.

[210] O. Scherzer, M. Grasmair, H. Grossauer, M. Haltmeier, and F. Lenzen. *Variational Methods in Imaging*, volume 167 of *Applied Mathematical Sciences*. Springer, New York, 2009.

[211] W. Schirotzek. *Nonsmooth Analysis*. Springer, Berlin, 2007.

[212] E. Schock. Approximate solution of ill-posed equations: arbitrarily slow convergence vs. superconvergence. In *Constructive Methods for the Practical Treatment of Integral Equations (Oberwolfach, 1984)*, volume 73 of *Internat. Schriftenreihe Numer. Math.*, pages 234–243. Birkhäuser, Basel, 1985.

[213] F. Schöpfer, A. K. Louis, and T. Schuster. Nonlinear iterative methods for linear ill-posed problems in Banach spaces. *Inverse Problems*, 22(1):311–329, 2006.

[214] F. Schöpfer and T. Schuster. Fast regularizing sequential subspace optimization in Banach spaces. *Inverse Problems*, 25(1), 2009.

[215] F. Schöpfer, T. Schuster, and A. K. Louis. An iterative regularization method for the solution of the split feasibility problem in Banach spaces. *Inverse Problems*, 24, 2008. 055008.

[216] F. Schöpfer, T. Schuster, and A. K. Louis. Metric and Bregman projections onto affine subspaces and their computation via sequential subspace optimization methods. *Journal of Inverse and Ill-Posed Problems*, 16(5):479–506, 2008.

[217] T. Schuster. The 3D Doppler transform: elementary properties and computation of reconstruction kernels. *Inverse Problems*, 16(3):701–723, 2000.

[218] T. Schuster. An efficient mollifier method for three-dimensional vector tomography: convergence analysis and implementation. *Inverse Problems*, 17:739–766, 2001.

[219] T. Schuster. A stable inversion scheme for the Laplace transform using arbitrarily distributed data scanning points. *J. Inv. Ill-Posed Problems*, 11(3):263–287, 2003.

[220] T. Schuster. *The Method of Approximate Inverse: Theory and Applications*, volume 1906 of *Lecture Notes in Mathematics*. Springer, Heidelberg, 2007.

[221] T. Schuster, J. Plöger, and A. K. Louis. Depth-resolved residual stress evaluation from X-ray diffraction measurement data using the approximateinverse method. *Z. Metallkd.*, 94:934–937, 2003.

[222] T. Schuster and E. T. Quinto. On a regularization scheme for linear operators in distribution spaces with an application to the spherical radon transform. *SIAM J. Appl. Math.*, 65:1369–1387, 2005.

[223] T. Schuster and F. Schöpfer. Solving linear operator equations in Banach spaces non-iteratively by the method of approximate inverse. *Inverse Problems*, 26, 2010.

[224] T. Schuster, D. Theis, and A. K. Louis. A reconstruction approach for imaging in 3D cone beam vector field tomography. *Int. J. Biomed. Imag.*, 2008. DOI:10.1155/2008/174283.

[225] T. I. Seidman and C. R. Vogel. Well posedness and convergence of some regularisation methods for non-linear ill posed problems. *Inverse Problems*, 5(2):227–238, 1989.

[226] R. E. Showalter. *Monotone Operators in Banach Spaces and Nonlinear Partial Differential Equations*. American Mathematical Society, Providence, 1997.

[227] V. P. Tanana. *Methods for Solution of Nonlinear Operator Equations*. Inverse and Ill-posed Problems Series. VSP, Utrecht, 1997.

[228] U. Tautenhahn. Optimality for ill-posed problems under general source conditions. *Numer. Funct. Anal. Optim.*, 19(3-4):377–398, 1998.

[229] U. Tautenhahn. Lavrentiev regularization of nonlinear ill-posed problems. *Vietnam J. Math.*, 32(Special Issue):29–41, 2004.

[230] U. Tautenhahn, U. Hämarik, B. Hofmann, and Shao Y. Conditional stability estimates for ill-posed PDE problems by using interpolation. *Preprint 2011-16*, 2011. Fakultät für Mathematik, Technische Universität Chemnitz, `http://nbn-resolving.de/urn:nbn:de:bsz:ch1-qucosa-72654`.

[231] U. Tautenhahn and Q. Jin. Tikhonov regularization and a posteriori rules for solving nonlinear ill posed problems. *Inverse Problems*, 19(1):1–21, 2003.

[232] A. N. Tikhonov and V. Y. Arsenin. *Solutions of Ill-posed Problems*. Wiley, New York, 1977.

[233] A. N. Tikhonov, A. V. Goncharsky, V. V. Stepanov, and A. G. Yagola. *Numerical Methods for the Solution of Ill-posed Problems*, volume 328 of *Mathematics and its Applications*. Kluwer, Dordrecht, 1995. Translated from the 1990 Russian original.

[234] A. N. Tikhonov, A. S. Leonov, and A. G. Yagola. *Nonlinear Ill-posed Problems. Vol. 1, 2*, volume 14 of *Applied Mathematics and Mathematical Computation*. Chapman & Hall, London, 1998. Translated from the Russian.

[235] G. M. Vaĭnikko and A. Yu. Veretennikov. *Iteratsionnye protsedury v nekorrektnykh zadachakh (in Russian)*. "Nauka", Moscow, 1986.

[236] C. R. Vogel. *Computational Methods for Inverse Problems*, volume 23 of *Frontiers in Applied Mathematics*. Society for Industrial and Applied Mathematics (SIAM), Philadelphia, PA, 2002.

[237] Z. Wang and J. Liu. New model function methods for determining regularization parameters in linear inverse problems. *Appl. Numer. Math.*, 59(10):2489–2506, 2009.

[238] J. Xie and J. Zou. An improved model function method for choosing regularization parameters in linear inverse problems. *Inverse Problems*, 18(3):631–643, 2002.

[239] Z.-B. Xu and G. F. Roach. Characteristic inequalities of uniformly convex and uniformly smooth Banach spaces. *Journal of Mathematical Analysis and Applications*, 157(1):189–210, 1991.

[240] Q. Yang. The relaxed CQ algorithm solving the split feasibility problem. *Inverse Problems*, 20:1261–1266, 2004.

[241] Q. Yang. On variable-step relaxed projection algorithm for variational inequalities. *Journal of Mathematical Analysis and Applications*, 302:166–179, 2005.

[242] C.A. Zarzer. On Tikhonov regularization with non-convex sparsity constraints. *Inverse Problems*, 25(2):025006, 13, 2009.

[243] E. Zeidler. *Nonlinear Functional Analysis and its Applications III*. Springer-Verlag, New York, 1984.

[244] J. Zhao and Q. Yang. Several solution methods for the split feasibility problem. *Inverse Problems*, 21:1791–1799, 2005.

Index

BV-space 89, 91
C^k-space 89
L^1-space 89
L^2-space 91
L^∞-space 89
L^p-space 55, 62, 88
Ω-minimizing solution 61
ℓ^1-space 89
ℓ^2-space 91
ℓ^∞-space 89
ℓ^p-space 62
η-condition 69
$\overline{x}$-minimum norm solution 52
q-coercive Bregman distance 70, 96, 105

a posteriori parameter choice 58, 88, 93
a priori parameter choice 58, 86, 93, 98, 99, 102, 103, 113
adjoint 29
adjoint operator 58
approximate inverse 235, 236
  semi-discrete 255
approximate source condition 66, 96, 97, 99
Asplund theorem 35
at-the-money options 17
auxiliary Banach space 17

Bakushinskii veto 58
Banach space 28
Banach's bounded inverse theorem 52
Banach–Alaoglu theorem 89
benchmark source condition 66, 67, 97, 98
Besov space 62
biadjoint operator 67
Black–Scholes formula 14
Black–Scholes function 14
Bregman distance 44, 66, 70, 90, 95
  properties 45
Bregman distance, estimates 110
Bregman domain 45, 61
Bregman projection 157, 158

coercive functional 60
compact set 55
concave index function 92, 93, 96
conditional stability estimate 56, 103
conditionally well-posed 55, 103, 105
conjugate exponent 28
convergence 57
convergence of regularized solutions 84
convergence rate 16, 57, 66, 92, 106
convergence rate result 92, 93, 96, 99, 101–103, 107
converse result 98
convex
  p 37
  spaces, properties 42
  of power type, space 37
  strictly, space 37
  uniformly, space 38
convex constraints 99
convex functional 60
convexity 31
CQ-algorithm 145, 178

degree of nonlinearity 73, 99, 106
differentiability
  Fréchet 30
  Gâteaux 30
directional derivative 61
discrepancy principle 87, 93, 156, 165, 199
  Morozov's 115
  modified 116
discretization 51, 89
distance function 66, 67, 96, 97, 99
domain
  effective 31

dual method 122, 135
  convergence rates 137
dual pairing 28, 50, 66
dual space 28
duality gap 87
duality mapping 34, 35, 43, 50, 100
duality of convexity and smoothness 42

enhanced source condition 102
enhanced-rate region 101
equivalent 30
error estimate 91, 95
error measure 90, 95
exact penalization 60
existence of regularized solutions 81
extremal problem 59

Fenchel conjugate function 93
fidelity term 57, 60
Fréchet derivative 16
function
  convex 31
  sparse 18
functional of sympathy 59

Gâteaux derivative 15, 62
generalized Young inequality 96
GPS data 21

Hölder convergence rate 97, 98, 103
Hölder source condition 98
Hahn–Banach theorem 43
Hausdorff distance 178
heuristic parameter choice 58
Hilbert space theory 51, 54, 58, 64, 92, 98

ill-conditioned 51
ill-posed 49, 51
  type I 51
  type II 51
ill-posedness phenomenon 49
index function 55
inequalities
  Xu–Roach 38
inequality
  Cauchy's 29
  Hölder's 29
  Young's 30

inner inverse 52
interior point 62, 100
inverse option pricing 13
iterative regularization 59
iteratively regularized Gauss–Newton method 212
iteratively regularized Landweber method 204

Josefson-Nissenzweig theorem 54

Kadec–Klee property 85
Kullback–Leibler divergence 90

Lagrangian functional 87
Landweber iteration 146
Landweber method 144, 199
Lavrentiev regularization 63
linear operator equation 50
link condition 64
local regularization 63
locally ill-posed 54
locally well-posed 54
logarithmic convergence rate 97
logarithmic decay rate 97
logarithmic source condition 65
logmoneyness 15
low-rate region 101
lower semi-continuous 60

mathematical finance 13
metric 90
metric projection 162
minimum norm solution 50
misfit term 57, 60, 96
modulus
  of convexity 38
  of smoothness 38
mollifier 236, 254
mollifier property 254
monotonicity
  of the duality mapping 35
  of the subgradient 34
Moore-Penrose inverse 51
multiplication operator 15
multiplicative variational inequality 69

Nemytskii operator 15
noise level 56
noisy data 56

non-reflexive Banach space 67, 88
nonlinear operator equation 52
nonlinearity condition 69, 72, 73, 96, 97, 101, 102, 106
norm convergence 85
norm-to-norm continuous 55
normal cone 100
null-space 29

objective a priori information 55
observation operator 253
operator
  adjoint 29
optimality conditions 32
orthonormal basis 91, 105

parameter identification in PDEs 9, 103
penalized minimization 59
phase retrieval 6
positively homogeneous functional 91
predual space 89
primal method 122, 124
  convergence rates 128
projected Landweber method 178
projected source condition 100
proper 31

quantities
  equivalent 30

Radon–Riesz property 85
range 29
range inclusion 65
reconstruction kernel 236, 237, 255
reflexive Banach space 53, 55, 62, 67, 100, 105
regularization parameter 16, 57
  choice 113
regularized solution 57
RESESOP 145, 157
residual method 59
Riesz Representation theorem 244

saddle point 87
saturation 65, 66
separable Banach space 89
separable Hilbert space 91, 105
SESOP 144, 157
smallness condition 16, 72, 102
smooth
  spaces, properties 42
  of power type, space 37
  p 37
  space 38, 102
  uniformly 38
Sobolev space 62
solution smoothness 64, 97
source condition 16, 64, 96, 101, 106, 108
  high order 109
  low order 109
source element 64
sparsity 18, 91, 105, 106
  almost 21
  constraints 18
split feasibility problem 145, 177
stability of regularized solutions 82
stabilizing functional 60, 105
stable approximate solution 56
starlike 61, 98
stopping rule 63
strictly convex functional 91
subdifferential 31
subgradient 31
  of compositions 33
  of sums 33
  of translation 33
subjective a priori information 59
sublevel set 53, 86, 92, 96
super-rate 105–107

three-point identity 47
Tikhonov functional 18, 31, 81
Tikhonov's theorem 55
Tikhonov-type regularization 20, 59, 81, 99

variable Hilbert scales 55
variational inequality 67, 92, 95, 98
variational regularization 59
volatility identification 13

weak convergence 82, 84
weak* compact 89
weak* convergence 89
weak*-to-weak continuous 89
weak-to-weak continuous 53, 55

weakly closed 53
weakly compact 53, 60
well-posed 49, 51

X-ray diffractometry 4, 248

Young's inequality 69, 71, 74

Bei Fragen zur Produktsicherheit wenden Sie sich bitte an:
If you have any questions regarding product safety,
please contact:

Walter de Gruyter GmbH
Genthiner Straße 13
10785 Berlin
productsafety@degruyterbrill.com